WULFRUN COLLEGE

Women, Crime, and Custody in Victorian England

LUCIA ZEDNER

CLARENDON PRESS · OXFORD

Oxford University Press, Walton Street, Oxford OX2 6DP

Oxford New York Toronto
Delhi Bombay Calcutta Madras Karachi
Kuala Lumpur Singapore Hong Kong Tokyo
Nairobi Dar es Salaam Cape Town
Melbourne Auckland Madrid

and associated companies in
Berlin Ibadan

Oxford is a trade mark of Oxford University Press

Published in the United States by
Oxford University Press Inc., New York

First Published 1991
This Paperback Edition 1994

British Library Cataloguing in Publication Data
Data available

Library of Congress Cataloging in Publication Data
Data applied for
ISBN 0-19-820264-3
ISBN 0-19-820552-X (pbk)

1 3 5 7 9 10 8 6 4 2

Printed in Great Britain
on acid-free paper by
Biddles Ltd.
Guildford and King's Lynn

ACKNOWLEDGEMENTS

The debts I have incurred in researching and writing this book are too numerous and too great to be repaid here. I owe enormous gratitude to Roger Hood, All Souls College, Oxford, for his able direction of the doctoral research on which this book is based and his careful criticism of the innumerable drafts through which it progressed. His vast knowledge of the history of crime and penology was matched only by his generosity of time, patience, and encouragement. John Goldthorpe, Nuffield College, Oxford, was an important source of inspiration, not least in providing invaluable insight into the possible fruits and dangers of applying sociological concepts and methods to historical material.

Many others read all or parts of this book: for their comments and criticisms I would like to thank A. H. Halsey, Brian Harrison, Kenneth O. Morgan, H. C. G. Matthew, Allison Morris, and Janet Howarth. In developing my ideas I was helped considerably by comments made by members of the following seminars to which I presented papers: at Oxford, the Feminist History Group, the Graduate History Group, the Centre for Criminological Research, the Nuffield Sociology Seminar, the All Souls History Seminar, and the History Workshop Conference on 'Women in Prison' (January 1988); and at the University of Edinburgh, the Centre for Criminology and the Social and Philosophical Study of Law.

The research and writing of this book was carried out whilst I was a student and later Prize Research Fellow at Nuffield College, Oxford. Together the Warden, Fellows, students, and staff at Nuffield form a superb environment for research. I was also greatly aided by the staff of all the libraries and archives in which I worked, in particular the staff of Nuffield College Library, the Bodleian, the Greater London Record Office, the Public Record Office, the Home Office Notice Office, and the Prison Service Museum.

I owe a more personal debt to my family and friends who provided much needed support as well as welcome distractions. I would like to express special thanks to my parents who, willingly accepting the role of 'intelligent laypersons', read and criticized the final drafts. This book is dedicated to them.

CONTENTS

INTRODUCTION

In Victorian England women made up a far larger proportion of those known to be involved in crime than they do today. During the second half of the nineteenth century over a fifth of those convicted of crime were women—today they make up only an eighth. And whilst a hundred years ago women made up 17 per cent of the daily average local and convict prison population, today the figure is less than 4 per cent. Unsurprisingly then, female criminality attracted considerable attention in the nineteenth century and preoccupied those trying to provide for women within the penal system. This book explores how the Victorians perceived and explained female crime, and how they responded to it—both in penal theory and prison practice.

In recent years there has been a spectacular growth in the history of crime and of the agencies and institutions set up from the late eighteenth to the mid-nineteenth century to control it. Many historians have suggested that the police and the prison are best understood as methods of social control, set up to maintain order in a society undergoing rapid urbanization or to impose the social discipline necessary to capitalism. Provoked but not wholly convinced by these accounts, this book examines several problems that have not yet been satisfactorily addressed by historians. A number of key questions have to be answered: What determines public perceptions of the social costs of crime? In what ways do attitudes vary according to the type of crime or offender? Why do prevailing concerns appear to change over time and what are the consequences for penal policy? And how far do penal institutions match up to the ideals of penal reformers?

This book addresses such questions primarily, though not exclusively, in relation to female offenders, and focuses particularly on the role of gender in determining attitudes and responses to criminality. The growth of feminist history has revealed the extent to which gender divided Victorian society. Gender, quite as much as class, was the focus of intense interest and elaborate theorizing which had the effect of amplifying

distinctions between the sexes. Notions of femininity and, no
less, masculinity influenced the ways in which crime was
viewed. The result was that criminality was perceived, judged,
and explained differently according to the sex of the offender.
Moreover, differential opportunities and needs were seen to
create sex-specific criminal characteristics and types of criminal
activity.

The book begins by examining the extent to which gender-
based ideologies influenced attitudes to female criminality.
This necessarily raises questions about the degree to which
criminal men were also responded to according to parallel conceptions of masculinity. Historians have not yet examined how
far notions of appropriate male behaviour coloured definitions
of male criminality or determined how men were treated within
the criminal justice system. Rather, the history of criminality
and its control has been documented with scant regard to the
sex of the offender, other than the recognition that crime was
something engaged in mainly by men. Until historical research
has been carried out on the place of gender norms in the explanation and treatment of male criminality, it will not be
possible to view responses to female criminality in proper perspective. It is clearly beyond the scope of this book to give equal
weight to male criminality. It can only open up the debate by
exploring how far gender norms informed understandings and
explanations of female crime.

By establishing what constituted or was constructed as normality, one can begin to understand responses to what, consequently, was seen as deviant. The prescriptive ideology of
femininity in Victorian England gave women an important
moralizing role: not least the responsibility for maintaining the
respectability of their family. As a result women's crimes
contravened not only the law but, perhaps more importantly,
their idealized role as wives and mothers. The high social costs
attributed to criminality in women attracted considerable
public anxiety and, as a result, female offenders were likely to
be severely stigmatized. Most strongly condemned were
women involved in so-called 'crimes of morality', particularly
offences relating to prostitution and alcoholism. Such women
were seen to betray the very ideal of womanhood as passive,
respectable, and virtuous. Moreover it was commonly feared

that women, once they lost the respect of society, would fall to even greater depths than the male criminal.

Literature about the nature, extent, or causes of female criminality is a fruitful source for social historians in that it reveals much about the way women were regarded more generally. Changing explanations of female criminality over the course of the century mirrored shifting views of women and prevailing social issues quite as much as trends in the nascent science of criminology. Although the range of explanations was extremely heterogeneous, a broad shift is discernible, from the moral analyses favoured around the middle of the nineteenth century, to a growing tendency to interpret criminality as indicative of biological or psychological disorder. Understanding why female criminality was so readily integrated into these medical interpretations is central to explaining the course of penal policy well into the twentieth century.

After examining explanations of female criminality, this book goes on to look at penal responses—both the penological theories which determined the structure of the prison regime and, more revealingly, the realities of prison life itself. Recent 'revisionist' historiography has claimed that the prison was 'reformed' by the 1860s (see Chapter 3 below). However, accounts of prison reform as the universal replacement of the disorder and squalor of eighteenth- and early nineteenth-century gaols by uniform, disciplined, and finely tuned penal regimes seem to have been overplayed. This book reveals that the process of reform was far more problematic and consequently less successful than most historians have allowed. The contest between possible prison regimes (most notably the 'silent' and 'separate' systems), the competing demands of reform versus deterrence, of economy (not least the doctrine of 'less eligibility') versus humanity, continued to create fundamental tensions within the penal regime. Added to this was the major problem of how women were to be fitted into the system. Although some historians have claimed that nineteenth-century prisons were designed solely with men in mind, policy makers at both national and local levels debated anxiously how to adapt the official regime to accord with their assumptions about women. Middle class women philanthropists were especially influential in developing regimes oriented towards that idealised model of femininity from which inmates had so obviously departed. Despite the difficulties facing female ex-prisoners (not least in finding respectable work), penal policy makers remained committed to developing a regime that could reform. Believing that women were more malleable than men, they permitted greater flexibility in female prisons in the hope that 'lady visitors' and prison matrons could persuade criminal women to repent. The resultant emphasis on establishing personal influence over each woman prisoner was in marked contrast to the quasi-militaristic, anonymous, and strictly uniform regime imposed on men.

Historians have tended to concentrate on the implementation of penal ideology in national prisons such as Pentonville and Millbank. However, in addition to these 'convict' prisons set up to contain serious offenders for sentences of penal servitude (of five years or more), hundreds of local prisons held the petty offenders sentenced to only a few days, weeks, or months. These local prisons constituted the prison experience of the vast majority of offenders and their importance is increasingly being recognized by historians. By examining attempts to put penal ideology into practice at both levels of the system, I suggest that nineteenth-century penal policy was far more complex, and in many ways less coherent, than historians have allowed. Moreover, in the Victorian mania for classification, the sex of the offender became a further basis for separation, and assumptions about gender the basis for fundamental differentiation in penal policy.

Whereas most penal histories have relied mainly on external reports, official papers, and the writings of penal reformers, this book also attempts to examine the internal social organization of the prison by looking at surviving day-to-day records. Rich archival material, in the form of prison regulations, officers' journals and reports, punishment books, prisoners' histories, memoirs, letters, grievances, and appeals, provides a counterbalance to the more usual sources. From this it emerges that the prison cannot be satisfactorily understood in terms of the intentions of the policy makers alone. Daily prison life was shaped largely by the inmates and warders who lived within its walls. Inside prisons the regime was constantly modified, and often subverted, by relations between staff and inmates. In local prisons the outside world continually intruded: women were mostly sentenced to very short terms of only a few days or weeks but returned again and again to the same prisons, thus blurring the divide between life inside and their own communities outside. Some very poor women even looked to the local prison as a refuge and thus, quite unwittingly, subverted its punitive purpose entirely. Even the more formal regime imposed in convict prisons (set up for those serving long-term sentences of penal servitude) was in no way a perfect realization of the intentions of the reformers. Although fewer internal records have survived from female convict prisons, there are sufficient to glean that order inside was maintained by a continual round of compromise and bargaining. Whereas male convicts went out each day to work in regimented gangs as labour on the 'public works', female convicts were confined within a regime whose character depended heavily on the personalities and personal relationships of inmates and warders alike. Regimes in both local and convict prisons were clearly differentiated according to the sex of their inmates. Looking behind the rhetoric of penal policy, this book attempts to unveil life inside the prison and to show how it often departed dramatically from official rules and regulations.

Towards the end of the nineteenth century, penal policy makers and administrators became progressively disillusioned as they saw that habitual offenders failed to respond to the reformative influences of the prison regime. Prisons seemed to have become a dumping ground for the socially inadequate, for confirmed alcoholics, and for the mentally deficient. Women actually outnumbered men in the class of most hardened—so-called 'habituals' who had ten or more convictions—and many worried contemporaries feared that they were simply immune to the pains of imprisonment. New scientific 'experts' proffered medical explanations for the apparent 'incorrigibility' of these recidivists, suggesting that their behaviour was the result of some pathological condition rather than any ill will. Ironically whilst this lessened moral condemnation of female offenders, it actually amplified the dangers associated with their condition. Wider fears about the declining health of the nation, coupled with heightened anxiety about the eugenic implications of female crime, created a new impetus to control those women whose freedom, it was thought, would lead to 'the rapid multi-plication of the unfit'. If such women were allowed to remain unprotected, or uncontrolled, in normal society their progeny could only contribute to the feared decline of the race.

Prompted by the interest of doctors and psychologists, new departures at the end of the nineteenth century focused on two main groups, who made up the core of the worst recidivists: 'habitual inebriates' (confirmed alcoholics) and the 'feeble-minded' (the mentally deficient). Women were seen to be the most serious social problem in both these groups.

Female alcoholics moved continuously in and out of the prison system: their addiction was unaffected by the short sentences they served and, as it worsened, they disrupted prison life more with each successive term. Outside prison, drunkenness was seen as an urgent social problem: its costs were particularly grave when female drunks gave birth to sickly children whom they neglected or even abused. The Eugenics Movement played on fears of the declining health of the nation and lobbied for special reformatories to contain, and it was hoped to cure, such women. The inebriate reformatory experi-ment which was launched as a result poses a particular puzzle for historians: it lasted only fifteen years (1898–1914), and it catered overwhelmingly for women, who made up about 80 per cent of admissions. It strove to reform alcoholics by placing them in a propitious rural environment and attempted to sub-ject them to a variety of positive moral influences. However, by the time women reached the state of chronic alcoholism which secured their admission to reformatories, many were barely susceptible to moral influence and some were so mentally impaired as to be incurable. Although the failure of the reform-atory experiment led to its abandonment, observation of the women who were admitted to reformatories furthered medical diagnoses of female deviance. Its failure to reform more than a tiny proportion of inmates raised serious questions about the usefulness of any penal or reformatory endeavour and suggested that, for the very worst cases, some form of benign containment was the only remaining solution.

The book ends by discussing a second major area of penal innovation around the turn of the century—the provision for those labelled as 'feeble-minded'. Many women prisoners were undoubtedly mentally weak, whether as an inherited condition or as a result of disease, malnutrition, or prolonged alcoholism. It is debatable, however, whether 'feeble-mindedness' constituted an actual medical condition or whether it was a construct, designed to explain the general failure of penal and welfare institutions to deal with their more recalcitrant inmates. Significantly, feeble-mindedness had no precise medical definition but was used as a generic term encompassing all those who appeared to be blighted by a mental incapacity for making 'correct' moral judgements. As such it provided a ready explanation of women's apparent imperviousness to the rewards and sanctions of penal discipline.

The fact that the majority of those labelled and confined as feeble-minded were women raises a number of questions: Were women actually more commonly mentally deficient or were they simply more likely to be diagnosed as such? What were the costs of feeble-mindedness in women, and why did contemporaries fear them so much as to demand their long-term or even permanent segregation? Eugenic concerns about the need to limit such women's ability to breed undoubtedly furthered the campaign for legislation to protect, to sterilize, to restrain, and to confine those identified as feeble-minded. However, psychiatric diagnoses of those who could be deemed to be feeble-minded were all too often confounded with moral judgements about women who seemed to be out of control. Whilst many women so defined were undoubtedly mentally impaired, others were incarcerated on very questionable grounds. Under the 1913 Mental Deficiency Act, for example, any unmarried woman in receipt of poor relief when pregnant or at the time of giving birth was automatically classified as feeble-minded. New specialist institutions set up outside the prison system abandoned moral reform and punishment in favour of medicalized treatment or, at their most pessimistic, the containment of those women deemed incurable for as long as they were capable of bearing children.

A principal theme of this study is the relationship between responses to female criminality and prevailing social values and concerns. It is not possible to understand the history of crime, or its control, in isolation. Changing views of womanhood and of woman's role in society informed the ways in which criminal women were perceived and treated at every juncture. Similarly the history of penal institutions remains mere antiquarianism unless it is related to their wider social and moral world. The contemporary values that these institutions endorsed and the anxieties they reflected are central to understanding changes in penal policy and practice. For the nature, quality, and purpose of penal responses to women were in a state of perpetual flux. Only one factor remained constant—the presumption that penal provision must be differentiated on the basis of sex.

At one level this book is a study of penal theory and policy as it related to women. At quite another it is an attempt to go beyond the rhetoric of the policy makers to get, in so far as it is possible, at the 'realities' of prison life as experienced by women. The picture it reveals must place in question the revisionist orthodoxy that the 'reformed prison' of the nineteenth century was in any real sense 'a triumph of rationalization'.

PART I
VICTORIAN VIEWS OF FEMALE CRIME

1 NORMAL AND DEVIANT WOMEN

EARLY VICTORIAN IDEALS OF FEMININITY

Nineteenth-century writings about criminal women can only be understood within the more general framework of Victorian morality. Criminal women were perceived and judged against complex, carefully constructed notions of ideal womanhood. This chapter begins by examining early Victorian models of femininity and exploring the relationship between these ideal types and women's roles at all levels of society. We explore the implications of deviation from these ideals and seek to provide a basis for understanding contemporary reactions to female crime.

Since the Middle Ages the mythology of woman as Eve-like, both corruptible and corrupting, had proved powerful and enduring. In the early years of the nineteenth century it was steadily overlaid with the carefully contrived model of woman as 'Mary'—the 'divine guide, purifier, inspirer of the man'.[1] Accordingly, by the beginning of the Victorian era two stereotypes of woman prevailed. The ideal of femininity was invested in the middle-class wife and mother whose asexual, morally uplifting influence was held as a vital bulwark against the sordid intrusions of industrial life.[2] Her antithesis was the epitome of female corruption—fallen from innocence, she had plummeted to the depths of degradation and contaminated all who came near her. Although Christianity taught that everyone was fallen from the original state of grace, the notion of the 'fallen woman' suggested a descent to far greater depths.

Just as in early modern Europe the image of the witch had stood as the negation of femininity, so in early Victorian England the woman who had fallen through sexual misconduct, drunkenness, or criminality became the subject of a mass of

---

[1] Charles Kingsley quoted in F. Basch, *Relative Creatures* (1974), 6.
[2] See Anon., 'The Mental and Moral Dignity of Woman', *The Female's Friend* (Jan. 1846).

literature.[3] To avoid being stigmatized as 'witch' or 'fallen woman', women were obliged to conform to the prescribed patterns of acceptable female behaviour of their time. To fail to do so lost them all claim to the respect of 'decent society'. Denied access to economic or political power, women gained coveted status through this respectability, which, as a result, was likely to be of considerable importance to them. Yet in attempting to maintain respectability, women, to a degree unknown by men, became subject to the sanctions of public opinion. These two stereotypes, of the respectable wife and mother and of the fallen woman, can be seen as symbolizing middle-class hopes and fears. The historian Leonore Davidoff asserts that these images were so powerful mainly because they were founded on the realities of women's lives.[4] I would argue that they were no more than ideal types, often bearing little relation to women's lives at any level of society but nonetheless intended to shape and direct them. For women, far more than men, were subjects of social definition.

Whilst this book is primarily concerned with the imagery and literature of the 'fallen woman', we need first to examine the feminine ideal who graced the pedestal from which she fell, and to consider what role this idealization fulfilled.[5] In order to do this we begin by locating femininity in the broader social and moral schema of early Victorian England. For it is clear from the literature that by the 1830s the definition of the ideal woman was tied closely to her role within the family and home.[6] In turn the family was central to middle-class morality, according to which it served as a sanctuary for the preservation of traditional moral and religious values.[7] The massive social

[3] For example, on prostitution see M. Ryan, *Prostitution in London* (1839); W. Acton, *Prostitution* (1857); H. Mayhew, *London Labour and London Poor* (1861–2), iv; G. P. Merrick, *Work among the Fallen as Seen in the Prison Cell* (1891).

[4] L. Davidoff, 'Class and Gender in Victorian England', in J. L. Newton *et al.* (eds.), *Sex and Class in Women's History* (1983).

[5] There are many good works on the Victorian ideology of femininity, see W. E. Houghton, *The Victorian Frame of Mind 1830–1870* (1957); C. Christ, 'Victorian Masculinity and the Angel in the House', in M. Vicinus (ed.), *A Widening Sphere* (1977), 146–62; O. Banks, *Faces of Feminism* (1981), especially ch. 6, 'The Ideal of Feminine Superiority'; D. Gorham, *The Victorian Girl and the Feminine Ideal* (1982); S. K. Kent, *Sex and Suffrage in Britain 1860–1914* (1987), especially ch. 1, 'The Sex'.

[6] Basch, *Relative Creatures*, pt. 1.

[7] Houghton, *Victorian Frame of Mind*, 341–8; A. Wohl (ed.), *The Victorian Family* (1978).

and economic transformations brought about by industrialization provoked severe anxieties about the security of all that
the newly emergent middle classes valued. The demands of
working-class political and trade union movements challenged
political stability; the disintegrating effects of urbanization
undermined hierarchical social relations; and the apparent
irreligion of the urban poor threatened the position of the
Church.[8] Growing state intervention concerning the condition
of the working-class family was not, then, simply the product of
humanitarian impulse. Concern about housing, sanitation,
and education all reflected a desire to bolster the family as a
source of stability in an otherwise fast-changing world.[9] As we
shall see, middle-class reformers' obsession with the condition
of the urban slum arose less out of alarm about its insalubrity
or the dangers of disease than from their perception of it as a
breeding ground of disorder. The importance of the family lay
in its ability to lessen the demoralizing effects of overcrowding
and, above all, to police its own members. For example, it
mitigated the worst effects of close cohabitation by trying to
keep unmarried adults segregated from one another, or at least
in separate beds. Even more importantly, it raised children in
as near an approximation to innocence as conditions allowed.[10]
Similarly, men pulled into competing in the market-place, and
seeing themselves as contaminated by the demands of business
and corrupted by their own self-interest, aggression, and
greed, held up women as the epitome and enforcer of the values
they missed in themselves.

Alongside the State, the Church also regarded the family as
both a source of moral values and a means of protecting them.
During the 1830s and 1840s a particularly important boost to
the position of the family, and of woman's place within it, came

---

[8] S. Delamont, 'The Domestic Ideology and Women's Education', in S.
Delamont and L. Duffin (eds.), *The Nineteenth Century Woman* (1978), 165.

[9] B. Harrison, 'State Intervention and Moral Reform', in P. Hollis, *Pressure from
Without* (1974); J. Lewis, 'The Working-class Wife and Mother and State Intervention
1870–1918', in J. Lewis (ed.), *Labour and Love* (1986).

[10] Though, as Jeffrey Weeks points out, the family was also of practical importance
to the working classes as a self-supporting 'mutual aid society'. The success of the
family is seen by Weeks not as passive working-class acceptance of a middle-class
model but as the product of 'negotiated redefinition'. J. Weeks, *Sex, Politics and Society*
(1981), 75.

from Evangelicalism.[11] Evangelicals were a powerful reformist group within the Church of England, centred around a collection of influential figures known as the Clapham Sect (a group which included members of the Thornton family based at Henry Thornton's house at Clapham and also figured such powerful individuals as William Wilberforce, Zachary Macaulay, James Stephen, and Hannah Moore). They were charged with a profound sense of the universality of sin, against which all must constantly struggle. They sought to ensure that religious observance pervaded every aspect of daily life by requiring continual attention to the state of one's soul.[12] Given the temptations and demands of the outside world, the domestic sphere became the one arena where the fight against sin could be hoped to prevail. As the poet William Cowper insisted:

> Domestic happiness, thou only bliss
> Of paradise that has survived the Fall[13]

Women were of central importance to the successful preservation of the family as a haven of morality. However, maintaining this role involved several contradictions. In order to act as guardians of the domestic sphere they were defined as occupying an elevated position of moral superiority. In this respect their role was considered to be absolutely vital, as the Revd Horsley's fervent declaration signified: 'Woe to that country in which men are not able to consider women as living lives on the whole more sober, righteous and godly than their own!'[14] Yet, although this superiority was supposed to be innate, it was generally thought that it could only be maintained by careful attention to exacting standards of propriety. Within Evangelical morality women were regarded as being naturally different from men. Said to be more fragile and more impressionable, their superior moral standards had to be continually reinforced. Moreover, although women gained considerable power within the limited sphere of their influence, in order to protect their own purity they were admonished to

[11] See C. Hall, 'The Early Formation of Domestic Ideology', in S. Burman and B. Harrell-Bond (eds.), *Fit Work for Women* (1979), 15–33.
[12] See B. Hilton, *The Age of Atonement* (1988), ch. 1, 'Evangelicalism in the Age of Atonement'.
[13] W. Cowper, *The Task*, quoted in Hall, 'Early Formation', 23.
[14] J. W. Horsley, *Jottings from Jail* (1887), 62.

leave the home as little as possible. Powerful in one sphere only, they were to be cloistered away from the potentially corrupting influences of the wider world.

Amongst the upper classes the moral code of the lady and the attributes of the ideal woman were lengthily elaborated. The means to attaining them became the subject of a mass of literature by writers such as Sarah Ellis, Elizabeth Sandford, Mary Anne Stodart, and Louisa Tuthill.[15] Advice books, etiquette manuals, and didactic fiction aimed primarily at the middle-class girl carefully detailed the virtues she was supposed to possess. Submissive, innocent, pure, gentle, self-sacrificing, patient, sensible, gentle, modest, quiet, and altruistic, the middle-class woman was to have no ambition other than to please others and care for her family. Having minutely catalogued her 'innate' qualities, this literature went on to suggest how she ought to conduct herself at every stage of her life and gave awful warnings of the dangers of deviating from the prescribed path.[16] Her friendships were intended to foster feminine qualities of empathy and generosity within the safe confines of the domestic environment. Later in life woman was to find fulfilment in marriage: establishing her own home, caring for her husband, and bringing up children. In the middle classes, such ideals were generally upheld by mothers, sisters, and friends, who were only too quick to condemn any contravention.

How far these prescriptions constrained the lives and behaviour of working-class women is less easy to ascertain. Although they did not have the time, inclination, or, in many cases, the ability to digest manuals on etiquette or household management, working-class women were not immune from the influence of their teachings. The canons of feminine conduct were soundly impressed upon young girls in elementary school, in

[15] I am grateful to Anne Gelling, Nuffield College, Oxford, for her unpublished paper 'The Didactic Moralists and the Ideology of Femininity' (1984). See also C. Dyhouse, 'Mothers and Daughters in the Middle-Class Home *c*.1870–1914', in J. Lewis (ed.), *Labour and Love* (1986).

[16] See, for example, E. Sandford, *Women in her Social and Domestic Character* (1831), S. Ellis, *The Women of England* (n.d.), *The Daughters of England* (1842), *Mothers of Great Men* (1861); and periodicals such as *British Mothers' Journal* and *The British Mother's Magazine*. A much fuller list is provided in B. Kanner, 'The Women of England in a Century of Social Change 1815–1914. A Select Bibliography', in M. Vicinus (ed.), *Suffer and Be Still* (1980), 178–82.

Church, and through the informal but nonetheless effective
channels of working-class morality. Judith Walkowitz has
traced the 'hierarchical female networks' which operated
amongst the urban poor to articulate and enforce social and
sexual norms.[17] Mothers socialized their daughters into fatalistic but secure dependence on men and taught them to shun
'bad women' at all cost. The enduring importance of such networks is testified to by the respondents to Elizabeth Roberts's
oral history of women between 1890 and 1940. Many of them
admitted to going to great lengths to keep up the appearances
of respectability, not least by 'keeping a distance' from their
neighbours and refraining from 'gossiping' on the doorstep.[18]

Such standards relied on a myriad of informal, but nonetheless severe, sanctions against their contravention. Community
self-policing and the way women, in particular, observed,
discussed, and condemned their neighbours' behaviour operated as powerful constraints against deviance. Of the slums of
Edwardian Salford, Robert Roberts remembered: 'Over our
community the matriarchs stood guardians, but not creators,
of the group conscience and as such possessed a sense of social
propriety as developed and unerring as any clique of Edwardian
dowagers.'[19] Unlike the police, they did not merely patrol
behaviour in the public sphere but extended into the home to
condemn the woman who kept a slovenly house, or neglected
her children, or became pregnant outside marriage. Elizabeth
Roberts interviewed many women who themselves experienced
the power of the Lancashire matriarchs; one respondent observed the plight of an unmarried mother—'[they] would talk
about it and she would probably know and be ashamed. It was
the shame that was the worst part.'[20]

Conformity to the idealized norms of femininity, at least so
far as circumstance permitted, was induced less by persuasion
than by veiled threat.[21] To avoid the fatal stigma of ostracism,
the working-class woman was expected to adhere rigorously to
certain standards of social conduct, if not to the minutiae of

[17] J. Walkowitz, 'Male Vice and Feminist Virtue', *History Workshop*, 13 (Spring 1982), 86.
[18] E. Roberts, *A Woman's Place: An Oral History of Working Class Women 1890-1940* (1984), 14-15 and 196.   [19] R. Roberts, *The Classic Slum* (1971), 26.
[20] Roberts, *A Woman's Place*, 79.
[21] On the use of negative images as warnings to middle-class girls, see Gorham, *Victorian Girl*, 37 and 52-6.

middle-class dress and parlour talk. Since in social standing she was so much closer to this social precipice, she needed to strive that much harder to maintain 'respectability'. Steven Marcus comments on the huge self-restraint factory girls needed to resist the temptation to earn from a few minutes' prostitution five times what they would normally earn from a whole day's drudgery.[22]

If attaining the ideal of femininity was difficult for the middle-class woman carefully closeted within the home, it was often impossible for poorer women.[23] Yet even they generally adopted the ideals of domesticity despite their patent inability to fulfil them. This dichotomy between the realities of their lives and the ideals to which they aspired created tensions in working-class families seeking to attain respectability, and inflicted guilt on women for whom work was a financial necessity. The working-class woman had to adhere, far more closely than her male counterpart, to an elaborate schedule of prescribed behaviour if she was to avoid the sanctions of public opinion. In turn the working-class wife and mother was charged with ensuring the good behaviour of her menfolk and offspring. The responsibilities weighed by these dual roles necessarily demanded that she be *seen to be* morally superior. A range of subtle measures—moral pressure, prescriptive teachings, religious sanctions—were employed to elevate women to this evidently artificial ideal. One might argue that, since women were already in a far weaker position than men, there was less need to employ the coercive agents of police and prison to secure their conformity. So pervasive was this ideal of the 'angel in the house', the little wife and caring mother, that only the very lowest stratum of society remained completely immune from its influence.

Although the ideals of femininity filtered down the social scale, the working-class woman had only limited powers to ensure the moral propriety of her family. Restricted in her movements and denied access to economic or political power, she wielded her beneficial influence mainly by her own example. Simply by being a model of chastity, altruism, and

---

[22] S. Marcus, *The Other Victorians* (1966), 145.
[23] See L. Jamieson, 'Limited Resources and Limiting Constraints', and E. Ross, 'Labour and Love: Rediscovering London's Working Class Mothers 1870–1918', both in Lewis (ed.), *Labour and Love*.

morality, she was supposed to induce men to raise themselves to her level of virtue.[24] Her beneficial influence was probably most important in the agonized sphere of sexuality. The heyday of what Frank Mort has termed 'moral environmentalism' in the 1830s and 1840s focused attention on sexuality as the key to working-class immorality.[25] The Victorians, whilst recognizing the potency of man's sexual appetite, sought to contain it by condemning its indulgence, and seeking to promote chastity. This preoccupation is clearly illustrated in the mass of literature on the dangers of sexual excess, and the debasement and debilitation it was believed to produce. As a result sexuality was suppressed and, where not suppressed, concealed and denied. Maintaining woman's supposed innocence demanded man's restraint before marriage, and woman's supposed lack of sexual appetite demanded the suppression of his own desires thereafter. Quite simply, the high price placed on female chastity was intended to limit male sexuality to procreation within marriage.

Notions of respectability furnished the working-classes with a code of accepted conduct that placed many restrictions on everyday behaviour. It added to the legal code numerous proscriptions and taboos. It would be wrong, however, to see respectability simply as a tool of middle-class repression. For in many ways it was also a positive moral force which the urban poor willingly strove to maintain. As many nineteenth-century feminists recognized, such standards could actually protect the factory girl or domestic servant from sexual exploitation; they helped to guard against incest in overcrowded homes, and kept men from squandering the family's income on gambling or alcohol.[26]

THE EXTENT OF FEMALE CRIMINALITY

We turn now from considering notions of femininity to examine those women who deviated from this ideal by committing

---

[24] Anon., 'Mental and Moral Dignity', 6–7.

[25] F. Mort, *Dangerous Sexualities* (1987), 11–63. For later changes in attitudes to sexuality, see P. Cominos, 'Late Victorian Sexual Respectability and the Social System', *International Review of Social History*, 8 (1963).

[26] See C. Rover, *Love, Morals and the Feminists* (1970); Kent, *Sex and Suffrage*, 36.

crimes. Given that, in every respect, criminal women represented the very negation of the ideal of femininity, assessments
of the extent and seriousness of female criminality are unlikely
to have been wholly dispassionate. Take, for example, the
completely unsubstantiated claims made by one observer at an
annual meeting of the Social Science Association:

One third of the convicts of the kingdom are women, but that is a
shallow calculation. Women are more often the accomplices of crime,
its aiders and abettors, than its actual perpetrators. Then also they
are the victims of crimes, and the seducers to crimes, which do not
come within the power of the law, while inflicting the deadliest
wounds on society; and over and above their own lives of crime, they
become the mothers of criminals.[27]

I shall go on to examine how far such reactions to female
crime simply mirrored the wider value structure, were driven
by stereotypical views of women, or were actually informed by
firsthand knowledge. But in order to gauge the relationship
between contemporary assessments and the reality of women's
involvement in crime, we first need to try to establish, as a
bench-mark at least, the amount of recorded crime perpetrated
by both sexes. Official figures were collected and published
annually under the heading 'Judicial Statistics'. These figures
are of extremely questionable reliability, not least given inconsistencies in the recording and enumeration of that crime
which actually came to the attention of the police (see the note
on the Judicial Statistics in the Appendix). However, they may
be put to some limited use to give a rough indication of trends
over time in reported crime. Major changes in the enumeration
of these statistics in 1857 and again in 1893 make comparisons
across these dates highly problematic and for this reason I will
confine my analysis to statistics for the intervening period.

Attempting to establish just how much crime men or women
committed is probably impossible. Nonetheless, the 'Judicial
Statistics' provide a table confidently headed 'The Criminal
Classes'. At best these offer an indication of the numbers of
those who were known to the police to have, at some time, been
involved in various broad categories of crime. Obviously they

---

[27] I. Craig, 'Emigration as a Preventative Agency', *The English Woman's Journal*
(Jan. 1859), 291. From an account of the meeting.

tell little of the level of crime being committed. These figures are divided by sex (and, up to 1893, by age). Ostensibly, they refer to the numbers of criminals 'at large' but, of course, actually comprise only those who came to the attention of the police. The numbers may, therefore, be distorted downwards by the fact that these figures mostly refer only to those with at least one previous conviction, so excluding those known to be involved in crime but not yet convicted. Or they may be distorted upwards by the fact that an individual may have been known to the police on the basis of a single conviction and have remained on the register thereafter but was not, for example, a confirmed thief (see Appendix, Table 1a). These figures should not be seen, in any sense, as a straightforward enumeration of the population involved in crime. First, the Victorian concept of the 'criminal classes' was founded on highly emotive notions of a predatory and dangerous portion of the population who not merely committed criminal acts but who were, by definition, vicious in character. Such perceptions are unlikely to have been conducive to careful, conservative estimates. Secondly, the exact method by which these figures were derived is not made clear but seems to have been based on assessments made by the police of known (and, after 1893, 'habitual') criminals. These designations were the product of subjective judgement and liable, therefore, to differ widely by police district according to the attitude of the officer charged with completing the returns. And the likelihood of inconsistency is compounded by the fact that there appears never to have been a Home Office circular advising individual police forces as to the exact scope or definition of each of the categories.

Bearing in mind these various caveats, what can we say of the 'criminal classes' and of women's part within them? Overall there was a considerable decline in those designated as the 'criminal classes'. Over the period 1860–90 they fell by more than a half. The number of women fell at roughly the same rate as men, remaining at around a fifth of the total in this category over the period. The relatively low percentage of women was, in part, a distortion produced by a decision in the late 1860s to exclude from the total figures the large numbers of prostitutes, male and female vagrants and tramps.[28] (When

[28] In order to ensure comparability over time I have excluded figures for prostitutes, vagrants, and tramps altogether.

these figures were still included, women were said to represent more than 40 per cent of the total known 'criminal' population. The exclusion of prostitutes, vagrants, and tramps had the effect of reducing the female population of the 'criminal classes' from 54,703 in 1860 to 11,445 in 1870. This disparity reveals how a seemingly minor alteration to category headings, or to the scope of data included in official statistics, can radically alter the resultant totals.) The remaining categories, recorded over the whole period, were: 'known thieves and depredators, 'receivers of stolen goods', and 'suspected persons'. The first two categories seem to offer a rather surer guide to the actual number of criminals, in so far as they refer to people known to have committed specific offences. Significantly, women had a slightly higher showing in these two categories than in the much vaguer category of 'suspected persons'. The latter category relied largely on subjective assessment and, in so far as any reliance can be given to it at all, probably reflects a tendency not to suspect women of crime. After 1893 the criminal classes table was redesignated 'Numbers of Suspected Persons at Large' (Appendix, Table 1*b*); the total numbers of known criminals were phased out, and instead figures were given only for the much smaller core group of those 'habitually engaged in crime'. In this last category women continued to figure in much the same proportions as in the 'criminal classes', that is around a fifth of the estimated total population.

The figures for the 'Criminal Classes' claimed to pertain to people still 'at large'. Statistics were also collected relating to the 'Character of Persons Proceeded Against', both summarily and by indictment (Appendix, Table 2). This was also intended to determine and describe the so-called criminal classes, but again was conducted largely on the basis of subjective and probably widely varying judgements. As the first report after the reorganization of the criminal statistics in 1857 admitted: 'What proportion these large numbers bear to the whole class which they represent, there are . . . little means of determining.'[29] Here again the headings—known thieves, vagrants and tramps, and suspicious characters—were used to categorize offenders. Unsurprisingly, men and women fell in very similar proportions to those given for the criminal classes at large. Of more interest are the headings: 'Prostitutes', provided

---

[29] 'Judicial Statistics', *Parliamentary Papers* (hereafter *PP*), 57 (1857–8), 392.

throughout the entire period; 'Habitual drunkard', introduced in the last quarter of the century; and 'Previous good character', giving details of those not previously known to the police. Though, of course, due to lack of standardization, inefficiency of record keeping, of tracking etc., these categories are also likely to have been very inaccurate before the turn of the century.

For most of the second half of the nineteenth century the largest single group of female offenders already known to the police were prostitutes. However, they fell considerably, if unevenly, from 28 per cent of all women proceeded against in 1857 to 12 per cent by 1890. This decline seems to belie the commonly voiced Victorian belief that all criminal women were 'immoral'. It may be, however, that this decline was produced by changes in recording rather than an actual fall in levels of prostitution. For example, it may be explained partly by the introduction of the new heading 'Habitual drunkard', to which many prostitutes may then have been more accurately ascribed. Women consistently made up over a quarter of those categorized as drunkards.

A far smaller proportion of women than men were said to be of previous good character. And although numbers of men and women in this category increased over the period, women remained, consistently, only about 12 per cent of the total. In 1857 36 per cent of men but only 17 per cent of women had no criminal record. The numbers of those without any previous criminal history steadily increased so that by 1890 53 per cent of men and 35 per cent of women coming before the courts were designated as first offenders. The smaller proportion of women in this class may reflect the tendency to see women as non-criminal. If a female offender had no previous criminal record, efforts may have been made to keep her out of the criminal justice system. Conversely, after women had been convicted once, the greater stigma attached to criminal women made them particularly liable to find themselves again in confrontation with the law. In the main, then, those women coming before the courts were already labelled as criminals.

Towards the end of the nineteenth century, observers discerned a general marked decline in female criminality. In 1894 Arthur Griffiths, Deputy Governor of Millbank Convict Prison and subsequently Inspector of Prisons, claimed of women criminals: 'the sources of supply are running dry.'[30] A number of articles in women's papers such as *The Englishwoman's Review* made similar claims.[31] In fact, their observations were based not on the total crime figures but, curiously, were extrapolated from the marked decline in the number of women convicted of more serious offences and sentenced to penal servitude. The suggestion that women were becoming less criminally inclined was more soundly based on the observation that the much higher rates of recidivism amongst women than men meant that the criminal statistics referred to a proportionally smaller number of women, each repeatedly reconvicted.[32]

EXPLAINING WHY WOMEN COMMITTED LESS CRIME

It is, perhaps, unsurprising that the Victorians had little difficulty in explaining the relatively low rate of female crime. Since they expected women to be morally superior to men, they expected their social conduct to be better too. Thought to be more honest than men, women were said to be less likely to commit perjury; thought to be less acquisitive and competitive, they were seen as less liable to steal or cheat; being naturally chaste, they would be less liable to commit sexual crimes. In short, being naturally good and sociable, women were thought not to be drawn to the evil and anti-social activities of criminality. The fact that women were more religiously inclined than men was also thought to act as a restraint against crime.

This view of most women as innately non-criminal was strongly propounded throughout the period, but arguments and explanations based upon it grew considerably towards the end of the century as commentators sought to explain the apparent decline in female crime. A typical example is Havelock Ellis, an influential writer concerned with criminal anthropology, who argued: 'there can scarcely be any doubt that the

---

[30] A. Griffiths, *Secrets of the Prison-House* (1894), ii. 4.

[31] See, for example, Anon., 'Women Convicts', *The Englishman's Review* (Oct. 1887), 473, and A.M.M., 'The Decrease of Crime', *The Englishwoman's Review* (Feb. 1890), 61–3.

[32] As recognized, for example, by C. E. B. Russell, 'Some Aspects of Female Criminality and its Treatment', *The Englishwoman* (Jan. 1912), 35–7.

criminal and anti-social impulse is less strong in women than men . . . Not only are women by their maternal function more organically tied to the social relations of life, but their affectability renders an anti-social and unusual course of life more organically difficult.' [33]

Commenting on the small number of women in prison, Sarah Amos went even further in praising the innate goodness which kept women from crime despite what she considered to be greater temptations for them to commit certain types of crime than men. She argued: 'The very goodness of women, their law abidingness, withdraws them from notice, and this in spite of the fact that women have more laws to obey than men have, while the temptation to disobey some laws that are equally encumbent on both sexes is, by consent and necessity, greater to women than to men.' [34]

In addition to her innate moral superiority, woman's primary function in life—bearing and bringing up babies—was believed to foster qualities of altruism, compassion, and a feeling of self-fulfilment, all of which mitigated against any tendency towards crime. Revd W. D. Morrison, a long-serving prison chaplain and penal reformer, neatly encapsulates such an assumption in *Crime and its Causes* (1891):

The care and nurture of children has been their lot in life for untold centuries; the duties of maternity have perpetually kept alive a certain number of unselfish instincts; these instincts have become part and parcel of woman's natural inheritance, and as a result of possessing them to a larger extent than man, she is less disposed to crime. [35]

More cynical late Victorian writers argued that woman's lack of physical strength and her natural passivity kept her from criminal activity. Such physical hindrances were said to prevent women from committing murder, assaults upon the person, or robbery with violence. Even Morrison accepted the view that: 'Undoubtedly the lack of physical power has as much to do with keeping down female crime as the want of will . . . where

[33] H. Ellis, *Man and Woman* (1894).

[34] S. Amos, 'The Prison Treatment of Women', *Contemporary Review* (June 1898), 803.

[35] W. D. Morrison, *Crime and its Causes* (1891), 152. See also V. Harris, 'The Female Prisoner', *The Nineteenth Century and After* (May 1907), 783.

the temptation is strong and the power sufficient, women are just as criminally inclined as men.'[36]

Although Victorian assumptions about female physical frailty provided a plausible basis for explaining the low level of female crime, sociological analyses were more common still. These recognized that women were not necessarily innately superior but that their apparently stronger sense of morality was largely dictated by their social situation. For example, the American sociologist Frances Kellor argued, 'the capacity for good or evil, which in general distinguishes the sexes, finds its explanation not so much in sex as it exists, but in the influences and conditions which have determined these characteristics.'[37] Confined for much of their time to the home and to childcare, women were simply denied the opportunity to commit many forms of crime. Their mobility was restricted, they spent little time in the public sphere, and they tended not to congregate in pubs or on the streets, as men did, enabling them to devise and commit crime. Those acts of deviance women did commit often took place in the privacy of their own homes and so rarely reached the public scrutiny of the criminal justice system. As long as women remained little involved in the struggle for existence, they would not, argued Kellor, be put under the same pressures as men to drink, gamble, or to compete in commerce—all of which encouraged criminality. In explaining the low level of female crime by reference to social circumstance, Kellor unwittingly supported the fear that, as women's position in society changed, as they moved more and more into the public sphere, their crime rate would necessarily increase. This view was strongly argued by many opponents to women's movement outside the home. Take, for example, the views expressed by the prison inspector, Vernon Harris:

There seems to be no doubt that women's usual environment shelters them from many and various evil tendencies which are the undoing of men. The conservative disposition of women is remarked the world over, and, being more domestic and sedentary in their habits, they less exposed than men to the varying and disturbing influence of

---

[36] Morrison, *Crime and its Causes*, 153. In fact, assault formed a higher proportion of female crime than it did of male and, even more strikingly, women made up 40 per cent of those tried for murder.

[37] F. Kellor, *Experimental Sociology* (1901), 162.

ordinary life . . . It is probable that as women come more to live with men in their occupations, in their struggles for existence, in their independence, and the like, they will suffer as men do, and exhibit similar signs of degeneracy and an equal tendency to criminality.[38]

Such arguments clearly resonate with those advanced in more recent years concerning the supposed rise of the 'new female criminal' as a corollary to the second wave of the women's movement in the 1960s and 1970s.[39]

A final, less readily verifiable explanation of the low showing of women within official crime statistics related to their treatment within the criminal justice system. As today, there was an implicit assumption, common to much nineteenth-century criminological literature, that women were kept out of prison as much by men's chivalry as by their own virtue. This suggested that women were less vigorously pursued than men, that judgements made upon those who were prosecuted were less severe, and that punishments meted out were less harsh. Through male sympathy, women's already low rate of criminality was thought to be artificially lowered yet further. For example, Harris argued, 'It must . . . be remembered that women are punished more lightly than men for a similar offence, that the public and police are less disposed to charge them.'[40]

The Judicial Statistics, pertaining to both summary and indictable offences, do provide reasonably reliable figures showing that a consistently higher proportion of women than men were discharged by the courts. Even so, for both sexes, proportions of those discharged declined over the second half of the century. For summary offences, discharges fell from 47 per cent of women and 34 per cent of men in 1857 to 23 per cent and 17 per cent respectively in 1890 (Appendix, Table 3*a*). Similarly for indictable offences, discharges fell from 55 per cent of women and 38 per cent of men in 1857 to 26 per cent of women and 18 per cent of men in 1890 (Appendix, Table 4*c*). One might explain the disparity purely by reference to notions of male chivalry which suggest that women were more likely to be let off, regardless of their actual culpability. However, this

---

[38] Harris, 'Female Prisoner', 783.

[39] Amongst others by F. Adler, *Sisters in Crime* (1975).

[40] Harris, 'Female Prisoner', 782.

begs the question of whether women were more or less likely to be proceeded against in the first place. Moreover, such an explanation sits uneasily beside the sorts of attitudes generally expressed about women who were thought to be criminal. Whilst chivalry may have operated in the favour of first offenders, it does not provide a plausible explanation in relation to women with previous convictions, towards whom attitudes were generally much less sympathetic.

The general decline in the number of discharges of both sexes is no more easily explained. It may be at least partially accounted for by the increasing efficiency of policing. This raised the chances that the correct suspect was apprehended and that sufficient evidence was gathered to ensure his or her conviction. Also new alternatives to imprisonment made possible by the extension of summary jurisdiction to a range of formerly indictable offences (so reducing the gravity of the charge and consequently the resultant tariff) may have made juries more willing to convict than before.

For most of the nineteenth century, criminologists accounted for the low level of female crime by referring to woman's supposed moral superiority. How far encouraging women to aspire to lofty standards of propriety and labelling them as non-criminal curbed any tendency to commit offences we can only guess. Similarly the assumption that women were not criminally inclined may, indeed, have made them less likely to be suspected of crimes or prosecuted even when they were actually under suspicion. The low level of recorded female crime may, therefore, be at least partly attributable to the pervasive designation of women as non-criminal.

DEVIANCE FROM FEMININITY

The ideology of femininity, so important in defining women's ideal conduct, was no less important in responding to their worst. The gravity of female criminality was measured primarily by their failure to live up to the requirements of this model. For example, Thomas Holmes, Secretary to the Howard Association (the prison reform society that preceded the Howard League), in *Known to the Police* (1908) recounted the case of a wife prosecuted for inebriety. He described her downfall not in terms of immorality but by reference to the woman's loss of her natural vocation: 'She had been a good wife and mother till late in life. Then her children had all dispersed, and great loneliness came upon her.'[41] Deprived of her maternal role she turned in desperation to alcohol. The proof of her inebriety was found not in the amount drunk but in her failure to fulfil her remaining wifely duties, as Holmes saw it: 'It was the old, old story of drink, neglect, waste, and dirt—no food provided, no house made tidy, no beds made, no washing of clothes.' An article published in the *Cornhill Magazine* (1866) made similar observations about the relationship between women's failure to keep up appropriate feminine standards of appearance and their criminality: 'those awfully wretched-looking creatures that lounge about or squat down at the entrance of the courts with dirty faces, hair uncombed, a kerchief tied over the half-exposed bosom . . . When a woman gets to be utterly careless of her personal appearance—personal cleanliness—you may be sure that she is careful for nothing else that is good.'[42] Such descriptions illustrate the tendency to assess female crime not according to the act committed or to the damage done but according to how far a woman's behaviour contravened the norms of femininity. Deviance from femininity alone, then, was grounds for suspicion and condemnation.

Victorian writers on female crime frequently abandoned objective assessment in favour of emotional outbursts and moral censure. The tone of any given author depended mainly on his standards of female propriety and his tolerance of those unwilling or unable to conform to them. Note, for example, the description of criminal women by the well-known journalist and social investigator Henry Mayhew, and his perceptive commentary on his own reactions: 'in them one sees the most hideous picture of all human weakness and depravity—a picture the more striking because exhibiting the coarsest and rudest moral features in connection with a being whom we are apt to regard as the most graceful and gentle form of humanity.'[43]

[41] T. Holmes, *Known to the Police* (1908), 60–3.

[42] M. E. Owen, 'Criminal Women', *Cornhill Magazine*, 14 (1866), 155.

[43] H. Mayhew and J. Binney, *The Criminal Prisons of London and Scenes of Prison Life* (1862), 464.

For the most part, male crime was lucidly examined and explained with reference to economic need or to motives which, if morally unacceptable, were nonetheless recognized as rational. Analyses of female crime, however, all too often lapsed into outraged condemnation. A good illustration of this latter tendency is provided by the heated discussions provoked by the subject of infanticide at the annual meeting of the Social Science Association in 1866.[44] Here a sharp division arose between those who considered that the woman who committed infanticide was an 'object of peculiar compassion and sympathy'[45] and those who felt that she represented the very antithesis of womanhood. The former group argued that it was 'the most modest women who were tempted to commit infanticide'.[46] They stressed that such women were motivated by the feminine quality of shame and the natural desire to save their child from a life of misery and themselves from possible further degradation by the need to turn to prostitution to survive. Less sympathetic participants, on the other hand, ascribed infanticide to a lack of chastity and characterized offenders as callous single mothers concerned only 'to get rid of an encumbrance'. Sir Thomas Chambers, a prominent Liberal lawyer, speaking for this latter group, warned of the dangers of treating infanticidal mothers sympathetically, arguing: 'The motives to female virtue should be re-enforced, instead of being weakened, and nothing but harm could ensue from mitigating the shame, inconvenience, and hardship of unchastity.'[47]

In these sets of arguments the woman was portrayed as representing the embodiment or the antithesis of femininity but, significantly, not as a murderess. A paper given by A. Herbert Safford at the same meeting provides a clue to this curiously slanted analysis. Having idealized women as non-criminal, it seems that men found it extremely difficult to conceive of them as criminal at all, let alone as a child murderer: 'That a mother should be capable of killing her infant is a fact that even the strong intellect of man cannot compass, and we consequently rarely find a jury that returns a verdict of wilful

---

[44] *Transactions of the National Association for the Promotion of Social Science (NAPSS) 1866*: discussion on 'What are the best means of preventing infanticide?', see especially 293–4. [45] Sir J. Eardley Wilmot, MP, ibid. 294.
[46] Mr C. H. Bracebridge, ibid. 293. [47] Sir Thomas Chambers, ibid. 294.

murder against a woman so accused.'[48] The central issue,
responsibility for the death of the child, is conspicuously underplayed in all these arguments.[49] For concern lay less with the
wrong done than with the moral circumstances of its commission. Condemnation of the infanticidal woman rested less on
the fact of murder than on her status: for example, whether she
was an unmarried mother or not. Similarly, appreciation of her
position, whether as the unprotected victim of seduction or
destitute of the means to bring up her child, allowed the more
sympathetically inclined to exonerate her act of murder. The
case of infanticide seems to be a particularly useful illustration
of a more general tendency to moralize about, rather than to
seek the substantive causes of, female crime.

For the most part women criminals were viewed not so much
as economically damaging, physically dangerous, or destructive
to property but as a moral menace. Consequently, less attention was paid to the costs of female theft or the harm done, for
example, by cruelty to children, than to the female offender's
moral condition. Descriptions of women's crime frequently
referred to past conduct, marital status, protestations of regret,
or shamelessness, and even to the woman's physical appearance. It would no doubt be instructive to study court cases to
ascertain how far this attitude affected judgements made about
women actually on trial, to establish to what extent courts
demanded information about female defendants' moral credentials, and how far they made judgements upon the basis of this,
strictly speaking irrelevant, information.

MID-NINETEENTH CENTURY CONCERN WITH
'CRIMES OF MORALITY'

This moralizing attitude was instrumental not only in responding to female crimes, but in actually designating what types
of crime were regarded as most serious when committed by
women. As the historian David Jones has noted, mid-nineteenth
century Britain witnessed a rising obsession with 'crimes of

---

[48] In a paper of the same title as the discussion, by A. Herbert Safford, ibid. 224.
[49] For further discussion of the question of responsibility in the case of infanticide,
see Ch. 7 below.

morality': sexual crimes, prostitution, drunkenness, vagrancy, and 'low' entertainments such as gambling and dancing.[50] The attention and publicity given to these largely behavioural, and generally victimless, activities was out of all proportion to their gravity. But to middle-class observers, the urban poor seemed to be turning their backs on the precepts of respectability, self-restraint and sobriety. Charitable organizations such as the Association Institution for Improving and Enforcing the Laws for the Protection of Women, set up by women 'to promote the purity and morality, the welfare and security of [their] own sex, and especially of the young, the friendless, and the exposed'[51] regarded women as being particularly at risk from this per-ceived decline in moral standards. Writers in *The Female's Friend*, the Association's journal, warned that the effects of moral degradation were much worse in women: 'she abandons herself to sensuality, drinks to drown her sense of shame, becomes unsexed in her manners, practises every vice for the sake of a living, and in her delirium of guilt and infamy spares neither men nor women.'[52]

Exposés of street life and the 'licentious' entertainments of the urban poor exacerbated such concerns. These in turn generated public scrutiny, policing, and in some cases the criminalization of deviant but hitherto not illegal behaviour (the campaign against betting houses is an example here). Growing publicity given to 'crimes of morality' in works such as Jelinger C. Symons, *Tactics for the Times* (1849), Henry Mayhew, *London Labour and London Poor* (1861–2), and James Greenwood, *The Seven Curses of London* (1869) may also have increased the seriousness with which such offences were re-garded by the courts, regardless of any actual increase in their commission.

This mid-century concern with crimes of morality became a veritable obsession in defining female criminality. J. C. Symons, a lawyer and educational expert, found the apparent rise in the proportion of female offenders during the 1840s not merely worrying but went so far as to suggest that it was indic-ative of the 'increasing demoralisation' of the country as a

---

[50] D. Jones, *Crime, Protest, Community and Police in Nineteenth Century Britain* (1982), 129.   [51] *The Female's Friend* (Jan. 1846), 2.
[52] Ibid. (Apr. 1846), 83.

whole.[53] Links were commonly made between female crime and sexual morality, not least because all female sexual activity outside the bounds of marriage was seen as an undesirable and particularly damaging form of deviance. This thinking had the effect of placing special emphasis on sexual misdemeanours committed by women. More curiously, it also tended to characterize all criminal women as sexually deviant, so that assessments of sexual conduct were used to measure the depth of their criminality. In the light of these concerns, the prostitute obviously attracted most attention for she epitomized what the historian Peter Cominos has labelled 'the tainted model'.[54]

In general, little distinction was made in early criminological literature on women between criminality and unchastity—the female thief and the prostitute tended to be grouped together as one. John Binney, one of Mayhew's collaborators on the *London Labour and London Poor* series, argued that 'the habitual crime of the female portion of the community is in most cases associated with prostitution',[55] though he later admitted that 'we found it impossible to draw an exact distinction between prostitution and prostitute thieves.'[56] In a subsequent study Mayhew himself worked from the assumption that 'it will, we believe, be found to be generally true that those counties in which the standard of female propriety is the lowest, or where the number of prostitutes is the greatest, there the criminality of the women is the greatest.'[57]

It was also commonly held that 'prostitution, besides being itself a vice, is also a fruitful cause of crime' and that 'prostitution is one of the special sources of crime as well as of vice.'[58] Opportunities for crime were seen to arise in the course of the prostitute's daily activities: petty theft from the person or burglary of their homes, assault or even murder of difficult clients, or the fencing of articles stolen by less honest ones. Prostitution as a motivational cause of female crime was explained with unwitting reference to the more pernicious effects of the moral code enforced against women. Mayhew, who

---

[53] J. Symons, *Tactics for the Times* (1849), 25.

[54] P. Cominos, 'Innocent *Femina Sensualis* in Unconscious Conflict', in Vicinus (ed.), *Suffer*, 166–8.   [55] J. Binney in Mayhew, *London Labour*, iv. 275.

[56] Ibid. 355.   [57] Mayhew and Binney, *Criminal Prisons*, 462.

[58] W. Tallack, *Penological and Preventative Principles* (1896), 45; Howard Association, *Annual Report* (1881), 19.

believed that women needed social approbation for their self-respect, claimed that prostitution necessarily led to crime, for having committed 'the one capital act of shamelessness', he argued, a woman gave up all right to respect, became reckless, and finally turned to crime.[59]

In mid-Victorian England standard works on female crime almost all document this downward progress from sexual experience to criminality. The prostitute whose original motive had been economic and who used her trade primarily for the purposes of petty theft was, nevertheless, seen as a sexual deviant without economic rationale.[60] Even when the sexual origin or content of a crime was not immediately apparent it still tended to be characterized as sexually motivated and made the subject of much pious moralizing. Such assumptions fitted into the wider view which proliferated in the decades around the middle of the century of crime as itself a 'moral disease'.[61] As we shall see, these views were to be gradually but almost completely superseded in the later nineteenth century by new interpretations of crime which derived from emerging schools of biological and psychological thought.

THE NATURE OF FEMALE CRIME

Clearly the moral role assigned to women in early Victorian England meant that concern about female deviance tended to concentrate on moral transgressions, and, as we have seen, particularly on sexual ones. How far did these notions of the nature of female deviance tally with the range of crimes for which women were actually proceeded against? For what types of offence were men and women convicted? What differences can one observe in patterns of crime by sex? And were there specifically feminine offences for which men were rarely or never convicted? To attempt to answer these questions we can only turn again to the annual publications of the 'Judicial Statistics'. As we have mentioned above, statistics given for both summary and indictable offences by sex provide only the

---

[59] Mayhew and Binney, *Criminal Prisons*, 456.

[60] Mayhew, *London Labour*, iv, section on prostitute thieves, 355–66.

[61] See, for example, E. Lettsom, 'What are the Principal Causes of Crime . . .?', *Transactions NAPSS* (1869), 324.

most general data about the actions of the courts and even
these figures are of questionable reliability, given the somewhat
haphazard fashion in which data were collected and returned.
In any case, such statistics obviously do not give a true indication of the frequency with which different types of offence were
actually committed. Their relation to the actual level of crime is
further attenuated by the filtering processes of detection,
arrest, and prosecution.

Summary Convictions

The only data available giving types of summary offence according to sex are for conviction (Appendix, Table 3*b*). Total
numbers by sex were given of those proceeded against and discharged but these were not divided by type of offence. Given
these limitations, this section will be confined to examining the
cases where conviction was obtained, the data for which are
only readily comparable between 1857 and 1892. The most we
can safely do is to establish an approximate idea of the types of
crime for which men and women were convicted and how these
changed over this period. Overall, women's crimes made up a
steady 17 per cent of all summary convictions. Strikingly, the
types of offences for which men and women were convicted
(apart from a number of sex-specific offences, most obviously
prostitution) were fairly similar. The largest single category of
offence for both sexes was drunkenness (or drunk and disorderly
behaviour). Over the period convictions for drunkenness grew
considerably from 19 per cent of all summary offences in 1857
to over 28 per cent in 1890. On average more than a fifth of
those convicted for drunkenness were women. A great many
more people were convicted in the years immediately after the
1872 Habitual Drunkards Act but this had only a short-term
impact: due to the lack of appropriate provision for treating
common drunkards the Act fell out of use and the figures
levelled out again. Drunkenness made up a consistently higher
proportion of all female convictions than it did of males and
showed a greater increase for women than for men over the
century (see Chapter 6 below).[62]

[62] Increasing from 22 per cent of all female convictions (19 per cent of male) in 1857
to 37 per cent (27 per cent of male) in 1890, with a peak in the years after the 1872 Act
(44 per cent of female convictions and 34 per cent of male in 1875).

The second most common summary conviction, for both sexes, was for 'common assault'. Women made up a fifth of the total convictions for this offence. Perhaps surprisingly, common assault made up a larger proportion of all female convictions than it did of male. Anecdotal evidence suggests that assaults committed by women were often drink-related, for example, brawls between women outside pubs, or assaults committed by prostitutes resisting arrest or seeking to defend their 'patch' from rival trade.[63] Fights between men may have been ignored by police as inoffensive 'manly' behaviour (unless they seriously threatened public order or the lives of those involved), but fights between women were much more likely to be deemed unseemly and stopped. Assaults were also often carried out by prostitutes on drunken clients, both to avoid giving services for which they had been paid and in order to rob them of all their valuables. Both sexes were convicted of proportionally fewer assaults as the period continued,[64] perhaps reflecting the increasing stability and respectability of late Victorian society. Convictions for prostitution (an offence against the Vagrancy Act) also fell by half over the second half of the century; though it may well be that many of those women prosecuted for drunkenness, for being drunk and disorderly, or for indecent behaviour were in fact prostitutes. Whatever the case, it does not seem that the proportion of female crime made up by prostitution was as high as Victorian literary sources would lead us to believe.

Larceny was the other single, large group of offences which brought both men and women before the courts. Any examination of trends under this heading is complicated by the considerable impact of the 1879 Summary Jurisdiction Act.[65] Up to 1879 women made up roughly a third of convictions for larceny under 5 shillings. This category of offence disappeared under the 1879 Act and the new, greatly enlarged category of 'simple larceny' (for all values of theft) was introduced. As a result women's share fell to little over a fifth of the total under this heading. Larceny (in all its forms treated summarily) never

[63] Jones, *Crime, Protest*, 105.

[64] Assaults showed a steady decline from 17 per cent of female offences and 13 per cent of male in 1857 to 10 per cent and 6 per cent respectively by 1890.

[65] See the note in the Appendix on the Judicial Statistics.

made up much more than 8 per cent of all summary convictions of either sex.

The mass of other offences treated summarily each formed only tiny proportions of the total convictions for each sex. However, many of those most commonly leading to convictions of both sexes may be seen as directly related to the conditions created by a poverty-stricken existence lived largely on the streets (offences such as begging, breaches of the peace, and offences against local Acts and by-laws). The only single category where convictions of women consistently outnumbered men was that of 'offences against the Pawnbroker's Act by persons unlawfully pledging or disposing'. This may be partly explained by the fact that charges were often brought when seamstresses and other sweated home-workers pledged the raw materials given them by their employers, in order to buy food, and then found themselves unable to redeem the goods in time to return them as finished goods or garments. The preponderance of women convicted of this offence tells us much of both the role of women in managing and attempting to eke out an inadequate household budget, and also of the place of petty crime in the economy of the urban female poor.

Indictable Offences

Unfortunately the Judicial Statistics do not provide comparable figures for convictions for indictable offences by sex. We are provided, however, with statistics, by sex, for apprehensions, discharge, bail, those committed for want of sureties, and those committed for trial (Appendix, Table 4*c*). Over the second half of the nineteenth century, women formed a declining proportion of those proceeded against by indictment (falling from 27 per cent of the total in 1857 to only 19 per cent by 1890). In attempting to provide data similar to that given above for summary convictions we can only turn to 'committals for trial' as the nearest available category. These are obviously not in any strict sense comparable with the figures for summary convictions since they give no indication of how many of those committed were subsequently found guilty. At best they are a very imperfect indicator of the types of indictable offence which brought men and women before the courts (Appendix, Tables 4*a* and 4*b*).

By far the majority of men and women, and a larger proportion of women than men, were tried for crimes coming under the general heading 'offences against property without violence'. The total for this class reached a high point in 1870, when 81 per cent of female committals and 68 per cent of male committals came under this heading, but declined towards the end of the century—by 1890 such property offences represented 70 per cent of female and 52 per cent of male committals on indictment. Within this class 'simple larceny' was the single most frequent category. It represented over a third of female commitments on indictment for most of the period and women made up over a fifth of all commitments for this offence. Other sorts of larceny also made up a large proportion of indictable crimes reaching court. 'Larceny from the person' represented 23 per cent of all serious female crime in 1857, though it declined towards the end of the century. Women formed a strikingly large, though declining, proportion of all those tried for this offence.[66] The most common form of larceny from the person committed by women seems to have been theft by prostitutes from their clients. The historian David Jones has studied prostitutes working in 'China', a slum area of Merthyr Tydfil, Wales, and has found that theft was a major part of their business. He argues that without it prostitution alone failed to provide a viable means of income.[67] 'Larceny by servants' also commonly brought women before the courts (women made up between 20 per cent and 30 per cent of all commitments)—though given the large numbers of women in domestic service it is perhaps surprising that they did not make up an even larger proportion.

Together various forms of larceny made up around 70 per cent of the offences for which women were tried on indictment, but only about half of male indictments. Two other crimes falling into the category of offences against property without violence also accounted for a significant proportion of women appearing on indictment. 'Receiving stolen goods' made up 5 per cent of all female committals on indictment in 1857,

[66] Women made up 46 per cent of the total in 1857, declining to 32 per cent by 1890.   [67] Jones, *Crime, Protest*, 107–8.

falling slightly to 3 per cent by 1890. Notably women made up
a third of all those tried for this crime around mid-century.[68]
That women apparently formed a relatively high proportion
of those acting as receivers indicates how constraints on
their mobility limited opportunities for female criminal activity
but did not prevent them from taking up the role of 'fence'
operating from home. Unlike most of the preceding offences,
committals in the class 'fraudulently obtaining goods by false
pretences and attempts to defraud' accounted for an increasing
proportion of female commitments on indictment, rising from
4 per cent in 1857 to twice that by the end of the century. This
apparent rise probably reflected women's increasing involvement in commercial and public life and, with it, greater
opportunities to commit fraud.

The second largest category of offences for which both men
and women were tried on indictment were 'offences against the
person'. These made up an increasing proportion of committals
by both sexes.[69] In the early part of the period, women formed
a striking 40 per cent of all those tried for murder. Since
murder is probably the least likely of all offences to go un-
detected, this figure raises interesting questions about the
amount of other serious female crime that was perhaps not so
readily detected and so was not always reflected in the official
crime statistics. In general, women were more likely to offend
in the private sphere and so were less likely to be detected or
suspected, and less likely to be committed for trial even if they
were. Murder was the single exception to these tendencies, and
perhaps gives a better indication of 'actual' levels of female
crime than any other offence. In the last decades of the nineteenth century, statistics for the murder of infants under one
year were separated out from the general murder statistics. As
a result the distribution of women in the statistics changed.
Less than a quarter of those tried for murder of victims over a
year old were women, but women made up nearly all those
tried in the new category of murder of an infant. This last
offence was distinguished from 'concealing the birth of
infants', under which many cases of suspected infanticide were

[68] Though again this fell to 24 per cent of the total by 1890.
[69] Offences against the person made up 8 per cent of female and 13 per cent of male
committals in 1857, rising to 13 per cent and 20 per cent respectively by 1890.

tried. Given the difficulty of establishing whether or not the baby was born dead, died during delivery, or was deliberately killed immediately afterwards, women were often found guilty only of failing to prepare responsibly for the impending birth of their child—obviously a much lesser charge than the capital offence of child murder. Not surprisingly nearly all those tried for 'concealment' of birth were women. The numbers of those tried for the other peculiarly female offence, 'attempts to procure miscarriage', remained very small, never much above a dozen a year. [70]

The next largest class of offences for which both sexes were tried was 'offences against property with violence'. This made up an increasing proportion of male offences, though a lesser proportion of female offences. [71] Of these, 'burglary and house-breaking' made up just 2–3 per cent of female committals. Other offences in this category formed very small proportions of female indictable crime and an insignificant part of the total offences against property with violence brought to trial. Those women who were charged were, by contemporary testimony at least, generally accomplices or 'look-outs' for male burglars. [72]

Offences in the category 'forgery and offences against the currency' made up roughly similar proportions of committals of both sexes—that is between 3 and 4 per cent. Most common in this category were those offences under the heading 'utter-ing, putting off, and having in possession Counterfeit Coin'. Perhaps surprisingly, women made up nearly a third of those tried for 'uttering' in the early part of the period (declining slightly to a quarter by 1890). Other currency offences seem to have been committed in far smaller numbers, but here again a large proportion of those coming to trial were women—around a third of those tried for forgery and up to 40 per cent of those tried for 'coining and having in possession implements for coining'. These were skilled crimes, requiring neither strength nor brutality, which women could pursue within the secrecy of their homes. Women's relatively high level of involvement in

[70] For a discussion of recourse to abortion, see P. Knight, 'Women and Abortion in Victorian and Edwardian England', *History Workshop*, 4 (Autumn 1977).

[71] Male committals rose from 13 per cent in 1857 to 18 per cent by 1890, female committals from 4 per cent in 1857 to just over 6 per cent by 1890.

[72] See, for example, J. Binney, 'Thieves and Swindlers' (and especially the section on 'Housebreakers and Burglars', 334–5), in Mayhew, *London Labour*, iv.

these activities where the usual hindrances did not apply seems rather to belie contemporary assertions about women's innate non-criminality.

There was, of course, a wide range of other offences for which men, and to a lesser extent women, were tried. However, each of these made up only a tiny proportion compared to those discussed above. The only other offence where committals of women consistently outnumbered those of men was that of 'keeping disorderly houses' or brothels, for which they were between 54 and 70 per cent of all those tried. Only in the last years of the nineteenth century did male pimps begin to outnumber 'madames'—in 1890 the proportion of those tried for this offence who were women fell to 40 per cent.

In the main, then, women tried on indictment were charged with financially motivated crimes. These were often planned, organized, and, in the case of currency offences at least, highly sophisticated ventures. Even taking into account their dubious reliability, such figures do seem to belie the widely held notion of female criminals as sexually motivated or driven by impulse to commit irrational, behavioural offences.

STIGMATIZING CRIMINAL WOMEN

Having examined exactly what types of crime brought women before the courts, we will next explore the sorts of responses their criminality provoked. We have seen how the idealization of femininity attempted to raise women to an ideal model of virtue. For negating this ideal, the criminal woman was condemned far more harshly than her male counterpart, since he was only seen to be enacting man's natural sense of adventure. Indeed, many attributes central to criminal activity were not far removed from Victorian notions of masculinity. Entrepreneurial drive, courage, physical vigour, and agility were all lauded characteristics which, though abused in the pursuit of crime, were thought to be appropriate male traits.

The criminal woman, on the other hand, offended against her very social role; her whole character repudiated the revered qualities of femininity. As a result, a range of negative imagery surrounded her. Just as the ideal woman was portrayed as the 'angel in the house'—the heart of a warm and loving home—so the criminal woman was condemned as the neglectful absentee from a cold and loveless one—if, indeed, she had anything that might be called a home. As the ideal woman was compassionate, self-sacrificing, and deferential, so the criminal woman was seen as hardened, self-seeking, and bold.[73] Above all, when the revered innocence of the ideal woman was discarded by her criminal counterpart, it was thought, as we have seen, to be replaced by sexual degradation. Typical is the casual remark by G. L. Chesterton, Governor of the House of Correction, Coldbath Fields, concerning one 'shameless and profligate woman' in his care, that she 'was one of those supremely vicious creatures who aspired to nothing apart from the most debasing sensuality'.[74]

It was commonly held that 'a bad woman is the worst of all creations.'[75] In the Judeo-Christian tradition the innate sinfulness of humanity could be especially dangerously exhibited in women. Thus Charlotte M. Yonge, the prominent High-Church author, asserting the greater sinfulness of women, recounted how, 'when the test came whether the two human beings would pay allegiance to God or to the Tempter, it was the woman who was first to fall, and to draw her husband into the same Transgression.'[76] Beneath the façade of the loving, honest woman there thus lurked a darker self, capable of all manner of evil. This was normally suppressed by rigid adherence to the tenets of femininity and the desire to win public approbation. However, if she lacked such self-discipline, woman would, it was thought, fall quickly to the very depths of depravity.[77] A good example of such thinking is provided by Henry Mayhew's assertion that criminal women lacked a sense of shame, 'so that the female criminal being left without any moral sense, as it were to govern and restrain the animal propensities of her nature, is really reduced to the same condition

[73] A view widely expressed in works as chronologically separate as Symonds, *Tactics*, 26, and H. Adam, *Woman and Crime* (1914), 3–4.

[74] G. L. Chesterton, *Revelations of Prison Life* (1856), i. 67.

[75] Harris, 'Female Prisoner', 783.

[76] Quoted in B. Heeney, *The Women's Movement in the Church of England 1850–1930* (1988), 7–8.

[77] See e.g. J. B. Thomson, 'The Psychology of Criminals', *Journal of Mental Science*, 17 (1870), 337–9.

as a brute, without the power to check her evil propensities.'[78] Moreover, the dichotomy between the high moral standards woman was expected to maintain and the paucity of moral powers she was thought to possess made the chances of her retrieval, once fallen, seem remote. The Directors of Millbank argued that: 'The gentler sex, as a whole, are superior in virtue to the sterner sex; but when woman falls, she seems to possess a capacity almost beyond man, for running into all that is evil . . .'[79]

The ferocity with which the fallen woman was condemned seems to suggest that, having repudiated every aspect of her assigned role, she was no longer considered to merit male respect. Perhaps it was also because such women blatantly demonstrated how very fragile the construct of femininity was that men responded by vilifying them. In this vilification, criminal women acquired a striking range of character defects. A few observers claimed to be so shocked by the traits they observed that they were loathe to describe them. Thomas Holmes (Secretary of the Howard Association) in discussing habitual criminals, declared himself especially horrified by criminal women: 'who dare plainly describe the female contingent of this dolorous army? I dare not!'[80] Mary Carpenter, not normally unsympathetic to the plight of women in prison, deplored the depths to which some women appeared to fall: 'The very susceptibility and tenderness of woman's nature render her more completely diseased in her whole nature when this is perverted to evil; and when a woman has thrown aside the virtuous restraints of society, and is enlisted on the side of evil, she is far more dangerous to society than the other sex.'[81] She saw criminal women as characterized by 'a very strong development of the passions and of the lower nature. Extreme excitability, violent and even frantic outbursts of passion, a duplicity and disregard of truth hardly conceivable in the better classes of society . . .'[82]

[78] Mayhew and Binney, *Criminal Prisons*, 467.
[79] Reports of the Directors of Convict Prisons (RDCP), Millbank for 1859, *PP*, 35 (1860), 486.
[80] T. Holmes, in Howard Association, *Annual Report* (1908), 81.
[81] M. Carpenter, *Our Convicts* (1864), i. 31–2.
[82] M. Carpenter, *Reformatory Prison Discipline* (1872), 68.

Many writers were unsparingly vitriolic. They condemned women as immodest, callous, cruel, and degraded. Not untypical is the following extract from an article in the *Cornhill Magazine* (1866):

The man's nature may be said to be hardened, the woman's destroyed. Women of this stamp are generally so bold and unblushing in crime, so indifferent to right and wrong, so lost to all sense of shame, so destitute of the instincts of womanhood, that they may be more justly compared to wild beasts than to women. . . . Criminal women, as a class, are found to be more uncivilised than the savage, more degraded than the slave, less true to all natural and womanly instincts than the untutored squaw of a North American Indian tribe.[83]

This highly unsympathetic portrayal highlights two aspects commonplace in much of the literature on criminal women. Firstly, that personality traits were subject to much closer attention than in similar literature on men. In highly moralizing tones, they were minutely catalogued and described, effectively attributing crime to personal failings. Emphasis on internal, rather than possible external, causes of female criminality laid blame primarily on defective personality. Arthur Griffiths, when Deputy Governor of Millbank Female Convict Prison, observed that criminal women reflected 'the seemingly indomitable obstinacy and perversity of the female character, when all barriers are down and only vileness and depravity remains'.[84] Secondly, at their worst, criminal women were portrayed as sunk beyond redemption and certainly beyond the supposedly reformative powers of the prison. At its most condemnatory, Victorian criminology represented criminal women as utterly depraved and corrupted beyond repair. As we shall see, one of the most significant changes towards the end of the century was the shift from this moralizing stance to psychological interpretations of the supposedly defective nature of criminal women.

It did not seem to occur to most observers that it was the very effect of these attitudes, in stigmatizing the female offender, that hindered her attempts to reform. Rejected by respectable society, she was repeatedly refused jobs which demanded any degree of trust, especially in domestic service. Consequently,

[83] M. E. Owen, 'Criminal Women', 153.
[84] A. Griffiths, *Memorials of Millbank* (1884), 117.

she was liable to be forced to return to crime or prostitution to survive. Rosamund Hill stressed this point at an annual meeting of the Social Science Association, arguing: 'women have much greater difficulty in obtaining employment than men, because they are unable to perform the rough out-door work for which no character is required. They must be domestic servants, or be employed in factories; or if they work at home, they must be trusted with property; all three modes requiring, as we know, a good character.'[85] Similarly, Revd W. D. Morrison observed: 'A woman's past has a far worse effect on her future than a man's. She incurs a far graver degree of odium from her own sex: it is much more difficult for her to get into the way of earning an honest livelihood.'[86]

Only towards the latter years of the nineteenth century did more sympathetic observers begin to recognize that it was partly their own value system that made it so much more difficult for criminal women to return to an honest life: 'the dual moral standard makes it more difficult for a woman who has once done wrong to regain her status, than for a man, and thereby tends to degrade her into an abandoned reprobate.'[87] Often it was those women who actually worked with criminal women who recognized their plight, as we can see in this observation by an Irish lady philanthropist: 'the world is unspeakably harder to a woman who falls than to a man . . . an amount of indignation mingles with the displeasure expressed, more vehement than is ever addressed to a man . . .'[88]

Whatever the cause of their reoffending there was undoubtedly a core of habitual female offenders whose persistent recidivism attracted much attention, especially towards the end of the nineteenth century. In one sense the habitual female offender was held up as the epitome of female depravity. Vernon Harris, an inspector of prisons, asserted that it was commonly known that 'a bad woman is the worst of all creations.'[89] Although women were, of course, much less likely than men to offend, once convicted their rates of recidivism were much

---

[85] R. Hill, 'A Plea for Women Convicts', *The English Woman's Journal* (Apr. 1864).

[86] Morrison, *Crime and its Causes*, 161-2.

[87] M. M. B., 'Men and Women as Habitual and Occasional Criminals', *The Englishwoman's Review* (July 1888), 307. Also J. Devon, *The Criminal and the Community* (1912), 158-9.

[88] Miss Todd, 'Prison Mission and Inebriate's Home', *The Englishwoman's Review* (July 1881), 248-9.   [89] Harris, 'Female Prisoner', 783.

greater. As the first Lady Inspector of Female Prisons, Mary Gordon, soon learned, the 'manufacture of the habitual offender, or recidivist, as she is called—is a very swift affair.'[90] Women sentenced for minor offences to local prisons seem to have been given slightly shorter sentences than men and could very quickly notch up a dozen or more sentences of a week or two. As Gordon noted:

elderly or aged women who began in this way and who have been coming to prison in short sentences practically all their adult lives, are to be found at all times in the prisons. There are women who have been convicted 20 or 30 times before they are 20 years old. There are women whose convictions run into hundreds, or of whose convictions all count has been lost.[91]

But even a possible disparity in sentence lengths could not alone explain the frequency with which these women returned to prison. Revd J. W. Horsley, writing in the *Pall Mall Gazette* about habitual criminals, asserted that 'men in this category are, at their worst, twice as good as women at their best.'[92] Even if this were a wild exaggeration, many writers provided local and national figures to show that, in the highest categories of recidivism at least, women were actually worse than men.[93] Figures for the turn of the century showed that women actually outnumbered men in the class of most hardened recidivists. Women made up 61 per cent of those convicted twenty times or more.[94] The huge number of reconvictions notched up by the older female habituals might have given observers cause for celebration that crime was, therefore, committed by a proportionally smaller number of women. Instead, around the turn of the century, attention focused on these apparently 'incorrigible women'.[95] The Annual Report of the Howard Association for 1906 portrayed the female 'habitual' as quite hopelessly beyond control. At worst she was

the heroine of a hundred convictions, whose life is a perpetual horror and a public scandal, who only lapses into decency when undergoing

[90] M. Gordon, *Penal Discipline* (1922), 24.   [91] Ibid.

[92] Quoted in M.M.B., 'Men and Women', 307.

[93] See e.g. figures for Scotland in Thomson, 'Psychology of Criminals', 337–8; figures for Liverpool in J. Nugent, 'Incorrigible Women: What are We do Do with Them?', *Transactions NAPSS 1876* (1877), 375–6; or national figures for 1900 in M. F. Johnston, 'The Life of a Woman Convict', *Fortnightly Review*, 75 (1901), 566–7.

[94] Johnston, 'Life'.   [95] See e.g. Nugent, 'Incorrigible Women'.

imprisonment, and who is only kept alive by repeated terms of imprisonment . . . She has flouted every warning, despised all advice, and refused every offer of help. To live the life of an un-controlled animal has been her one object and desire.[96]

Typically the nature and cause of such recidivism was thought to be innate. Some saw it simply as an unbreakable 'habit of wrong-doing';[97] others, more dramatically, saw it as the manifestation of 'fatal perversity';[98] others, most damningly of all, as signifying 'moral and spiritual death'.[99]

In order to explain how the idealized moral exemplar—woman—could repudiate every positive quality attributed to her, it became necessary to construct notions such as 'moral insensibility'. This condition was used to explain why the female habitual offender, in complete repudiation of the femin-ine ideal, seemed to be devoid of self-respect and completely uncaring of social disapproval. Placing the blame squarely with the criminal woman, it ignored the possible role of stigma in denying her any chance to regain her reputation or secure honest employment, or even lodgings.[100] By the end of the century, early psychological studies began to suggest that hazy constructs like 'moral insensibility' were better understood as indicators of mental incapacity, or, as it was then known, 'feeble-mindedness'. As a result the habitual female offender came increasingly to be seen less as 'bad' than as 'mad'. As we shall see below in Chapter 7, the recognition that many of the very worst female recidivists were in fact merely mentally inadequate was to have significant consequences for their treat-ment within the penal system in the early twentieth century.

THE COSTS OF FEMALE CRIMINALITY

By denouncing female criminality so strongly, the Victorians ostracized female offenders. They thus made it more difficult

[96] Howard Association, *Annual Report* (1906), 24.

[97] Harris, 'Female Prisoner', 782.

[98] J. A. Bremner, 'What Improvements are Required in the System of Discipline in County and Borough Prisons?', *Transactions NAPSS 1873* (1874), 281.

[99] T. Holmes, *London's Underworld* (1912), 109.

[100] Significantly, women philanthropists actually involved with the provision of after-care for female prisoners recognized the social barriers faced by women trying to reform—not least in the importance of 'character' for gaining many forms of female employment. See Craig, 'Emigration', 291, and Hill, 'A Plea', 131–2.

for them to reform; for a core it was made all but impossible. Denunciations were, as we have seen, partly provoked by the degree to which criminal women seemed to transgress norms of femininity. But the costs of female criminality were seen to be far wider than the ruin of the offender herself. Woman's role as an agent of moral influence over the rest of society was also negated by her criminality. The criminal woman had not only repudiated her responsibility as a benign moral influence but, infinitely worse, had become a source of moral contagion. And the ill effects of this were compounded by the fact that the force of woman's influence for good or ill was thought to be considerable.

The costs of female crime reappeared on the agenda of Social Science Association meetings. At the 1863 annual meeting, one delegate, Miss Rosamond Hill, stressed that the 'comparatively small numbers' of female criminals belied the extent of the damage they did, for 'the conduct of the female sex more deeply affects the well-being of the community. A bad woman inflicts more moral injury on society than a bad man.' [101] The rationale behind such an assertion is clear from the argument put forward at the 1874 annual meeting, where David Gibbs lamented: 'There is domestic purity and moral life and example in a good home, and individual defilement and moral ruin in a bad one.' [102] This view that women's influence within the home exacerbated the costs of female criminality appears to have been widely held. For example, J. C. Symons argued that female criminality, though quantitatively insignificant, was a serious problem: 'for female crime has a much worse effect on the morals of the young, and is therefore of a far more powerfully depraving character, than the crimes of men . . . the influence and example of the mother are all powerful: and corruption, if it be there, exists in the source and must taint the stream.' [103] Thus female criminality figured prominently in explanations of juvenile delinquency, which was often traced back to the evil influence of a corrupt mother or the neglect of an uncaring one. The delinquent mother, it

[101] Hill, 'A Plea', 134.

[102] D. Gibb, 'The Relative Increase of Wages, Drunkenness and Crime', *Transactions NAPSS 1874* (1875), 334.

[103] Symons, *Tactics*, 25. See also J. Devon on the evil influence of criminal mothers, especially on daughters: *Criminal and the Community*, 22.

was suggested, left her children to play in the gutter or to fend for themselves on the streets whilst she indulged in debauchery, drunkenness, or theft. By courting imprisonment in this way, she risked depriving them of maternal care, however inadequate. The children, if they survived at all, grew up sickly, ill-nourished, and even as alcoholics themselves. Denied a 'proper' mother's loving restraint they quickly ran beyond control, became rebellious, and, in time, became incapable of settling to the discipline of school or work.[104]

More fiercely condemned still were those women who actively led their children into crime by sending them out to beg or steal. Mary Carpenter in *Our Convicts* (1864) recounted numerous cases of women who encouraged and even instructed children in varieties of petty crime. One woman, 'S.J.', with no children of her own but entrusted to care for her sister's, 'induces all the lads and girls she can to bring plunder to her house, and harbours them therein'. Another, 'E.T.', herself a mother of five, 'keeps a notorious house for young thieves, who bring their plunder there. Several have been transported from her den. Her own children steal.'[105] Such case studies provoked especial moral outrage, for the woman had not only offended against her supposedly natural maternal instinct but she had repudiated her responsibilities in bringing up the next generation.

It was not only juvenile delinquency that was attributed to the evil influence of the criminal woman. Even crime and immorality amongst grown adults were traced back to this source. It was commonly held that 'one bad woman drew after her many criminal men'.[106] Automatically linking female criminality with prostitution, many writers characterized woman's evil influence over men as primarily sexual. A single prostitute became 'the seducer of virtue, not in one, but hundreds of young men; (she) robs them of their strength, their money, their character; and then confederates them in those

---

[104] On the social consequences of women who failed to fulfil their maternal duties see, for example, Tallack, *Penological and Preventative Principles*, 32–7; J. W. Horsley, *How Criminals Are Made And Prevented* (1913), 186–209; M. Carpenter, *Juvenile Delinquents, their Condition and Treatment* (1853).

[105] Carpenter, *Our Convicts*, i. 69–70.

[106] A. Lewis during discussion on 'The Prevention of Crimes Act 1871', *Transactions NAPSS 1875* (1876), 327.

bands of midnight robbers, and those hordes of swindlers and gamblers . . .' [107] At its most extreme this literature attributed to the corrupt woman pseudo-mystical powers by which she exercised an almost hypnotic influence over any man foolish enough to allow himself to fall within her grasp. She entranced weak-willed men with promises of sexual favours and so inveigled them unthinkingly into any immorality she might choose to instigate. This portrayal presented the man as more likely to be perceived by the courts as the instigator, so that he was liable to be convicted of the crime, whilst the woman, the real culprit, was set free. One contributor to the *Quarterly Review* clearly had such instances in mind when he warned: 'We venture to prophesy that our Pentonvilles, be they ever so multiplied, will never cease to be furnished with cargoes of living vice. We may never cease to hope for empty cells and maiden assizes so long as, when the thief's punishment has expired, his paramour is waiting at the gate.' [108]

This quasi-biblical interpretation of the sinful woman leading men astray, repeated so often in the criminological literature, was evidently pervasive, for male criminals often claimed to have been corrupted in this way. The Revd J. Kingsmill, the Chaplain of Pentonville Prison, in *Prisons and Prisoners* (1854) quoted numerous cases where convicts questioned about the origins of their criminal career attributed it to a woman's influence. To cite just a few examples from the hundred convicts he questioned: one asserted that 'a bad wife was the first cause of all my trouble'; another blamed 'bad company, particularly female'; a third explained that he had frequented singing rooms where 'one female so persuaded me to adopt her life, that, in order to gratify her wishes, and have an opportunity of seeing the plays, I was led to steal'. [109] Since we can presume that these men expected their excuses to be taken seriously, the attribution of blame in these accounts further corroborates the common view of criminal women as corruptors. We can see, then, that even when women were not actually convicted of crime themselves, they were readily blamed for the corruption and criminality of others. As

[107] J. Kingsmill, *Chapters on Prisons and Prisoners and the Prevention of Crime* (1854), 64.
[108] J. Armstrong, 'Female Penitentiaries', *Quarterly Review* (Sept. 1848), 368.
[109] Kingsmill, *Prisons and Prisoners*, 285 ff.

William Acton observed of fallen women: 'though we may call these women outcasts and pariahs, they have a powerful influence for evil on all ranks of the community.'[110] The perceived costs of female criminality were amplified far beyond the scope of offences committed by women to attract a quite disproportionate level of concern.

As we will see in later chapters, although condemnation of criminal women was, over the course of the nineteenth century, increasingly mitigated by growing understanding of the reasons for their offending behaviour, such explanations did little to calm fears about its costs. Indeed, as biological and medical explanations tended to replace moral interpretations of female deviance, they actually drew an even more dramatic picture of its potential for social harm.

[110] Acton, *Prostitution*, 73.

2 EXPLAINING FEMALE CRIME

The attempt to define and evaluate women's crimes naturally led investigators back to consider the root causes of female delinquency. Interpretations of these causes can be slotted into various broad groups, for example, ecological, moral, economic, biological, and psychological. None was exclusive, each overlapped and occasionally merged with other interpretations. Equally the views and findings of one body of investigators often contradicted and occasionally refuted those of another. To treat them, therefore, as discrete and coherent schools of thought would be to underestimate the intricate intertexture of nineteenth-century criminological theory. At the risk of overstating their heterogeneity, I present them here under separate headings for the sake of convenience.

It is not possible to determine an exact chronology of trends in prevailing schools of thought, for explanations of female crime were not discrete intellectually or temporally. In order to understand the rise and decline of types of explanation, and why certain interpretations persisted whilst others fell out of favour, one must set criminology against more general trends in contemporary social thought.

EARLY VICTORIAN ECOLOGICAL ANALYSIS

In the period 1830–60 the predominant interest was in regional and territorial studies (for example, by Joseph Fletcher, Henry Mayhew, and James Greenwood) which revealed that crime rates varied enormously by area. Certain locations, particularly inner city slums, were identified as veritable 'crime areas' of dense, frenetic criminal activity.[1] Much effort was devoted to exposing the 'promiscuous herding' and 'scenes of profligacy', the 'polluting language', and the 'vicious abandonment'

[1] Yale Levin argues that in mapping 'crime areas' such men acquired an 'amazing ecological knowledge'. In Y. Levin and A. Lindesmith, 'English Ecology and Criminology', *Journal of Criminal Law*, 27 (1936–7), 810.

they engendered. The specific social characteristics of these areas were catalogued and evaluated to determine which were the crucial causal factors. Overcrowding, destitution, poor sanitation, and filth were just a few of the more salient traits identified as demoralizing the urban poor. As we have seen, a home was thought to be essential to the preservation of femininity. Yet in the slum, domestic order and cleanliness was impossible to achieve and as a result woman was thought to be unavoidably degraded by the 'want of order, tidiness, and those decencies of life which incapacitate the female from fulfilling those duties to which she is likely to be called as a mother or servant'.[2] A woman's attempt to maintain a decent home, the strain of multiple pregnancies, the anxieties of trying to feed and clothe her often sickly offspring compounded the miseries of female destitution. As a result she risked turning to less desirable occupations and amusements in the dangerous world outside the home. Little wonder, more sympathetic observers remarked, that she sought in the pub the light, warmth, and conviviality she so badly lacked in her own poor home.

In many areas, gross overcrowding reduced the home to conditions of squalid demoralization. Overcrowding often meant that a room was shared by whole families, and even unmarried adults, with obvious and, to contemporary observers, horrifying consequences: 'congregated together in masses too frightful to contemplate, and almost beyond belief,—men and women, boys and girls, relations, and strangers mingled together in most promiscuous intercourse . . . herded together in one common room, in utter disregard of every sense of decency'.[3] The effects of overcrowding were thought to be especially harmful to women: 'The indiscriminate herding together of the poor of all sexes and ages in their cramped, comfortless and inconvenient houses has a necessary tendency to sap that instinctive modesty and delicacy of feeling which the Creator intended to be the guardian of virtue, and which are

---

[2] Greater London Record Office (hereafter GLRO) MA/RS/1/742. Reports of the Governors and Chaplains of the Houses of Correction, essay by Foster Rogers, Assistant Chaplain, House of Correction, Westminster, 1850, 46.

[3] Ibid., essay by G. H. Hine, Chaplain of the House of Correction, Westminster, 1850, 31. See also essays in the same collection by G. L. Chesterton, E. A. Illingworth and J. Williams, and A. F. Tracey.

more especially necessary for the defence of the female
character.'[4] The ideal of feminine innocence and purity,
always at variance with the cruder realities of working-class
life, could not possibly hope to prevail in such conditions. In
part, these exposés reflected the terror felt by the middle classes
of the spreading urban slums which sucked in the masses of
workers who flowed to the cities in search of employment. And,
in part, such writings fed that insatiable demand for sensation-
alism exploited by the popular press, who employed investigators just to provide such lurid 'pictures of the underworld'.

London, larger, filthier, and, most importantly, inescapably
under the noses of those in power, epitomized the very worst of
slum conditions. It was avidly studied, described, and deplored.
Andrew Mearns, a Congregationalist minister in Chelsea,
wrote a pamphlet which gained instant fame, lamenting: 'The
low parts of London are the sink into which the filthy and
abominable from all parts of the country seem to flow. Entire
courts are filled with thieves, prostitutes and liberated convicts.'[5] He also recognized that the appalling conditions of
urban life were, in turn, major factors in dragging the poor
down. Forced to associate so closely with criminals, vagrants,
and prostitutes, the 'honest' poor were very likely to be corrupted. Mearns demanded: 'Who can wonder that every evil
flourishes in such hotbeds of vice and disease? Who can wonder
that young girls wander off into a life of immorality, which
promises release from such conditions?'[6]

If overcrowding and especially the degrading effects of close
confinement with others still more abjectly miserable were seen
as major sources of corruption, what horrified middle-class
observers even more was the utter filth of the urban slum. They
deplored the diseases it engendered, suggesting that these were
not only physical but moral too. The sickliness of the underfed,
puny urban poor found its counterpart in the 'contagious
diseases of the moral world'. The statistician Frederic Mouat,
speaking to a meeting of the Social Science Association in 1881,
posed the evidently rhetorical question:

[4] Ibid., essay by Foster Rogers, 1850, 45.
[5] A. Mearns, *The Bitter Cry of Outcast London* (1883), 10. See also J. Greenwood, *The Seven Curses of London* (1869); J. London, *The People of the Abyss* (1902).
[6] Mearns, *Bitter Cry*, 9.

How can clean water come out of a dirty vessel? The undue pressure of population upon limited areas, the struggle for mere existence of multitudes born into this world of civilisation and progress, and the exigencies of a community placed in such unnatural circumstances, deprived of pure air and denied the bright light of heaven, are all important factors in the great questions of pauperism and crime, twin children of unnatural parents.[7]

Women were considered to suffer particularly in the filth of the urban slum. Dirt not only reduced them to the level of the gutter but corrupted the very ideal of womanhood, for physical filth simply could not be separated from moral impurity in women.[8] Only much later did writers such as Mary Higgs, Regional Secretary of the National Association for Women's Lodging Houses, more sympathetically and more practically recognize that dirt could actually be an indirect cause of female crime. Since female propriety demanded cleanliness, dirty clothing or appearance inevitably designated the offender as being outside respectability. She stressed: 'It must be remembered that to a woman, for respectable existence, cleanliness is an absolute necessity . . . a woman must "look tidy" or no one will employ her. Therefore conditions destructive to cleanliness are for her equivalent to forcing her lower and lower into beggary and vice.'[9] Whereas men could seek rough but respectable work whatever their physical appearance, women seeking employment generally needed to be presentable. Domestic service was a major source of female employment which demanded very high standards of dress. A dirty appearance was cause for suspicion about a woman's character and, in any case, was simply unacceptable—note for example the common assumption that 'the greater the outward dirt of a woman's personal appearance, the broader and coarser are her mental perceptions generally.'[10] If women were unable to keep clean or their clothes became ragged and stained, they were all but unemployable. In desperation some inevitably turned to prostitution or to crime in order to survive.

[7] F. Mouat, 'Address on the Repression of Crime', *Transactions NAPSS 1881* (1882), 259.   [8] S. K. Kent, *Sex and Suffrage in Britain 1860–1914* (1987), 39–40.
  [9] M. Higgs, *Glimpses into the Abyss* (1906), 250.
  [10] R. Smith Baker, 'The Social Results of the Employment of Girls and Women in Factories and Workshops', *Transactions NAPSS 1868* (1869), 542.

The environment of the urban slum was a massive obstacle to preserving any degree of propriety for the female poor. Recognizing that this was so, observers did not exonerate women's recourse to crime so much as adopt the following fatalistic view: the slum environment was degrading and demoralizing; women inhabiting that environment were already outside respectability; and, if in the least susceptible to its influence, they were doomed to crime.

HIGH VICTORIAN MORAL ACCOUNTS

From the middle of the nineteenth century a growing body of wealthy philanthropists supported empirical research into the problems of Victorian society with a view to seeking their solution. 'Ameliorism', as this school of thought was known, suggested that most social problems were caused by a combination of two main factors: the strains of certain social situations, and individual moral weaknesses. The conjunction of these two lay behind all the major social problems of the day: not least destitution, sexual immorality, and inebriety.[11] These premises led to the investigation of crime both in terms of its social context and the moral character of those who, in this environment, became criminal.

Many writers, and especially those concerned with delinquency amongst young women and girls, stressed the devastating effects of 'bad upbringing'. They thought women were all the more lastingly damaged, believing that 'woman not only receives impressions more easily, but retains them more tenaciously than man.'[12] Whereas men could rise up out of the slum and overcome the disadvantages of their early years, women were considered to be less equipped to do so. They were seen to be irrevocably dragged down by the corrupting influence of 'ungodly and drunken parents' who failed to provide the necessities of a religious education, moral training,

[11] This view was epitomized by the work of the National Association for the Promotion of Social Science (NAPSS). See P. Abrams, *Origins of British Sociology* (1968), 33–52; L. Radzinowicz and R. Hood, *A History of English Criminal Law and its Administration from 1750* (1986), vol. v, ch. 3, 'The Ameliorative Creed'.

[12] M. E. Owen, 'Criminal Women', *Cornhill Magazine*, 14 (1866), 152. See also G. L. Chesterton, *Revelations of Prison Life* (1856), i. 71–6.

or in some cases even a proper home.[13] Even greater harm was
thought to be done by parents who were not merely neglectful
but a positively evil influence on their young daughters. Their
profane language, bullying, or even violence drove young girls
out on to the streets to beg, to steal, and to prostitute themselves. For example, it was claimed of the Birmingham poor
that 'it is a fairly common practice among the mothers of quite
young girls in the terribly over-crowded slums to force their
daughters to sleep with men lodgers; and that women similarly
situated, and perhaps a shade more degraded, compel their
little girls to go on the streets soliciting.'[14]

However depraved or drunken the parents, or dissolute the
home, the girl who was abandoned, orphaned, or otherwise left
with no home at all, was obviously at greater risk. The girl
whose upbringing was entrusted to the workhouse and whose
only education came from the district school was believed to
need tremendous inner moral resources to withstand the influence of such an environment. The Revd J. W. Horsley, in
Jottings from Jail (1887), examined a wide range of causes of
female crime and found the demoralizing effects of a workhouse
upbringing perhaps the most pernicious and yet the most often
ignored. Horsley demanded: 'When will people understand
that no list of the causes of hereditary pauperism, of crime, and
of prostitution, would be complete or honest which did not
include the workhouse training of the young, and especially of
girls?'[15] The education provided there was inadequate, the
company corrupting, and the influence of a good mother totally
absent. He could only conclude that 'the huge barrack schools
are utter ruin for pauper girls.'

Perhaps the most sympathetic and certainly the most an-
guished portrayal of a young woman forced into crime by the
misfortune of homelessness was that of the fictional criminal
and prostitute, Nancy, in Charles Dickens's *Oliver Twist*.
Recounted in the first person, the descriptions of her early life
gain an added poignancy. Introducing herself to the angelic
Miss Maylie, Nancy laments: 'I am the infamous creature you

[13] Reports of the Directors of Convict Prisons (RDCP), Woking for 1878, *PP*, 35
(1878–9), 540.

[14] H. M. Richardson, 'The Outcasts', *The Englishwoman* (Sept. 1909), 4.

[15] J. W. Horsley, *Jottings from Jail* (1887), 47.

have heard of, that lives among the thieves, and that never from the first moment I can recollect my eyes and senses opening on London streets have known any better life, or kinder words than they have given me, so help me God!' She exhorts Miss Maylie to be grateful 'that you were never in the midst of cold and hunger, riot and drunkenness, and—and—something worse than all—as I have been from my cradle. I may use the word, for the alley and the gutter were mine, as they will be my deathbed.'[16]

Nancy, it is revealed, was raised by Fagin as a child thief and prostitute, trained from an early age and forced on to the streets to earn a living as 'the miserable companion of thieves and ruffians, and fallen outcast of low haunts, the associate of the scourings of the jails and hulks, living within the shadow of the gallows itself'.[17] This portrayal of Nancy was remarkably sympathetic in its characterization of woman as the victim both of circumstance and of male abuse. Perhaps it is significant that this telling and relatively rare antidote to the usual condemnatory portrayal of the criminal woman as outside any claim to respect or sympathy was a fictional heroine, albeit one drawn from Dickens's own crusading work amongst prostitutes and thieves.

Another heroine, more common to both literary and documentary works, was the innocent virgin ruined by the worldly male seducer who abandoned her pregnant, unprovided for, and friendless.[18] The seduced and abandoned woman was doubly exposed. She had lost all right to reputation, was barred from respectable employment and decent company, and so was left with no lawful means of supporting herself. Moreover, the consequences of seduction itself often led to those peculiarly female offences catalogued here by William Acton: 'The results, then, of seduction and desertion are to force the mother to take to prostitution, and to tempt her to make away with her child either by her own hand or by means of baby farmers, thus giving rise to an amount of crime fearful to contemplate.'[19]

[16] Charles Dickens, *Oliver Twist* (1837–9), 362.   [17] Ibid. 361.

[18] Even in works as chronologically separated and as ideologically disparate as W. Acton, *Prostitution* (1857), A. Bebel, *Woman in the Past, Present and Future* (1885), and H. Adam, *Woman and Crime* (1914).

[19] Acton, *Prostitution*, 285.

The woman was called to account for the perpetration of her crimes whilst the original transgressor, the male seducer, escaped not only unpunished but without even the stigma of moral sanction.

Male responsibility for female criminality was also seen to operate in a more general way. Whilst women were supposed to be able to rely on the protection of a 'good man', their downfall was often attributed to the corrupting influence of 'vicious' men. In studies of 'the underworld', women were commonly portrayed as an integral part of the criminal community—instructed by the male professional in the techniques of petty theft, shoplifting, and pickpocketing, acting under his influence and direction.[20] In such circumstances, women were recognized to be less culpable, as Hargrave Adam noted in 1914: 'Very properly the law makes merciful allowance in dealing with the female offender where it can be apparent that she has been impelled to commit breaches of the law under masculine influence . . .'[21]

A phenomenon which occasionally thwarted those trying to reform criminal women was their 'invincible, unreasoning, and self-sacrificing devotion to a most brutal ruffian of a man'.[22] In *Oliver Twist* Dickens portrays Nancy as unfailingly loyal to the villainous Bill Sykes. Despite her desire to be saved by Miss Maylie, she refuses to go with her, giving the clichéd lament: 'among the men I have told you of there is one: the most desperate among them all: that I can't leave; no, not even to be saved from the life I am leading now.'[23]

The common perception of the criminal woman as a helpless victim was based on this image of the weak-willed, dependent female stuck limpet-like to her man, however vicious a criminal or brutal a lover he was. It seems somewhat ironic that this stereotypically feminine devotion to the 'stronger' sex should also have been recognized as cause of her descent to crime—the very negation of femininity. Recognizing this risk, writers such as Mary Higgs warned against the dangers of engaging in 'imprudent' relationships.[24] Precocious and hasty marriage

[20] See H. Mayhew, *London Labour and London Poor* (1861–2), iv. 334–5 and 355–66 for accounts of such collaboration.

[21] Adam, *Woman and Crime*, 4.

[22] Ibid. 5; see also F. Scougal (pseud.), *Scenes from a Silent World* (1889), 69–81.

[23] Dickens, *Oliver Twist*, 364.   [24] Higgs, *Glimpses*, 317–20.

amongst the poor, Higgs argued, led to unhappiness, separation, adultery, and profligacy. Becoming a mother before she
was old enough to have developed her maternal instincts fully,
a woman would be unable to manage her meagre budget or
keep a comfortable home, and would turn in desperation to
drink. In sum, she 'may be placed in circumstances of destitution in pursuit of the *ideal* life'.[25] The lauded middle-class
institution of marriage, in this perverted form, became only an
additional cause of female immorality.

This gap between the ideals of middle-class morality and the
realities of working-class life is all the more apparent in the
failure of middle-class observers to recognize that what they
saw as the immediate causes of crime were themselves the
product of environment. Drunkenness is a prime example
here. Throughout the nineteenth century, social observers both
in and outside the Temperance Movement saw public houses,
spirit shops, and dancing saloons, and the drunkenness that
accompanied them, as sources of corruption, degradation, and
ultimately of criminality. All too often they failed to realize that
it was the appalling misery of slum life that drove men and
women to seek solace in drink.[26] Nonetheless, they deplored
the amount of crime committed either when drunk or as an
indirect result of inebriety, for example to raise money for
alcohol. In 1834, the Select Committee of the House of Com-
mons on Drunkenness declared that alcohol aggravated 'all the
worst passions of the heart' and led to an increase in crime of
'every shape and form'.[27] Drunkenness as direct or indirect
cause of crime later became common in explanations otherwise
widely differing in perspective and purpose. Some writers
identified drunkenness as a specific factor in motivating particular crimes, not least violence. Others made more general
claims about alcohol's effects on moral standards: 'drink quiets
remorse and shame; more drink leads to reckless abandonment
and disorderly conduct, the police-court, and the prison.'[28]

[25] Ibid. 319.

[26] The relationship between alcoholism and crime will be discussed at length in
Ch. 6 below. Also see B. Harrison, *Drink and the Victorians* (1971) for a wider analysis of
Victorian attitudes to alcohol.

[27] Report of the Select Committee of Inquiry into Drunkenness, *PP*, 8 (1834), 318
and 320.

[28] S. W. Fletcher, *Twelve Months in an English Prison* (1884), 404. See also W. C.
Sullivan, *Alcoholism* (1906), 154–62.

Criminals also cited drink as the original cause of their downfall, responsible for their unhappy home, absent parent, or spouse. Such pleas were received with varying degrees of sympathy in mitigation of their criminality. One female prisoner in Tothill Fields House of Correction reported of her fellow inmates: 'nine in ten, and I think a larger proportion, owe their imprisonment solely to drink.'[29] The more sceptically inclined questioned the glib attribution of crimes to drunkenness, and W. C. Sullivan, Medical Officer at Holloway Prison, set about trying to calculate exactly what proportion and what types of crime could reasonably be explained by drink alone.[30] Estimates of the proportion and types of crime attributable to drunkenness, based mainly on subjective analyses of contributing factors, varied enormously in their respective findings. Yet underlying much Victorian criminological writing was an unquestioning acceptance that drunkenness weakened self-control, provided temptations to crime, and, in more serious cases, acted as a catalyst, destroying morals and multiplying transgressions. A downward spiral of recidivism was painted only too graphically in discussions of the 'habitual offender'.[31]

Prostitution was similarly linked with drunkenness. As the historian Frances Finnegan has observed, girls 'fell' or were first seduced whilst drunk, alcohol broke down resistance and excited sexual passion in both men and women, and drinking halls provided a trading ground for potential clients.[32] Nearly every witness questioned by the Select Committee on Drunkenness (1834) was invited to agree that rising prostitution was a direct consequence of the increasing number of gin-shops.[33] Most concurred, arguing that not only were they notorious places of assignation but that women who frequented them inevitably lost all sense of shame, that great regulator of womanhood. Pushed to the margins of society, they were lost irretrievably to 'immorality'.[34] Once launched upon a life of

[29] Fletcher, *Twelve Months*, 323.

[30] See for example, W. C. Sullivan, 'The Criminology of Drunkenness', in T. N. Kelynack (ed.), *The Drink Problem* (1907), 190.

[31] See E. Orme, 'Our Female Criminals', *Fortnightly Review*, 69 (1898), 792.

[32] F. Finnegan, *Poverty and Prostitution* (1979), 143–51.

[33] Select C. on Drunkenness, Minutes of Evidence, *passim*.

[34] Anon., 'Intemperance and the Social Evil', *Temperance Spectator* (Jan.–Dec. 1860), 61–2.

prostitution, drinking became a necessary part of a life which contemporaries could only conceive as being horribly painful to woman's innate sense of virtue. According to Dr Norman Kerr: 'by drink, the "unfortunates" deadened their conscience and stifled the stirrings of remorse, thus fortifying themselves to ply their hideous calling.'[35] The Revd G. P. Merrick, Chaplain to Millbank Prison, conducted a massive study of 14,110 prostitutes whom he interviewed in prison. He ascribed even greater importance to alcoholism in overcoming woman's natural resistance to such a 'wicked and wretched mode of life', asserting: 'They loathe it, and their repugnance to it can only be stifled when they are more or less under the influence of intoxicating drinks.'[36] How far this supposed hatred was genuinely felt, merely professed for the benefit of a pious prison chaplain, or blithely ascribed by observers insensitive to the equal misery of alternative forms of employment is unclear. It may well have been that, given the location and nature of their calling, prostitutes drank regularly and heavily as part of their trade and gradually became confirmed drunkards more by habit than psychological need.

The social life of the urban poor, and especially its effect on the moral welfare of women, appeared frequently on the agenda of annual meetings of the Social Science Association. At the 1858 meeting the Revd Thomas Carter, Chaplain of Liverpool Borough Gaol, presented a paper denouncing casinos and dancing saloons as sources of female corruption; he argued that 'they would be bad enough if only visited by females of known profligate and abandoned character; but they are worse, from the undeniable fact that here, too, numbers of young women occupying positions of respectability . . . first acquire a fatal familiarity with vice.'[37] In such establishments women were said to take up drinking, fall in with 'loose immoral characters', keep late hours, and learn to enjoy 'unnatural excitement'.[38]

[35] See N. Kerr, *Female Intemperance* (1880), 9. See also G. P. Merrick, *Work among the Fallen* (1891), 28–31.
[36] Merrick, *Work among the Fallen*, 29.
[37] T. Carter, 'On the Crime of Liverpool', *Transactions NAPSS 1858* (1859), 351.
[38] GLRO MA/RS/1/742. Essay by G. H. Hine, 1850, 23. See also criticisms of the State's irresponsibility in the licensing of music-halls and pubs in Fletcher, *Twelve Months*.

To summarize thus far: overcrowding, filth, the lack of a decent home or proper upbringing, seduction, desertion, premature and imprudent marriage, low amusements, entertainments, and above all, alcohol figured prominently in the middle-class picture of the urban slum. These elements were central to the ecological and to the moral analyses of crime which flourished around the mid-nineteenth century. Such studies indicate humanitarian concern for the plight of the poorest section of society, but, even more strikingly, reveal a grave and unsentimental concern with the costs of urban growth.[39]

THE ECONOMIC COSTS OF INDUSTRIALIZATION

Concern with the social wastage consequent on industrialization prompted investigations into the profound changes taking place in the economic structure. For industrialization had revolutionized patterns of employment: it had forced many working-class women out of the relative safety of traditional employment in domestic service or home-based industries and into 'unnatural' roles in the factory, shop, and even in the coalmine, and iron works.[40] Country girls seeking employment in factory towns or cities were seen to be most at risk. Outside the protection of the home they were liable to temptation or depredation. Whereas no reaction of surprise or moral censure met young men leaving home in search of employment or adventure, the young girl who launched herself into the perilous outside world provoked great consternation. The Directors of Brixton Prison clearly believed that leaving the safe confines of home had been the moral downfall of many of those in their custody: 'With few exceptions female prisoners are extremely ignorant, and have been brought up in crime. They have been long accustomed to vice, dissipation, and idleness. Many of them ran away from home as the first step

---

[39] For the relationship of the growth of social investigation to concern about urbanization, see O. R. McGregor, 'Social Research and Social Policy in the Nineteenth Century', *British Journal of Sociology*, 8: 2 (1957), 146 ff.

[40] See the papers given under the heading 'The Social Results of the Employment of Girls and Women in Factories and Workshops', *Transactions NAPSS 1868* (1869).

to ruin, and consequently they fell into bad company and contracted habits of the worst kind.'[41]

Whilst the single woman was seeking work, finding a respectable place to live was a major problem. One of the few available sources of cheap accommodation were lodging houses, yet these were condemned as overcrowded, filthy, and in every way inadequate as a 'home'. Denied privacy or access to proper washing facilities, propositioned by male staff and lodgers alike, the young woman's virtue was gravely at risk. The contaminating effects of overcrowding in lodgings were compounded by the character of their tenants. Thieves, prostitutes, and habitual beggars all used lodgings not merely as places of rest but as bases from which to carry out their business. As a result, such houses came to be identified as veritable schools of crime. As late as 1906, Mary Higgs revealed in her penetrating study of the homeless that accommodation for poor single women was appallingly inadequate. With a liberal-mindedness far in advance of her day, Higgs argued that rented rooms in slum tenements, filthy lodging houses, and Magdalen Homes (which she recognized as well-intentioned but stigmatizing) should be replaced by decent lodgings for all women whether or not they were respectable themselves: 'There should be large, well-ventilated, well-provided women's lodging houses, open even to the prostitute, but under the care of wise, motherly women.'[42] Such provision would, she argued, provide 'a safe place of refuge . . . to arrest, if possible, a downward career'.[43] The expense involved was justified by the view that woman was a moral exemplar to the rest of society—as Higgs declared: 'women's lodging-houses—and what can be more needful for the morals of the community?'[44]

How did it arise that so many women were unmarried, living alone, and seeking employment outside the home? Certainly, the freedom of city life, of financial independence, and escape from the confining strictures of a closed rural community, attracted many young women away from home. A more intractable problem was the steadily increasing imbalance of the sexes in the population. In 1851 the excess of women over men

---

[41] RDCP, Brixton for 1861, *PP*, 25 (1862), 622.
[42] Higgs, *Glimpses*, 223–4.   [43] Ibid. 233.   [44] Ibid. 280.

already stood at 365,159; by 1901 the 'surplus' had risen to 1,070,617.[45] The essayist W. R. Greg noted, 'there is an enormous and increasing number of single women in the nation, a number quite disproportionate and quite abnormal; a number which, positively and relatively, is indicative of an unwholesome social state, and is both productive and prognostic of much wretchedness and wrong.'[46] This large and steadily increasing number of 'surplus' women, who could not hope to fulfil their 'natural' vocation as wife and mother, were both pitied and feared as 'a floating population of women who fall an easy prey to wrong conditions'.[47] They could not hope to take on their prescribed role and remained therefore, by definition, outside femininity. Unsurprisingly, one of the major preoccupations of the early feminists was to try to delineate a socially acceptable and productive role for these otherwise 'redundant women'.[48]

An over-abundant supply of women seeking work enabled employers to pay miserably low wages, leaving women economically distressed and, in times of illness or slump, poverty-stricken. The situation was poignantly described by one contributor to an annual meeting of the Social Science Association in 1858: 'The dire lack of employment, and consequent debasing struggle for the bare necessities of life, had told frightfully on the social condition of the humbler women of this country. The most terrible phase in the criminality of the country is the number of its female criminals'.[49] In order to earn even a pitiful wage many women had to work extremely hard for long hours. The most readily available, but worst paid, work was in the home, making hats, dresses, or doing plain machine sewing. As Greg observed: 'in great cities, thousands . . . are toiling in the ill-paid *métier* of semptresses and needlewomen, wasting life and soul, gathering the scantiest

[45] Figures from M. Vicinus, *Independent Women: Work and Community for Single Women 1850–1920* (1985), 293.

[46] W. R. Greg, *Why are Women Redundant?* (1869), 5.

[47] Higgs, *Glimpses*, 248.

[48] Vicinus, *Independent Women*, explores some of the fields opened up to single women as respectable forms of employment in the second half of the 19th cent. Another solution was emigration, as suggested, for example, in I. Craig, 'Emigration as a Preventative Agency', *The English Woman's Journal* (Jan. 1859), and in the far less sympathetic work by Greg, *Why are Women Redundant?*

[49] Craig, 'Emigration', 290.

of subsistence, and surrounded by the most overpowering and insidious temptations.'[50] Lowest paid of all was making match-boxes: Jack London in his exploration of the East End of London found an old woman 'supporting herself and four children, and paying three shillings per week rent, by making match boxes at 2¼d. per gross'.[51] Such work saved the employer the cost of providing workrooms, kept women in the home, and kept their exploitation out of sight and away from the public conscience. As a result the scandalous conditions of sweated home-work long continued unchecked. As late as 1912 Thomas Holmes drew this horrifying picture: 'Midnight! and thousands of women are working! One o'clock, and thousands are still at it! Two o'clock, the widows are still at work! Thank God the children are asleep. Three o'clock a.m., the machines cease to rattle and in the land of crushed womanhood there is silence if not peace.'[52]

The burden of poverty generally fell hardest on the woman of the family. It was she who shouldered the responsibility of feeding and caring for children, for husbands, for relatives who were ageing, invalid, or dying, and for friends. On the presumption that the husband, as the main breadwinner, must be fed, it was she who often went hungry when money was scarce. Small wonder then that, worn down by self-denial, overwork, and want, some sweated home-workers turned to petty theft or to prostitution as an easier means of subsistence. As Dr Sheridan Delephine, Director of the Public Health Laboratory, Manchester, observed: 'A poor woman with a large family in the midst of sordid poverty, from which there seems no escape, must ultimately lose hope for better things, become reckless in her actions as a drowning man in his movements.'[53] More sympathetic observers recognized that their situation offered the difficult choice between the semi-slavery of home-based work and the relative freedom of prostitution. The question was less one of moral propriety than of simple survival. The Chaplain of the House of Correction, Westminster, G. H. Hine, found that London's needlewomen, a large and particularly poorly paid group, 'when closely questioned, have

---

[50] Greg, *Why are Women Redundant?*, 6.   [51] London, *People of the Abyss*, 126.
[52] T. Holmes, *London's Underworld* (1912), 121.
[53] Quoted in P. Snowden, *Socialism and the Drink Question* (1908), 56.

been obliged to admit that they have recoursed to the streets in order to eke out the scanty pittance they receive from their employers'.[54] And William Acton's important study, *Prostitution* (1857), confirmed these findings. Although he also attributed women's descent into prostitution partly to vanity, to idleness, and to the love of drink, he admitted that 'by far the larger proportion are driven to evil courses by cruel, biting poverty. It is a shameful fact, but no less true, that the lowness of the wages paid to workwomen in various trades is a fruitful source of prostitution; unable to obtain the necessities of life, they gain, by surrendering their bodies to evil uses, food to sustain and clothes to cover them.'[55] Destitution was recognized, therefore, as a prime factor in forcing women to turn to prostitution and often also to crime.

By working from home, the sweated piece-worker did not contravene woman's separate sphere. But the woman who moved outside the home to work in the new industries in factories and workshops was viewed less sympathetically. Even in the first half of the century, the new woman worker was identified as a glaring social problem. Working in a factory or, worse still, down a mine, she represented the very antithesis of femininity. Moreover, her presence there was seen to be attended by risks to her moral purity which caused far more concern than the physical damage or hazards to health risked by overwork or dangerous conditions. As the historian Sally Alexander has pointed out, the abhorrence expressed towards female factory work did not simply reflect concern for physical working conditions. The conditions in which sweated needle-workers worked were often as bad but attracted less attention because they remained safely within the home.[56]

Young female workers were seen to be particularly vulnerable. It was argued that the majority of female criminals were recruited from their ranks. Away from the family's good influence, they found no substitute supervision from uncaring employers. Even during working hours they were exposed, for example, to 'the risk of their being brought into the companionship of those who are of indifferent or bad character . . . Of their losing their feelings of modesty and purity, through

[54] GLRO MA/RS/1/742. Essay by G. H. Hine, 1850, 28.
[55] Acton, *Prostitution*, 180.   [56] S. Alexander, *Women's Work in London* (1976).

being obliged to work amongst men and boys, or under immoral or irregular masters or foremen.'[57] Henry Mayhew
shocked his readers by claiming, on what possible foundation it
is not clear, that one in three 'female operatives' in London
was 'unchaste'.[58]

The corrupting effects of work were not confined to working
hours. Precociously acquainted with the very worst aspects of
urban life, working girls had little to restrain them from
drinking or from sexual promiscuity. Young women, finding
themselves with free time, money to spend, and no guidance as
to how to enjoy either, rejected the sober restraints of Church
or home for more alluring diversions, with the result that:
'At fifteen the London factory-bred girl in her vulgar way has
the worldly knowledge of the ordinary female of eighteen or
twenty.'[59] If she was liable to corruption whilst in work, the
dangers once out of it were much greater. Liable to dismissal at
a moment's notice, the migrant female worker was especially
vulnerable. If she had no friends or family nearby to fall back
upon, without wages she would soon lose her lodgings. The
Revd Joseph Kingsmill observed: 'The condition of both sexes
at such a crisis is appalling enough. That of young women is, of
course, the worst.'[60]

Whilst women working in factories were seen to be most
likely to fall into crime or prostitution, even those in domestic
service were not considered immune from risk. Though they
remained within the home, albeit someone else's, and did not
therefore trespass beyond the female sphere, they were still
seen to be prey to corruption. Note the fatalistic tone of the
following warning which opened a mid-century homily headed
'Friendly Hints to Female Servants': 'every day female servants
are convicted of dishonesty; few weeks pass without cases of
gross indecorum; thousands in the course of a year, lose their
character and enter on vicious and destructive courses . . .'[61]
Often the mistress of the house was blamed for failing to super-

---

[57] 'Social Results of the Employment of Girls and Women', contribution by R. S.
Bartleet, 603. Though, in a companion paper under the same heading, R. Smith
Baker argued that women corrupted their fellow workers more than men. With the
result, so he claimed, that 'sisterhoods of vice' operated in many factories (ibid. 541).
[58] Mayhew, *London Labour*, iv. 255.   [59] Greenwood, *Seven Curses*, 15.
[60] J. Kingsmill, *Chapters on Prisons and Prisoners and the Prevention of Crime* (1854), 411.
[61] 'Friendly Hints to Female Servants', *The Female's Friend* (Jan. 1846), 10–11.

vise the girls in her charge, for not allowing them time off to attend Church and yet giving them free evenings to visit theatres and music-halls. Neglectful of their religious duties and exposed to corrupting influences outside the house, female servants were thought to become flighty, unreliable, and careless of their own safety. Mayhew warned dramatically: 'There is no class more essential to the well-being and comfort of society, and none, it is to be feared, more exposed to dangers and temptations than domestic servants.'[62] Their mistresses, having no maternal interest in their welfare, failed to impose the tight moral discipline of a 'proper' home, and did little to prevent them flitting from one post to another. Some moved gradually downwards through progressively less respectable houses; others surrendered themselves to the charms of male members of the household, only to be denounced and dismissed if they became pregnant.

Whilst many writers criticized women's failure to guide or to restrain their female servants, few condemned the damage done by the casual sexual activities of male servants, sons, or masters of the house. Yet, for the female servant, loss of 'character' was disastrous—without it her chances of gaining any respectable employment were extremely limited. Young, pregnant, and friendless, the likelihood of her turning to crime or prostitution was extremely high.

THE CHANGING POSITION OF WOMEN IN LATE VICTORIAN SOCIETY

As we have seen, the economic transformations brought about by industrialization increasingly took women outside the home and into workshops and factories. During the second half of the century the sphere of working-class female activity grew. At work, commuting to and from the workplace, and even enjoying urban amusements with her own wages, she was both more seeing and more seen. Though middle-class woman's activity in the public sphere increased to a lesser extent, it was, perhaps, more shocking still. By the last quarter of the nineteenth century, women were much less cloistered in the home

---

[62] Mayhew, *London Labour*, vol. iv, p. xxvi.

but enjoyed wider social movement and took a growing interest in the affairs of public life. As a result, maintaining the image of the ideal woman as wholly innocent, as the 'angel in the house' unaware of the evils of the world, became more and more difficult. Instead, in the face of knowledge and temptation, it was argued that women's chastity would have to be secured by a more positive sense of personal responsibility. Women were still believed to possess special traits which protected them from temptation but they were now seen to need to resist these greater challenges far more actively. The later nineteenth-century ideal was of women as active moral agents —going out into the world to rescue their less fortunate sisters from sin, and to chastise and restrain their menfolk.

Negating this ideal was 'the girl of the period' (a description first coined in 1868),[63] who was derided as the 'new' or 'modern' woman. She was depicted as fast, irreligious, assertive, outspoken—'bold' was the term most commonly applied. This characterization of the 'new woman' narrowed the traditional divide between the middle-class woman and the fallen woman (prostitute or criminal), for this new image of female deviancy applied at all levels of society. In her excellent study of middle-class murderesses in nineteenth-century England and France, Mary Hartman suggests that the heightened public visibility of the new woman allowed critics to identify 'hitherto irreproachable women with their polar opposites, the adulteresses and whores'.[64] As Hartman's case histories show, the result was that whereas, previously, middle-class women cited in murder cases might have been viewed as above suspicion, they were now pursued and investigated as possible murderesses.

The 'new woman' quickly became a scapegoat, held responsible for the decline of the family. The belief that woman's emergence from her traditional sphere was an increasing cause of female immorality peaked in the last years of the nineteenth century and the early years of the twentieth.[65] Yet as early as the 1870s, the eminent criminal historian L. O. Pike voiced a

[63] The phrase was coined by E. L. Linton, 'The Girl of the Period', *Saturday Review*, 25 (1868), quoted in M. Hartman, *Victorian Murderesses* (1985), 132.

[64] Ibid. 133.

[65] The heyday of this view corresponded with the height of suffragette activity in the campaign for the vote for women, and, significantly, of opposition to it. See below for discussion of the criminalization of the militant members of the suffrage movement.

commonly held fear that woman's emancipation could only lead to her corruption. He argued that 'every step made by woman towards her independence is a step towards that precipice at the bottom of which lies a prison.'[66] As woman moved outside the protection of the home, as she became schooled in adversity, grew self-reliant, and acquired male courage and independence, so, argued Pike, would she become more likely to commit crime.

Towards the end of the century, criminological research began to explore this claim that women's emancipation directly increased the rate of female crime. Comparative studies of crime rates in different areas were used to demonstrate how the extent of women's movement into the public sphere correlated directly with the number of crimes they committed. In 1890, Havelock Ellis claimed that England's 'lead in enlarging the sphere of woman's work had created 50 years of rising female crime rates'.[67] The prison chaplain Revd W. D. Morrison, using a similar argument, claimed that women in the south of England committed less crime than those in the north because they still led a more cloistered life. Taking his comparison to the international level, he argued that the relatively low rate of female crime found in Italy was attributable to the continuing seclusion of women within the home in that country.

Woman's participation in public life—whether at the economic, political, or social level—was believed to reduce her capabilities as a mother and was used as a forcible argument for resisting her movement out of the home. Morrison, for example, insisted: 'One thing at least is certain, that crime will never permanently decrease till the material conditions of existence are such that women will not be called upon to fight the battle of life as men are, but will be able to concentrate their influence on the nurture and education of the young.'[68] By 1914 the arguments remained little changed but their tone was noticeably more resigned to woman's invasion of the public sphere, and, according to her critics, desertion of her maternal role. Hargrave Adam claimed: 'Women have continued, and

[66] L. O. Pike, *A History of Crime in England* (1876), ii. 529.

[67] H. Ellis, *The Criminal* (1890), 217.

[68] W. D. Morrison, *Crime and its Causes* (1891), 157. This view was widely shared, see also the paper by Dr R. C. R. Jordan given under the heading 'The Social Results of the Employment of Girls and Women', *Transactions NAPSS 1868* (1869).

do still continue, to drift farther and farther from the important and responsible duties of maternity, to embark on the, to her, demoralising activities of the prominent and sordid affairs of the world. As a result she is becoming, either directly or indirectly, more and more concerned in crime.'[69] By the time Adam was writing this, women were already gaining entrance to many sectors of the public sphere but this did not prevent him from desiring, and seeking, to force them back into the home: 'we are . . . labouring to restore woman to the home, the child to its mother . . . We wish to call a halt in the path of demoralisation down which the loosening of the family tie is leading the human mind.'

Changes in the position of women were not only seen to affect their relationships with their children but also, more damagingly, with men. Mary Higgs, writing in 1906, perceived the advent of what she called an 'erotic age', in which marriage changed from sensible, business-like partnerships to 'love-matches'. A woman who had already left the home before marriage was more likely to be economically independent and less to be sexually innocent. She would marry not for security and protection but, hastily and without forethought, in passion. Such a marriage was, so Higgs argued, less stable and, therefore, less likely to survive. Woman, whose very nature demanded the satisfaction of marriage and motherhood, would then risk enjoying neither. If her marriage broke down and she was left to fend alone, without the skills to earn an honest living, she would be prey to demoralization. Higgs drew a painful picture of the single, middle-aged woman deserted by her husband (he in turn became a dangerous divorcee), struggling to earn an honest living, and concluded: 'What wonder if she gives up the hard struggle and strays from this path.'[70] The failed love-matches of the so-called 'erotic age' would thus become yet another fruitful source of female criminality.

The worldly and assertive stereotype of the 'new woman' did force some to reappraise traditional assumptions about the need to shelter women protectively from all possible risk. Just how far perceptions of women had changed over the preceding half-century is clear from Higgs's adamant assertion that 'it is perilous and unwise to keep up the old conventional ideas as to

---

[69] Adam, *Woman and Crime*, 300.   [70] Higgs, *Glimpses*, 322.

"innocence" and "purity" being fostered by ignorance.'[71] Instead she proposed that women be educated for marriage and motherhood—a radical departure from traditional efforts to preserve her indefinitely in a state of exalted innocence.

One final aspect of women's movement out of the home and into the public sphere caused unparalleled public outcry and political upset. It also created an entirely new set of female offenders. This was the fight by women to gain access to the political sphere. Many writers roundly condemned the growth of women's involvement in public affairs and their attempts to gain political power.[72] Women's political leagues were criticized as 'militant' and, therefore, unfeminine. W. D. Morrison claimed that the combative nature of political activity was 'undoubtedly tending on the whole to lower the moral nature of women' and that knowledge and skills acquired, for example, in the 'shady tricks of electioneering' could only be corrupting to them.[73]

In the early years of the twentieth century the view that, by political activity, women moved outside femininity allowed opponents of women's suffrage to launch vitriolic attacks on those formerly respectable middle-class women who became suffragettes. When, in 1906, the Women's Social and Political Union embarked on the more militant phase of their campaign, leading activists committed often violent, criminal acts in its pursuit. Opponents of women's suffrage were quick to label them as unbalanced, hysterical, extremist, and highly dangerous.[74] After 1907 they were repeatedly denied the status of political prisoner, despite their claim that their offences were acts of political protest and not ordinary crimes.[75] These middle-class 'ladies', who had formerly represented the very

---

[71] Ibid. 321.

[72] See B. Harrison, *Separate Spheres: The Opposition to Women's Suffrage in Britain* (1978) for a full account of opposition to women's involvement in political life.

[73] Morrison, *Crime and its Causes*, 156.

[74] See discussion in Harrison, *Separate Spheres*. Also B. Harrison, 'The Act of Militancy: Violence and the Suffragettes, 1904–1914', in his *Peaceable Kingdom: Stability and Change in Modern Britain* (1982).

[75] The archives relating to the imprisonment of the suffragettes (including their own letters and petitions) are closed records held under the Public Record Office reference HO144. Inspection is subject to the approval of the Home Office Notice Office. See also L. Radzinowicz and R. Hood, *History of English Criminal Law* (1986), v. 439–60.

ideal of womanhood, offended as much by their political activism as by their violence. As a result they occupied prison cells alongside the common thief and the prostitute.

THE IMPACT OF THE EMERGENCE OF FEMINISM

As the traditional ideology of femininity led suffragettes to be stigmatized as deviant, the very evolution of the feminist movement led to a reappraisal of criminal women. In feminist eyes, the deviant or fallen woman ought not to be automatically condemned for being outside respectable society, but could often be more sympathetically recognized as the victim of that society.[76] Conventionally, marriage was proffered as the only really respectable prospect for woman. And yet hindrances to marriage, such as the unequal numbers of the sexes and economic difficulties for men in maintaining a wife and children, made marriage impossible for many women. Recognizing both the constraints of marriage and the hindrances to attaining it, many feminists rejected this extremely narrow prescription of 'correct' behaviour.

Since the enactment of the Contagious Diseases Acts in the 1860s, feminists campaigning against this legislation had publicly challenged the prevailing view of the fallen woman as a moral pollutant and social pariah.[77] The Acts empowered the police in designated port and garrison towns to pick up and register women suspected of prostitution and have them medically examined. Labelled as a 'queen's woman' under the Act, the prostitute was even more firmly ostracized. By their participation in such a campaign, respectable middle-class women such as Josephine Butler and the members of the Ladies National Association consciously sought to override class difference, and stressed, instead, a common bond of sisterhood. They tried to erase the stigma attached to the fallen woman by deliberately identifying with her plight. She became a paradigm for woman's condition—politically and economic-

---

[76] See e.g. arguments put forward by Craig, 'Emigration'.

[77] See J. Walkowitz, 'Male Vice and Feminist Virtue', *History Workshop*, 13 (1982), 80 ff. Also F. K. Prochaska, *Women and Philanthropy in Nineteenth Century England* (1980), 205–9.

ally powerless, without independent social status, and painfully vulnerable to male exploitation.

By refusing to join in condemning the fallen woman, many female philanthropists such as Josephine Butler, Catherine Booth, Frances Power Cobbe, Ellice Hopkins, Louisa Twining, and Mary Steer began to undermine the very ideology hitherto used to restrain and confine women. Many women's journals ran regular pieces on prostitution, female vagrancy, and criminality. These were not intended to satisfy prurient curiosity but to feed the active interest and concern expressed by many readers. For example, during the 1870s *The Englishwoman's Review* ran a series of investigative reports on management and discipline in female prisons in Britain and abroad, carried out almost exclusively by female correspondents. They not only advocated a more sympathetic attitude towards fallen women but stressed that it was the duty of all women to aid their less fortunate sisters. Similarly, Rosamond Hill in *The English Woman's Journal*, argued that it was the duty of '*women* to stretch forth a helping hand to our unhappy sisters sunk in crime'.[78]

In feminist analyses, then, criminal women and prostitutes were seen less often as evil pariahs but more commonly as somewhat pathetic victims of the social structure, of personal circumstance, or of men's brutality. Note, for example, this description of the fall of one girl: she was said to be 'tossed like a hapless piece of driftwood on the waves of this troublesome world;—beaten, bruised in body and mind; ill-treated, knocked down by drunken men;—dragged, resisting and half-intoxicated, to prison by the police . . .'[79] And Mary Steer summed up the view of many feminists when she applauded a fellow philanthropist who demanded, 'Call them *knocked-down* women if you will, but not fallen.'[80]

Prostitution existed, according to the German Marxist feminist, August Bebel, as an inevitable corollary to marriage.

---

[78] R. Hill, 'A Plea for Women Convicts', *The English Woman's Journal* (Apr. 1864), 134. See also L. M. Hubbard, *The Englishwoman's Year Book for 1881* (1881), esp. 59–60, and Higgs, *Glimpses*, 323.

[79] Scougal, *Scenes*, 126. See also Craig, 'Emigration', 290–1.

[80] M. Steer, 'Rescue Work by Women among Women', in Baroness Burdett-Coutts (ed.), *Woman's Mission* (1893), 157. Also note Miss Todd's view that 'the world is unspeakably harder to a woman who falls than to a man', in her 'Prison Mission and Inebriates' Home', *The Englishwoman's Review* (July 1881), 248.

'Prostitution', he argued, 'becomes a necessary social institution, just as much as the police, the standing army, the church,
the capitalist, etc., etc.'[81] Outside the respectability of marriage
limited employment opportunities and low wages all too often
forced women to seek subsistence in prostitution. A series
headed 'The Outcasts', published in *The Englishwoman* just.
after the turn of the century, sympathetically examined the
plight of a whole range of 'outcast women' and found their
condition to be produced mainly by adverse circumstances.
H. M. Richardson, for example, argued that prostitution was
caused mainly by poverty and that the prostitute was scarcely
more ostracized than the sweated home-worker.[82] Women
turned to prostitution not, as it was often asserted, as a result
of some diseased sexual mania but as a means of escape from
restricted social and economic opportunities.

Feminist analyses strongly decried the double standard
applied in judging behaviour. They argued that women were
condemned as deviant, and even as criminal, for activities
which went unpunished when committed by men. Certainly
the most obvious and extreme case was that of prostitution—
the woman alone paid the high moral and legal price; her male
customer remained unknown and unjudged. They recognized
that this unequal stigmatization had devastating consequences
in a society which demanded, above all else, that women be
chaste. Identifying such double standards did not promote
demands for sexual freedom for women equal to that for men
but, on the contrary, demands for male sexual continence. For
uncurbed male sexuality, it was feared, would only condemn
women to seduction, corruption, and abandonment. Bebel
cited sexual licence, particularly amongst young men, as the
pre-eminent source of all social evil: 'we see that vice, depravity,
error, and crime of all kinds, are bred by our social conditions.
The community is kept in a state of permanent unrest. But it is
the women who suffer most from these things.'[83] Feminists
found themselves in a curious alliance with the social purity
movement, demanding a single standard of chastity for both
sexes.[84] Only through such chastity, both parties argued, could

---

[81] Bebel, *Woman*, 92.   [82] Richardson, 'Outcasts'.
[83] Bebel, *Woman*, 4.
[84] E. J. Bristow, *Vice and Vigilance* (1977), 5; O. Banks, *Faces of Feminism* (1981),
63–8.

woman be spared the perils of seduction and the degradation of prostitution.

The arguments put forward in support of demands for women's suffrage suggested to some writers that female criminality and prostitution was not just the product of women's unequal position in society but the direct result of her political powerlessness. H. M. Richardson went so far as to suggest of prostitution and female crime: 'It is in some ways an expression of women's eternal revolt, now taking new and noble forms, against the State which has been created by men and for men.'[85] According to Richardson such proto-political revolt would continue until woman was herself in a position to 'enforce recognition of her claim to be considered arbiter of her own fate'.[86]

CESARE LOMBROSO AND THE HEYDAY OF BIOLOGICAL EXPLANATION

Towards the end of the nineteenth century, physicians and biologists took a growing interest in social problems. As they did, they sought to replace traditional conceptions of crime as the product of moral choice with more deterministic interpretations. Both Auguste Comte and Herbert Spencer argued that it was possible to derive principles of social organization and action on the basis of biological models.[87] A number, particularly among doctors working in the prison service, argued that it was possible to develop laws of criminality parallel to those of natural science. They even argued that judges evaluating moral responsibility should be replaced by human science experts qualified to decide whether or not an individual was innately criminal. Attention turned to the internal attributes which, it was believed, made people criminal. As a result, explanations focusing on external causes of crime were, to a great extent, overridden by a rising preoccupation with the pathological characteristics of the offender. This mode of enquiry was seen as particularly applicable in trying to understand female criminality. To understand why this was so we

[85] Richardson, 'Outcasts', 4.   [86] Ibid. 8.
[87] Their views were contested, for example, by August Weissman and the Scots biologist Patrick Geddes. See discussion in J. Conway, 'Stereotypes of Femininity in a Theory of Sexual Evolution', *Victorian Studies*, 15: 1 (1970).

need to examine pre-existing male assumptions about the constitutional make-up of women.

The Victorians assumed that woman was biologically inferior to man. The reproductive cycle was seen as such a massive tax on her energies that she had little surplus for other activities or exertions.[88] More damaging still was the view that woman's constitution was fundamentally pathological. Puberty, menstruation, pregnancy, and menopause were seen as so debilitating that woman was left barely fit for 'normal life'. Whilst all women were seen as inherently weak and prone to illness, prostitutes were perceived as fundamentally diseased. They were seen as the very physical (as well as moral) refuse of society. Together with criminals, drunkards, paupers, and beggars, they were stigmatized as the 'effluvia' of urban life. This imagery emanated from an organic view of society as the 'Body Politic'[89]—a body of interdependent parts with the middle-class male as its head, the middle-class female as its heart, and the working class as its hands. Prostitutes and criminal women represented various unmentionable and distasteful bodily functions. Consequently, two interrelated, if somewhat contradictory, biological images of the criminal or fallen woman arose—one as pathological, rotten, and corrupting; the other as a necessary drain of society's effluvia protecting the purity of middle-class wives and daughters at the expense of her own. Whichever one prevailed, the end result was the same—to set criminal women and prostitutes apart as a distinct, debased group within society.

Since mid-century the notion of an entirely distinct criminal class had gained considerable currency. Middle-class fears consequent on the ending of transportation gave strength to this idea of a recognizable 'criminal class'. Since it was no longer possible to embark shiploads of social undesirables to the colonies, they could only accumulate in prisons and slums in England. In 1851 Mary Carpenter identified these groups as the 'perishing and dangerous classes'—those who were liable

---

[88] A view advocated most strongly by Herbert Spencer, see L. Duffin, 'Prisoners of Progress: Women and Evolution', in S. Delamont and L. Duffin (eds.), *Nineteenth Century Woman* (1978), 60. See also Conway, 'Stereotypes', 48–9.

[89] L. Davidoff, 'Class and Gender in Victorian England', in J. Newton *et al.* (eds.), *Sex and Class in Women's History* (1983), 19 ff.

to become criminal and those who already were. Her observations on the origins of convict women illustrate how these
criminal classes were caricatured and stigmatized as distinct
from the rest of humanity: 'convict women usually spring from
a portion of society quite cut off from intercourse with that in
which exists any self-respect, and they are entirely lost to
shame or reputation'; they 'belong to a pariah class, which
exists in our state as something fearfully rotten and polluted'.[90]
Carpenter argued that, whereas male convicts came from all the
lower strata of society, convict women came exclusively from
a debased class, marked 'by a very strong development of the
passions and of the lower nature'.[91] The criminal class was
considered to be not merely culturally distinct but almost a
separate species; members were distinguished not only by their
character, mode of dress, language, or values, but also by their
physical constitution and appearance. And the latter two were
thought to be even more distinctly identifiable in women than
in men. A crude form of physiognomy was applied to identify
and describe criminal women, even by those who actually
worked with them: 'the sinister and hardened look of almost all
. . . the bruised and battered face, the broken down constitution and the loathsome disease, testify but too plainly of the
grossly immoral and abandoned lives they have led.'[92] Massive
migration by what was often wrongly assumed to be healthy
rural stock to the inner city was feared to be sapping the life-
blood of the country. Theories of hereditary urban decline
promoted the fear that the vast and ever-increasing urban
population would sink with each successive generation.

Gareth Stedman Jones has plotted changing conceptions of
this degenerating group or 'residuum' in London in the latter
part of the nineteenth century.[93] Up to the 1870s, though the
residuum was a cause of grave concern, it was thought to be
containable and possibly even eradicable by philanthropic
effort. By the 1880s, however, severe cyclical depression in
London created a deepening social crisis. Growing awareness
of the truly appalling state of the urban poor made middle-class
observers much less sanguine about the possibility of reform.

[90] M. Carpenter, *Our Convicts* (1864), ii. 208.   [91] Ibid. 209.
[92] RDCP, Millbank for 1872, *PP*, 34 (1873), 114.
[93] G. Stedman Jones, *Outcast London* (1971).

Note, for example, this assessment of the population from which criminals were supposed to be recruited: 'The majority are scrofulous, stunted in mind and body, the raw material of the idiots, epileptics, and others with less manifest evidence of physical defects, who fill our asylums with the blind, the halt, and the consumptive and people our prisons with their chief tenants.'[94] The residuum were no longer objects of pity but came to be feared as a degenerate, brutish threat to decent society. Women of the underworld were found especially repulsive and beyond all hope of redemption for, unlike men, they were said to have neither the physical strength nor the mental agility to reverse the tremendous pull of degeneracy.

The common initial reaction on identifying the residuum was one of horror. Middle-class social investigators drew breath and retreated to their suburban homes to write descriptions so fiercely condemnatory as to appear paranoiac. Only towards the end of the century did they return, armed with scientific tools of measurement and analysis, and fortified by the knowledge that the condition of those whom they ventured to study was not contagious but the product of inheritance. Scientifically minded observers, like prison doctors G. Wilson, J. Bruce Thomson and David Nicolson, began to investigate and to describe in minute detail the physical characteristics of criminals.[95] They believed that in doing so it was possible to identify a distinct criminal type.

Havelock Ellis, borrowing heavily from the research of Cesare Lombroso and Enrico Ferri in Italy, claimed that women guilty of infanticide were endowed with excessive down on their faces, that female thieves had underdeveloped teeth, and that criminal women in general went grey more quickly, were uglier, and exhibited more signs of degeneracy (especially of the sexual organs) than ordinary women.[96] Having identified this 'criminal type' he offered an extraordinary explanation of the low levels of female crime, based on a crude model of sexual selection: 'Masculine, unsexed, ugly abnormal women —the women, that is, most strongly marked with the signs of

[94] Mouat, 'Repression of Crime', *Transactions NAPSS 1881* (1882), 259–60.

[95] See Radzinowicz and Hood, *History of English Criminal Law*, vol. v, ch. 1, esp. 3–11.

[96] Ellis, *The Criminal*, *passim*. See also V. Harris, 'The Female Prisoner', *The Nineteenth Century and After* (May 1907), 795 ff.

degeneration and therefore the tendency to criminality—would
be to a large extent passed by in the choice of a mate, and would
tend to be eliminated.'[97] Ellis presented his work as the scientific
observation of a small but very distinct group of women who
were innately criminal. In fact, the attributes he claimed to
have discovered, though partly the result of mental deficiency,
inadequate diet, and physical defect, were also dictated by the
negative image of the ideal woman. The so-called scientific
process of observation was as culturally bound as any other
explanation of female criminality.

Ellis's tentative explorations into criminal anthropology
were somewhat overshadowed by the publication in 1895 of the
English translation of *The Female Offender* by Cesare Lombroso
and William Ferrero. This massive work elaborated the conception of female criminality as biologically determined.
Female crime was portrayed as non-cognitive, non-rational
and, instead, physiological in origin. Lombroso cemented the
view of a criminal woman as atavistic. According to him she
was even more biologically primitive than criminal man. And
yet when Lombroso actually investigated the physical characteristics of women criminals he found it extremely difficult to
confirm his hypothesis. Instead of abandoning his theory as
unfounded, Lombroso invoked the common belief that women
were, in any case, biologically inferior to men: if all women
were to a degree atavistic, he argued, it would obviously be
more difficult to discern atavism in criminal women because
the distinguishing signs would be relatively less marked. He
made an important exception of the prostitute, whom he found
to be the epitome of primitive woman and abundantly endowed
with atavistic traits masked only by the plumpness of her youth
and elaborate make-up.

Closer investigation of the traits Lombroso claimed to observe
reveals how far his supposedly scientific objectivity was, in fact,
determined by a culturally bound conception of what constituted normality, and what deviance, in woman. Moreover, he
evidently found it impossible to confine his explanation to the
supposedly inherent, pathological qualities of his subjects and
referred constantly to their social background and status. The
disinterested status of Lombroso's research was further marred

[97] Ellis, *The Criminal*, 217.

by his patent desire to seek confirmation in biology of his view of woman as in all respects inferior to man. He asserted that 'the normal woman is deficient in moral sense, and possessed of slight criminal tendencies, such as vindictiveness, jealousy, envy, malignity, which are usually neutralised by less sensibility and less intensity of passion. Let a woman, normal in all else, be slightly more excitable than usual, or let a perfectly normal woman be exposed to grave provocations, and these criminal tendencies will take the upper hand.'[98] In appearance he 'found' criminal women to be masculine, even virile, in appearance, coarse-voiced, and unusually strong. His sense of distaste, even horror, was barely concealed in these descriptions of criminal women.

In this scarcely dispassionate frame of mind, Lombroso conducted his famous anthropometric measures. Claiming impartiality and scientific rigour, he was, in fact, severely hampered by the difficulty of testing his measurements of criminal woman against any 'control', or 'normal', women, 'it not being too easy to find subjects who will submit to the experiment'.[99] This was hardly surprising since most women, inculcated with a deep sense of modesty, not unreasonably objected to Lombroso's requests to take measurements of their neck, thigh, and leg. In the end he was obliged to confine his sample of 'normal women' to only fourteen—a tiny number which rather belies the weight often attributed to his lengthy and detailed findings. Undaunted, Lombroso presented his catalogue of supposedly anomalous characteristics as certain proof of atavism in the criminal woman. For example, he claimed that whereas normal women had markedly different skulls from men, the skulls of criminal women more closely resembled those of ancient man. This, argued Lombroso, proved that criminal woman was a throwback to an earlier, less civilized age.

In line with the then current vogue for classification, Lombroso subdivided criminal women into various, supposedly distinct groups: the 'occasional', the 'hysterical', the 'lunatic', the perpetrator of 'crimes of passion', and the 'born female

[98] C. Lombroso and W. Ferrero, *The Female Offender* (1895), 263.
[99] Ibid. 56.

criminal'. Only the relatively small class of born female criminals, 'whose criminal propensities (were) more intense and
more perverse than those of their male prototypes',[100] clearly
fitted Lombroso's conception of degeneracy. The born female
criminal was the very antithesis of femininity: 'excessively
erotic, weak in maternal feeling, inclined to dissipation, astute,
and audacious.'[101] She was even endowed with masculine
characteristics, such as her love of violence and vice, and
'added to these virile characteristics [were] often the worst
qualities of woman: namely, an excessive desire for revenge,
cunning, cruelty, love of dress, and untruthfulness.'[102]
Lombroso condemned the born female criminal as 'completely
and intensely depraved', as 'more terrible than any man'.
That his condemnation derived from an ideology based on
double standards, which demanded a higher morality of
woman than of man, is only too evident from Lombroso's most
damning exclamation: 'As a double exception, the criminal
woman is consequently a monster.'[103] She had not only flouted
the criminal law but, more heinously still, had contravened the
norms of femininity.

Even at the height of Lombroso's influence in Continental
Europe, criminal anthropology was not without its critics,
particularly in English-speaking countries.[104] Apart from Ellis,
most respected commentators were extremely critical of his
claims. In America, the sociologist Frances Kellor reapplied
Lombroso's anthropometric measures on a much larger scale
to every aspect of the female body.[105] Unable to confirm his
findings, she severely criticized the conclusions he had derived
from such tests and refuted his interpretation of female criminality as biologically determined. By 1914 Hargrave Adam, in
England, felt able to repudiate the master criminologist and the
whole doctrine of positivist criminology as 'an utter fallacy'.[106]
He rejected the impressive apparatus of scientific testing and

---

[100] Ibid. 147.

[101] Ibid. 187.   [102] Ibid.   [103] Ibid. 152.

[104] For example, see C. Mercier, *Crime and Criminals* (1918). He was also criticized
by such eminent figures as Arthur Griffiths, Edmund Du Cane, Evelyn Ruggles-
Brise, H. B. Simpson, and James Devon. For the relative ease with which they
demolished his claims, see J. Devon, *The Criminal and the Community* (1912), 18–23.

[105] Using 55 white students, 60 white and 90 negro female criminals as subjects.
F. Kellor, *Experimental Sociology* (1901).   [106] Adam, *Woman and Crime*, 20.

appealed instead to simple commonsense: 'what nonsense it all is. As if we do not know that prostitutes are of all sizes and shapes, from the very thin to the very fat.'[107] His resounding refutation of all things Lombrostan concluded scathingly: 'the mystery is how such a fallacy ever came to be taken seriously . . .'[108]

EARLY TWENTIETH-CENTURY PSYCHIATRIC INTERVENTION

Though Lombroso's influence in England was limited by such widespread criticisms, in the case of female criminality it was obstinately enduring. In the early twentieth century, criminology became dominated by scientific and pseudo-scientific data collection and analysis. Analyses of body type or skull dimension were largely replaced by studies of the psychological characteristics, the aberrations, and motivations of the female offender. As Lombroso had stigmatized woman as physically diseased, so criminologists came to see the criminal woman as mentally deficient or sick. This trend, more recently identified by historians like Robert A. Nye, Roger Smith, and Ruth Harris as the 'medicalization' of deviance, abandoned traditional forms of moral judgement and notions of personal responsibility. Instead it suggested that female crime was preeminently pathological, with the result that judgement became less pertinent than diagnosis.[109]

In order to explain the increasing influence of the medical profession in the understanding of deviance during the latter part of the nineteenth century and the beginning of the twentieth, we need to examine the arena in which they operated. The links and interrelations between the lunatic asylum, the prison, and the inebriate reformatory were considerable. Inmates were liable to be transferred from one to the other, or, over the years, to find themselves incarcerated in more than one type of institution. Staff often served in various different institutions during their careers and would certainly be considered for

[107] Ibid. 23.   [108] Ibid. 28.

[109] R. Harris, *Murders and Madness* (1989), ch. 4, 'Legal Procedure and Medical Intervention'.

employment more favourably if they had previous institutional experience. Many prison doctors had been asylum medical officers or, indeed, held more than one such medical post simultaneously. Janet Saunders, in a local study of the 'asylum system', found that every doctor in Warwickshire prisons and asylums had held more than one institutional post in the course of his career.[110] They were regarded as well qualified, therefore, to make psychiatric assessments about prison inmates and to judge their level of responsibility. After 1865, prison doctors were required to make regular medical inspections of local prison inmates. For the first time they revealed that large numbers of prisoners, though not so bad as to be certifiably insane, were certainly mentally defective. However, assessing the mental state of offenders remained problematic: as late as 1904 the eminent institutional surgeon, Sir William Collins, reporting cases of women moved repeatedly from one institution to another, admitted 'even now the principles which should determine the class of case suitable respectively for a reformary [sic], an asylum, and a prison, are of the haziest description.'[111]

Roger Smith has examined the intervention of medicine in the criminal courts especially in relation to women defendants.[112] He points out that the lack of information concerning medico-legal relations prior to the nineteenth century severely limits the possibility of measuring what degree of change occurred. He also argues that whilst a direct measure of the medicalization thesis—the insanity defence in criminal trials—did become more widely used, it did not become much more successful. By narrowly restricting his analysis of the perceived relation between insanity and criminal responsibility to the case of the criminally insane, Smith fails to consider the far wider reinterpretation of criminal activity as the product of mental defect. He fails to consider the possibility that, as medical explanations

[110] J. F. Saunders, 'Institutionalised Offenders: A Study of the Victorian Institution and Its Inmates', Ph.D. thesis, University of Warwick (May 1983), 51.

[111] W. J. Collins, 'Institutional Treatment', *British Journal of Inebriety* (Jan. 1904), 102.

[112] R. Smith, 'Medicine and Murderous Women in the Mid-Nineteenth Century' (unpublished paper, 1977); R. Smith, 'The Boundary between Insanity and Criminal Responsibility in Nineteenth Century England', in A. Scull (ed.), *Madhouses, Mad-doctors and Madmen* (1981); R. Smith, *Trial by Medicine* (1981).

came to be applied across the whole range of deviance, neat distinctions between the sane and the insane were increasingly difficult to uphold. That is, once all criminals were liable to be diagnosed as mentally defective, the criminally insane no longer constituted so clearly distinct a category. Therefore, counting the number of criminal insanity verdicts alone will neither prove nor disprove any expansion of psychiatric influence.

A more convincing strand of Smith's analysis is his insistence that the 'medicalization' of deviance be integrated with the more general history of science. He argues that the trend towards psychiatric interpretation is only explicable as part of the far wider encroachment of science into public discourse.[113] 'Objective' scientific investigation propagated a deterministic analysis of deviance as the product of the mental inadequacies of the offender. This deterministic discourse was, according to Smith, endemic in late nineteenth-century British culture and by no means exclusive to the vocabulary of the psychiatric profession.

More recently, David Garland has examined the extent to which the nascent discipline of criminology was founded upon close examination of offenders within the prison.[114] This, in turn, drew attention to the peculiarities of the individual's physical and mental make-up. Here, again, the search for hard 'scientific' evidence concerning the aetiology of crime was strongly influenced by psychiatry. Garland, even more than Smith, sees the intervention of the psychiatric profession as vital. Increasingly prestigious itself, it provided criminology with quasi-medical status. In its early years criminology borrowed heavily from psychiatry the vocabulary, theories, categories, and strategies of what was then commonly known as the medical specialism of 'alienism'. Criminological interest in psychiatric interpretation is indicated by the growing numbers of medical men within the penal system who entered into debates in psychology journals.[115] It was from this discourse that

---

[113] Smith, 'Boundary', 364–75.

[114] D. Garland, 'The Criminal and his Science', *British Journal of Criminology*, 25: 2 (1985); D. Garland, *Punishment and Welfare* (1985).

[115] See, for example, the articles by J. B. Thomson, Resident Surgeon to the General Prison for Scotland at Perth, 'The Psychology of Criminals', *Journal of Mental Science* (Oct. 1870), and David Nicolson, Senior Assistant Surgeon, Convict Prison

categories of mental deficiency, of which 'feeble-mindedness' was to become most important, were drawn (see Chapter 7 below).

The relationship between criminology and psychiatry was by no means one-way. By informing and directing the development of criminology, psychiatry sought to expand its influence into the judicial and penal arena. Both Garland and Smith ask to what extent psychiatric intervention should be seen as empire building by a group of new professionals anxious to assert their status.[116] Psychiatrists certainly did not see themselves in that way but rather as impartial observers, elevating criminological understanding with their techniques of observation and classification. How the growth of psychiatric intervention in deviance control should be interpreted remains the subject of much historiographical controversy.[117] Should one see psychiatrists at the turn of the century merely as unscrupulous entrepreneurs? As innovative scientists? As guardians of the moral order? Or, as Vieda Skultans suggests, as agents of social control?

Despite the considerable heterogeneity of psychiatric discourse, a broad shift is discernible over the second half of the century. As we saw at the beginning of this chapter, up to the mid-nineteenth century the predominant criminological approach was moral. It involved examining 'character' to determine the individual's failings, and establishing the means, through external management or self-discipline, to strengthen moral fibre. In the second half of the nineteenth century, however, attention moved from deficiencies of character to the perils of biology, and it was in women that these new interpretations found their fullest, readiest confirmation. Roger Smith points out that, for the Victorians, 'conceptions of women's social position were integrated with naturalistic description of disease types and deterministic explanatory schemes. It was relatively easy to objectify women as part of physical nature.'[118] As we saw in the previous section, all

Department, HM Civil Service, 'The Morbid Psychology of Criminals', *Journal of Mental Science* (July and Oct. 1873).

[116] See Smith, 'Boundary', 375, and Garland, 'Criminal and his Science', 114.
[117] In addition to the preceding refs., see V. Skultans, *Madness and Morals* (1975), and P. McCandless, 'Liberty and Lunacy', *Journal of Social History*, 11: 3 (1978).
[118] Smith, *Trial by Medicine* (1981), 143.

women were seen to be closely bound to nature, the prisoners
of their biology. In turn, woman's psyche was thought to be
intimately connected with the reproductive cycle, the health or
pathology of which directly determined her mental health.[119]
So that even in the normal woman, the round of biological
crises—from puberty through menstruation, pregnancy, and
labour to menopause were fraught with danger. Being particularly vulnerable at each of these points, she was liable to fall
prey to a variety of mental illnesses, such as those catalogued
by Henry Maudsley in his *Body and Mind* (1870): from the
hysterical melancholy of pubescent girls to the recurrent mania
provoked by menstruation.[120]

The degree to which a woman could be seen as prisoner of
her biology directly lessened her culpability, as Maudsley later
affirmed: 'cases have occurred in which women, under the
influence of derangement of their special bodily functions, have
been seized with an impulse, which they have or have not been
able to resist, to kill or to set fire to property or to steal.'[121]
Deviant behaviour could also be explained away as the product
of delicate nerves, emotional disorder, or mental defect—all
directly related to woman's biology. Havelock Ellis argued that
women were so much victims of the monthly cycle that 'whenever a woman has committed any offence against the law, it is
essential that the relation of the act to her monthly cycle should
be ascertained as a matter of routine.'[122] Manifestations of
mental disorder, or even of insanity, were not merely unsurprising but almost to be expected in a constitution innately
predisposed to upheaval and crisis. Nor was this view of female
psychology tied only to her reproductive functions; it found
even greater currency in relation to her sexuality. Middle-class
norms of propriety set the female sexual appetite at zero and
established complete passivity, bordering on indifference, as
the healthy sexual state for women. Any deviation from this
state of sexual apathy could be seen, therefore, as indicative of

[119] Skultans, *Madness and Morals*, 4.
[120] H. Maudsley, 'Lecture Three—On the Relations of Morbid Bodily States to
Disordered Mental States', in his *Body and Mind* (1870), 79–89. See also Nicolson,
'Morbid Psychology', 401–2.
[121] H. Maudsley, *Responsibility in Mental Disease* (1874); this view became the
accepted orthodoxy, see e.g. Edward Fry, *The Problem of the Feeble-Minded* (1909), 22–3.
[122] H. Ellis, *Man and Woman* (1904), 293.

a disordered mind. Almost any expression of sexual desire by a woman could be interpreted as pathological and clinically described as nymphomania. Henry Maudsley saw women as being entirely at the mercy of the state of their reproductive organs: 'Take, for example, the irritation of ovaries or uterus, which is sometimes the direct occasion of *nymphomania*—a disease by which the most chaste and modest woman is transformed into a raging fury of lust.'[123] Pre-marital sex, infidelity, or even open expressions of sexual desire might also be seen as symptoms of this dangerous sexual mania and so justify drastic psychiatric or even surgical intervention.[124] With the ascendancy of 'scientific' discourse around the end of the century, outraged vilifications of the sexually active woman were replaced by increasingly clinical discussion of her sexual exploits as symptomatic of cerebral disorder.

By 1914 Hargrave Adam, putting aside Victorian prudery, was candidly exploring the relation of 'the sex question' (prostitution) to mental defect.[125] Denying the traditional assumption that prostitutes were victims of male lust, perverted by modern life, or driven solely by economic need, he argued that the majority were quite simply nymphomaniacs. And this notion of deviant women as driven by sexual mania was carried over to explain more serious female crimes. Poisoners, murderesses, and vitriol throwers were, according to Adam, primarily sexually motivated. 'Sexual mania' was, he suggested, so widespread a cause of female deviance that he called for research into 'the intimate connection which exists between nervous disorders and crime'.[126]

The most female of crimes—infanticide—particularly lent itself to psychiatric diagnosis. In the previous chapter we observed the rather surprising tendency to try to exonerate women who committed infanticide. Since it was commonly held that after childbirth all women were seriously mentally debilitated, public opinion was primed to excuse the new mother as not fully responsible for her actions. In addition, specific psychiatric disorders were attributed to the infanticidal

---

[123] Maudsley, 'Lecture Three', 82.

[124] E. Showalter, 'Victorian Women and Insanity', in Scull (ed.), *Madhouses*, 328, for discussion of the practice of clitoridectomy.

[125] Adam, *Woman and Crime*, 30.   [126] Ibid. 34.

woman. Immediately after labour women were said to be liable
to 'puerperal mania', a psychosis directly triggered off by the
massive physiological upheaval of giving birth. And if the
woman did not attack her baby until some later period, 'lactational insanity', closely correlated to the stresses of breast-
feeding, was invoked to explain away her behaviour. These
specifically female manias have been interpreted by historians
as legal constructs developed as a means of defending women
accused of infanticide.[127] However, they may well have
reflected the recognition of actual conditions or illnesses. For
example, lactational insanity was very probably the product of
malnutrition in mothers who persisted in trying to breast-feed
their babies though their own diets were wholly insufficient.

Efforts to deny the responsibility of the infanticidal mother
were primarily motivated by the desire to save her from the
death penalty (infanticide was not distinguished from murder,
which was a capital offence). Crucial to the success of such
defence was a diagnosis which ascribed psychological disturbance to social circumstances. Mental weaknesses, seen to be
latent in every woman, were finally exposed by the trauma of
childbirth. Moreover, the stresses of labour were seen to be
increased if there was evidence of poverty, insanitary or miserable surroundings, or cruelty or neglect by the husband. As
C. A. Fyffe observed, the 'abnormal condition of mind attendant on child-birth doubtless passes away within a few days; but
the crushing trials of poverty last through sad years. Hunger,
desertion, broken-heartedness, religious despair, drive women
to take the lives of their children at many ages.'[128] Thus
psychiatric diagnoses were built on to traditional, exculpatory
legal discourse to provide a formidable case for acquitting the
infanticidal mother, even when she was clearly the perpetrator
of the murder in question.[129]

Roger Smith proffers some reasons why Victorian juries
almost invariably sought to acquit 'murderous mothers'.[130]
Firstly, he argues that such women were not seen as a serious

[127] For example, N. Walker, *Crime and Insanity in England* (1968), i. 125.
[128] C. A. Fyffe, 'The Punishment of Infanticide', *The Nineteenth Century* (June
1877), 590.
[129] For further discussion of these points see Smith, *Trial by Medicine* (1981), and
Walker, *Crime and Insanity* (1968), i. 125–30.
[130] See Smith, *Medicine and Murderous Women*.

threat to the social order. Their violence, directed only at babies, was isolated to those who were scarcely full members of society, and it did not threaten the adult population. Secondly, since such women were not considered fully responsible for their actions, they were held to be inappropriate objects of punishment. In 1849 the last woman was hanged for murdering her child; thereafter it became increasingly clear that public opinion would no longer tolerate the hanging of infanticidal women. Perhaps the more remarkable achievement, not considered by Roger Smith, was that the medical profession succeeded in persuading lawyers of the validity of this psychiatric exculpation, effectively replacing traditional legal discourse with that of psychiatry. A third, possible explanation is that the very high rates of infant mortality pertaining in this period inevitably reduced the value of a child's life, so lessening the perceived seriousness of the mother's crime. Often the actual circumstances of the child's death were so unclear that it was impossible to distinguish between natural causes and murder, and so many women could be charged only with 'concealment of birth'.

Psychiatric discourse was employed not only to explain the clearly criminal but to establish standards of mental normality. It was particularly influential in defining the margins between acceptable behaviour and that which, by contravening accepted norms of propriety, could be seen as abnormal. Although psychiatrists set themselves up as objective arbiters of mental normality, they continued to operate entirely within the bounds of their moral world. Their resulting definitions were almost as much a product of contemporary values as were earlier explicitly moral and social interpretations. The prevailing belief that women were mentally, as well as physically, weak ensured, therefore, that they were more readily and enduringly integrated into the psychiatric model.[131]

[131] See Ch. 7 below for further development of this argument in relation to innovations in penal policy around the turn of the century.

PART II

WOMEN IN PRISON: REGIME AND REALITY

3 WOMEN AND PENAL THEORY

This central section examines how the perceptions and explanations of female criminality described in the previous two chapters influenced penal responses to women who broke the law. We begin in this chapter by looking at existing histories of the prison, at developments in penal thinking and policy over the nineteenth century, and at how ideas about the appropriate treatment of women were fitted into wider programmes of prison reform. The following two chapters will go on to examine the two main tiers of penal provision: local prisons and national convict prisons.

RECENT HISTORIOGRAPHY

The traditional 'Whig' history of the prison was presented as a success story.[1] The process of prison reform during the late eighteenth and the nineteenth century was described in terms of gradual progress motivated by humanitarian idealism. By individual effort, public campaign, and the promotion of legislation, middle-class reformers were seen to replace barbarism with benevolence, neglect with sophisticated mechanisms of manipulation, and chaos with uniformity. The professed good intentions of reformers were accepted by many historians at face value as honest, uncomplicated statements of intent. Most importantly the translation of penal ideas into practice was seen to be so straightforward as not to warrant any detailed investigation into their actual realization. In Whig histories, penal theory was all too often assumed to mean just what it said, and penal practice to constitute the perfect realization of theoretical ideals, marred only by the simple flaws of human incompetence. Many such accounts were official administrative

---

[1] e.g. see J. R. S. Whiting, *Prison Reform in Gloucester, 1775–1820* (1975); E. Stockdale, *A Study of Bedford Prison, 1160–1877* (1977).

histories written by retired penal administrators anxious to justify and to celebrate their life-long careers.[2]

More recently social historians and historically minded sociologists have reexamined this optimistic account of progress through reform. Their more analytical attention has focused particularly on the central question of why the prison emerged as the pre-eminent penal solution. They have considered how the prison was conceived, justified, and rationalized, and have examined its purposes, both ostensible and ulterior. In so doing they have sought to establish the wider relationship between the prison and the emergence of other formalized mechanisms for the maintenance of social discipline. All these questions have naturally drawn attention to the literature of penal theory and particularly the rhetoric of reform as expressed in philosophical works, such as those by Jeremy Bentham, and in the policy reports of government officials and prison inspectors. Deeply sceptical of simple notions of humanitarian idealism as the sole impetus behind prison reform, and alert to the problematic and often apparently contradictory nature of much of this literature, recent works have promoted renewed debate in what has become known as the 'revisionist history' of the prison.

The American historian David Rothman was amongst the first to identify a substantial gap between the high-flown conceptions of the reformers and the realities of penal practice. This raised the critical question of whether the prison system was simply failing to meet the aims of its proponents, or whether these avowed aims masked more sinister, hidden intentions.[3] Rothman suggested that either penal policy must be reassessed as inherently flawed, its deep internal contradictions manifesting themselves in the transition from theory to penal practice, or that prisons themselves were far more resistant to reform than traditional histories supposed. He argued perceptively that it was necessary to focus not merely on the rhetoric of reform but also on its implementation; as a result, he concluded that the prison was no more than a costly and irreversible failure. Yet in conducting this study Rothman retained a

---

[2] E. Ruggles-Brise, *The English Prison System* (1921); L. W. Fox, *The English Prison and Borstal Systems* (1952); J. E. Thomas, *The English Prison Officer since 1850* (1972).

[3] D. Rothman, *The Discovery of the Asylum* (1971).

disturbing preoccupation with ideology. In so far as he looked inside the prison it was primarily to assess the realization of ideology and not to consider the life of the prison itself. Consequently, he largely ignored the less high-flown, but nonetheless important, aspects of living conditions and daily routine, the details of human interrelation, frustration, and resignation. Rothman's was a study of the rational in which human emotion and irrationality have no part. Having said this, among revisionist histories, his study remains probably the most sensitive to the complex disparities between penal ideology and prison practice.

Subsequent works, similarly disillusioned with Whig historiography, also critically reevaluated nineteenth-century prison reform. Their tone and purpose can be gleaned by looking at two particularly influential works: Michel Foucault's *Discipline and Punish* (1975) in France and Michael Ignatieff's *A Just Measure of Pain* (1978) in England. Both delved beneath the avowed intentions of the reformers to elicit the 'real' purpose of the prison.

Foucault carried the vogue for concentrating on penal ideology to its logical extreme. He denied that the prison is explicable in relation to the demands of the economic system or even to the aims of social reformers. In doing so he effectively denied that the prison can be assessed in simple terms of success or failure at all. His lofty construction of the 'power/knowledge spiral' rests on a theory of total surveillance as the ulterior purpose of the prison. Complete knowledge of the prisoners is the key to total control. In Foucault's analysis the prison becomes a conceptual apparatus for the imposition of a discipline which extends throughout the whole social body. Endowed with total power, it knows no exceptions, failings, or inconsistencies; it has no need or reason for change. Once reformed, the prison is, he implied, immutable. In Foucault's work the reality of continuing administrative chaos and human error is lost. His cavalier treatment of evidence, the gap between his idealized account and the realities of most nineteenth-century prisons, and, above all, his failure to account for change over time combine to restrict the value of his work for historians. And yet Foucault's thesis undoubtedly forced historians to re-examine incarceration, to question its purposes—both avowed and covert.

On a rather more empirically sensitive level, Michael Ignatieff's *A Just Measure of Pain* analysed prison reform not as
the outcome of benevolently intentioned humanitarianism, but
as the work of the middle classes fearful of the social disorder
produced by industrialization. Looking nostalgically back at
the supposed stability of the old rural order, where religious
faith, deference, and community kept each in his place, prison
reformers were, according to Ignatieff, motivated primarily by
fear. They sought to reassert order by segregating criminals as
the 'dangerous classes'. They then triumphantly persuaded the
poor that imprisonment was legitimate punishment, 'a just
measure of pain'. Ignatieff thereby reclaimed the prison as a
success, though it was serving a quite different function from
that traditionally assumed. For in his view the prison was a
means of disciplining the masses as a whole, not just those few
within its walls. If, in his early work, Ignatieff was sceptical of
the purportedly benign intentions of reformers, he was no less
convinced of the impact of reforming ideology than his Whig
predecessors. He claimed that the ideal of the reformed prison
achieved 'a triumph in the 1840s'.[4] Largely confining his
analysis to the model prison at Pentonville he asserted that it
was 'an enormous success exciting emulation'.[5] Significantly,
he failed to examine just how far down the prison system this
influence extended and neither did he consider whether or not
this initial surge of success was actually maintained.

In an admirably self-critical article, published only three
years later, Ignatieff, describing himself as 'a former, though
unrepentant member of the revisionist school', recognized that
the revisionist histories of the prison had over-schematized
what was in fact a far more complex story.[6] He showed that the
eighteenth-century prison was not so squalid or incompetently
administered, and nineteenth-century reform not so efficiently
or widely carried out, as the reformers had claimed.[7] Taking

[4] M. Ignatieff, *A Just Measure of Pain* (1978), p. xiii.   [5] Ibid. 197.

[6] M. Ignatieff, 'State, Civil Society and Total Institutions: A Critique of Recent
Social Histories of Punishment', in M. Tonry and N. Morris (eds.), *Crime and Justice:
An Annual Review of Research*, 3 (1981); see the reprint in S. Cohen and A. Scull (eds.),
Social Control and the State (1983), 78.

[7] In developing this critique of his earlier assertions, Ignatieff cites, in particular,
the works of M. DeLacy and J. Innes (see Bibliography) as providing evidence as to

up some of the questions raised by this so-called 'counter-revisionist' critique, I will examine whether penal theory ever achieved the degree of unanimity or coherence of practice that this 'triumph of the separate system' perspective would suggest. Evidence from establishments other than the few model penitentiaries suggests that, well after mid-century, the operation of penal policy remained multi-faceted, often contradictory, and always problematic.[8]

Merely to criticize revisionist history is, perhaps, misleading. For it is the very power of the revisionists' highly analytical, schematic reappraisal that has generated renewed, and often intellectually rigorous, debate.[9] As a source of inspiration and interest in this field of history these works are undoubtedly important. And yet they risk over-schematizing; they pay in-sufficient attention to plain fact; and they tend to render history static. By ignoring the existence of conflict or even dissension they deny the possibility of change. Most importantly, for the purposes of this study, the revisionists take no account of gender. They fail to recognize how notions of appropriate male and female roles figured in the development of penal theory; how far penal policy was directed towards one sex; and how, in practice, the very presence of women in prison generated major anomalies. In fact, gender distinctions affected the whole of the penal system. As Patricia O'Brien, one of the few revisionist historians to deal adequately with the question of gender, rightly concludes: 'Sex remained the basis for the difference in institutional response for most of the 19th century.'[10] I will begin by examining how ideas of femininity slotted into the major themes of penal theory, and go on to consider how concern about the appropriate treatment of women generated theories and policies specific to their sex.

the more limited impact of prison reform. Ignatieff in Cohen and Scull, *Social Control*, 77–83.

[8] As M. DeLacy has shown of Lancashire penal administration, *Prison Reform in Lancashire, 1700–1850* (1986).

[9] See, for example, the essays in Cohen and Scull (eds.), *Social Control*; S. Cohen, *Visions of Social Control* (1985); M. Ignatieff, 'Total Institutions and Working Classes', *History Workshop*, 15 (1983); V. Bailey (ed.), *Policing and Punishment in Nineteenth Century Britain* (1981).

[10] P. O'Brien, 'Crime and Punishment as Historical Problem', *Journal of Social History*, 11:4 (1978), 516.

Very little history has been written about women's prisons in England.[11] Either they are dismissed as unimportant because the numbers of women in prison were relatively small, or the question of women's imprisonment has been considered an irrelevant diversion from the major philosophical issues of nineteenth-century penology. The more precise reasons are unclear. Until recently the only history to examine prisons for women in any depth was Ann Smith's *Women in Prison* (1962). Although it is carefully researched and provides a detailed account of administrative change, it makes little attempt to set penal reform within any wider social context or to discern the impact of reforms on daily prison life. Based almost exclusively on parliamentary papers and published sources, it tells the official story, accepting the rhetoric of contemporary reformers in traditional Whig fashion. Sweeping from the Middle Ages to the mid-twentieth century, it sees the history of women's imprisonment as one of incremental but steady improvement. Smith makes only a limited attempt to account for differences in the treatment of women within the criminal justice system. In so far as she does address the reasons for this, she observes that in all past societies the lawmakers have been men. They have prescribed the social code for women and determined for what offences women should receive punishment. Smith recognizes that this has led to different standards in the definition of deviant behavior in women and men. However, she does not address how far this discrepancy was reflected in the response of the criminal justice system to women, or to what degree it led to differential treatment within prisons.

More recently two works have provided more analytical accounts. Both are by sociologists primarily interested in the contemporary treatment of women in prison. These are Frances Heidensohn's *Women and Crime* (1985) and *The Imprisonment of Women* (by Dobash, Dobash and Gutteridge). Heidensohn's primary purpose is to provide an overview of research and writing on women and crime today. Her examination of the history of women's prisons is, therefore, tangential to her main task and necessarily brief. Using mainly

[11] There are, however, a growing number of good monographic studies of women's prisons in America, for example, E. Freedman, *Their Sisters' Keepers* (1981), and N. Hahn Rafter, *Partial Justice* (1985), as well as a myriad of shorter pieces.

secondary sources, her account reiterates the common view
that there was relatively little interest in women's prisons in the
nineteenth century and seeks to explain this assumed lack of
interest. This premise causes Heidensohn to overlook much
interesting evidence. Despite this limitation the book provides
a succinct and incisive interpretation of the development of
penal policy as inseparable from Victorian values and beliefs
about women. Effectively employing historical examples
Heidensohn shows that the aims and ethos of female prisons
are explicable only in the wider context of contemporary
values. In this sense she uses history primarily to illuminate
sociological analysis rather than 'for its own sake'. Significantly
her work also highlights the extent to which the modern-day
penal system is, and is seen to be, dominated by the legacy of
Victorian penal thinking.

In the work of Dobash *et al.* the historical section serves
mainly as a background to later analysis of 'the development of
the imprisonment of women today'. It serves, in particular, as
an introduction to their study of the new prison for women at
Cornton Vale in Scotland.[12] Surveying women's prison history
from the eighteenth century to the present day, their account
leaps curiously from the turn of the century to the 1970s, ignoring any continuing development in between. Like Heidensohn
they, too, are seeking to enhance understanding of contemporary female imprisonment, but their work also has a more
ambitious historical purpose. They aim to offer evidence where
'none has been widely available before' by providing an extensive account of women's imprisonment in the nineteenth
century based on some primary research. Unfortunately, they
barely differentiate between the terms of a few weeks or months
in local prisons, to which the mass of women prisoners were
committed, and the long years of penal servitude served by
more serious offenders in the national convict prisons (see
Chapters 4 and 5 below).

These few studies apart, most historians have ignored the
question of women's imprisonment and nearly all have chosen
to overlook the extent to which penal policy was differentiated
by sex. The latter failing has serious implications for the wider

---

[12] R. P. Dobash, R. E. Dobash and S. Gutteridge, *The Imprisonment of Women*
(1986), 11.

history of imprisonment in that it has led to a general portrayal of the prison as a place untouched by issues of gender. That prisons tended to be masculine environments, in which confrontation and aggression were common modes of interaction, has been downplayed. And that the policy makers themselves responded with tough, militaristic discipline and physical punishment has not generally been recognized as indicative of their predominantly male culture. Even where historians have commented on the masculine culture of male prisons, they have assumed that this was universal—an assumption which, as we will see, is not substantiated by historical research. A recent work by the historian Philip Priestley asserts: 'Prison was a man's world; made by men, for men. Women in prison were seen as somehow anomalous: not foreseen and therefore not legislated for. They were provided with separate quarters and female staff for reasons of modesty and good order—but not otherwise dealt with all that differently.'[13] Given that women made up around a quarter of all those sent to local prisons and around an eighth of all those sent to convict prisons up to the 1880s,[14] to suggest that they were simply 'not foreseen' is patently implausible. Moreover, given the pervasive importance of gender divisions in Victorian society discussed in previous chapters, it would be unwise to assume that women, however aberrant, were regarded and treated like men. Priestley is understandably surprised that women did not receive more attention from penal theorists and administrators, but he is wrong to suggest that they received none. Throughout the penal and criminological literature deviant women were clearly perceived as quite different creatures from men, requiring differentiated treatment appropriate to their sex and a reform programme orientated towards their eventual return to femininity.

To test Priestley's assertion that there was no 'positive policy' for the prison treatment of women in the nineteenth century, we need only examine the sources and nature of that policy. This raises the obvious question—why indeed does he make this assumption? It may be because Government Select

[13] P. Priestley, *Victorian Prison Lives* (1985), 69.

[14] After the 1880s the percentage of women committed to convict prisons fell dramatically to 4 per cent by the end of the century—source 'Judicial Statistics'.

Committees, Royal Commissions, and even Prison Inspectors, were not detailed to deal specifically with women's issues. As a result there is no ready body of conveniently indexed literature on penal policy relating to women. This does not mean that women were ignored or that there is no evidence to be found of 'official' opinion. There is in fact a mass of observations and suggestions concerning women scattered throughout Parliamentary Papers, Home Office, and Local Record Office archives (a perennial problem of researching women's history is not any paucity of information but the difficulty of locating it). The special requirements and purposes of the prison regime in relation to women figure even in the most general works, not to mention the numbers of books and many articles which deal specifically with the problems of imprisoning women.

PRISON REFORM

What were the purported aims of the prison as expressed by the penal theorists?

In the first half of the nineteenth century the impetus for prison reform came from a variety of theorists whose anxiety to ensure the ascendancy of their own beliefs and consequent policies greatly stimulated penal debate. Utilitarianism was one of the major philosophical forces in the development of penal reform. Jeremy Bentham, probably its greatest exponent, argued that punishment must have a positive purpose other than simply inflicting pain. If it did not limit, deter, or reform, then to punish was only to add another evil to that of the existing crime.[15] Prison should not be merely custodial but 'general prevention ought to be the chief end of punishment as it is its real justification.'[16] Prison labour therefore gained a central role as a means of deterrence. Hard labour would serve the complementary functions of deterring the lazy and creating hard-working citizens, conditioned to the discipline of industrial life. The prison, it was hoped, would become a hive of productive industry whose profits would render it self-supporting or, at

[15] J. H. Burns (ed.), *The Collected Works of Jeremy Bentham* (1970). Bentham's philosophy is discussed at length in S. McConville, *A History of English Prison Administration*, i, *1750–1877* (1981), 111–22.

[16] J. Bentham quoted in McConville, *History*, 114.

least, offset the heavy costs of containment: 'Saving the regard
due to life, health, bodily ease, proper instruction, and future
provision, economy ought, in every point of management, to
be the prevalent consideration. No public expense ought to be
incurred, or profit or saving rejected, for the sake either of
punishment or of indulgence.'[17]

Perhaps the most powerful and enduring legacy for penology
was Bentham's creation of a totally new form of prison design.
'The Panopticon' was a prison in which individual cells were so
arranged as to be observable by warders positioned in a central
circular tower. The warders would not be visible to the inmates,
who would live, therefore, with the sense of being continually
under the inspection of their captors. Largely due to Michel
Foucault's exposition in *Discipline and Punish*, Bentham's
Panopticon has attracted most attention for its conceptualiza-
tion of the power of total surveillance. Less often remarked on,
but equally significant, was Bentham's intention that the
Panopticon should operate much like a factory, the management and work of the prison devolved to a subcontractor who
would be encouraged to run it as a profitable enterprise.

Though many prison architects took up his ideas, Bentham's
vision was never built. Nor were prisons ever made self-
financing, let alone profitable. The inculcation of diligence
remained, nonetheless, an integral facet of penal policy.
Indeed, the common use of the treadmill, sometimes to grind
corn or raise water but more often for no other purpose than
to impose deterrent penal labour, underlines the importance
attached to obliging prisoners to work. Interestingly, Bentham
made no special provision for women—they were to be subject
to the same surveillance and the same hard labour as men. In
fact, as we shall see, prison administrators considered hard
labour wholly inappropriate for women and taxed their imagination to find some suitable substitute.

Another important source of penal theory was Evangelicalism, a powerful influence, as we saw in the first chapter, on
early Victorian social policy. Playing on fears about the growing irreligion of the urban poor, the evangelical perspective
focused on the prisoner as a sinner, a subject, therefore, for

[17] Ibid. 115.

redemption. Moral reform was the prime aim of evangelical reformers, and many of them joined together to form the Society
for the Improvement of Prison Discipline, whose committee
included influential Quakers such as Joseph Fry, Samuel
Gurney, Thomas Hancock, and Samuel Hoare.[18] In their
vision of the ideal prison, solitude, prayer, and reflection under
the spiritual guidance of the chaplain became pivotal to the
reformative process. By appealing to the prisoner's conscience
and evoking a sense of guilt for wrong deeds done, the chaplain
sought to elicit genuine penitence. Prison officials were to do all
in their power to establish a Christian presence in the prison,
not least by the example of their own conduct. Significantly,
some criminals were considered so sinful as to threaten to
contaminate would-be reformers.

Despite doubts that certain classes of prisoner could ever be
redeemed, the evangelical impetus to moral reform proved to
be highly influential. The status and influence of the Society
for the Improvement of Prison Discipline was further enhanced
by the involvement of such eminent figures as the Duke of
Gloucester, Lord John Russell, Lord Calthorpe, and the Earl
of Derby. Such influential patronage ensured that the Society's
aims gained wide acceptance, so that by the 1830s the importance of moral reform was generally accepted as a central
purpose of imprisonment.

Although the importance to be accorded to each of the various purposes of the prison caused controversy, early penal
theorists shared many common beliefs and aims. Central were
the twin aims of deterrence and reform. Many insisted on the
need to evolve a regime in which the two might coexist without
negating the effects of the other.[19] Mary Carpenter argued
that, in treating female prisoners: 'The two objects, the deterring from crime and reforming of the criminal ought always to
be associated together; no system is good in which both are not
combined, and no theories of punishment are of any value
unless they can be tested by results.'[20] The legacy of optimism

[18] See W. J. Forsythe, *The Reform of Prisoners 1830–1900* (1987), for a much broader
discussion of the goal of reform in 19th-cent. prisons, particularly chap. 2, 'The
Address to the Spirit 1840–1865'.
[19] See the writings of A. Machonochie, Sir Walter Crofton, Matthew Davenport
Hill, Joshua Jebb, William Crawford, etc.
[20] M. Carpenter, *Our Convicts* (1864), i. 89.

created by the Enlightenment, with its belief in the perfect-
ability of man, suggested to early reformers that it might be
possible to reform even the most badly deformed character.
This would be achieved by gradual retraining through varying
balances of regimentation and hard work, inculcation of habits
of obedience and sobriety, moral and religious influences, and
instruction (though as we shall see in subsequent chapters this
faith was severely dented by the 1860s).

These various means by which moral reform would be ef-
fected comprised the core of a penitentiary ideal which was
lengthily discussed and developed in the rhetoric of penal
reform.[21] The very aim of moral reform reflected an implicit
faith in the exhaustive disciplinary powers invested in the
prison, and particularly in the possibility of exercising total
control over malleable subjects in a completely closed environment. It saw no contradiction in trying to effect reform within
an environment that was utterly removed from that in which
the original causes of crime were located. Or rather it averted
attention from this irresolvable contradiction by focusing on
the character of the offender as the sole subject of reform. The
wider aetiology of his or her offence could then be ignored.

It would be misleading, however, to represent early nineteenth-century penology as an uncontroversial endeavour to
reform the offender. For others, many of whom were deeply
sceptical of the potential for reform, deterrence was paramount.
The aim of deterrence was to be reflected in the grim, high,
crenellated walls of the prison and by the severity of the regime
within. Less controversially, order would be reflected in the
regimentation of barrack-like buildings, careful ventilation,
attention to hygiene, and cleanliness. Just how far the twin
aims of reform and deterrence were compatible within the
prison remained a source of much dissension.

THE 'SILENT SYSTEM'

Whilst this competition between differing priorities aroused
debate, a far deeper rift was generated by competing views of

---

[21] Forsythe, *Reform of Prisoners*, provides detailed references and a critical appraisal
of this literature.

the systems around which the very structure of the prison was to be organized. Throughout the first half of the nineteenth century the greatest single schism lay between advocates of the 'silent system' (also known as 'silent association'—in which prisoners were allowed to live communally but not to talk to one another) and the 'separate system' (in which prisoners were isolated in individual cells and contact between them was reduced to a minimum). These two models originated in regimes evolved in two English prisons at Gloucester and at Southwell, but were far better known because of the rivalry in America between the silent associated system enforced at Auburn Penitentiary, New York, and the separate or solitary confinement evolved at the Western Penitentiary, Philadelphia. Both regimes shared common elements—isolation of inmates from wider society; prevention of communication; a programme of moral reform embodied in a severely disciplined routine. [22]

Advocates of the silent system, of whom perhaps the most notable was George Laval Chesterton, Governor of Coldbath Fields House of Correction, Middlesex, argued that the main purpose of imprisonment was reform. [23] This was most readily achieved by developing powers of self-control and an unquestioning obedience to discipline. Although prisoners were to work 'in association', that is, together, communication was forbidden. Silence was enforced by a large supervisory staff who could employ a wide range of punitive measures to discipline transgressors. The system was based on a profoundly pessimistic view of the offender as corrupt, defiant, and lacking self-control. [24] The system therefore allowed for immediate, severe punishment of any transgression of the silence rule, and so, as Chesterton saw it, forced prisoners to recognize the pains of indiscipline. The silent system also had other advantages—it required no structural alterations to prison buildings and was, therefore, relatively cheap to introduce. More importantly, it was relatively humane, allowing prisoners to associate with one another whilst minimizing corruption.

In England its detractors—amongst whom were prison inspectors William Crawford, Whitworth Russell, Frederic Hill,

[22] For further details see Rothman, *Discovery of the Asylum*, 79 ff.
[23] G. L. Chesterton, *Revelations of Prison Life* (1856), 2 vols.   [24] Ibid. ii. 40.

and John Perry, and Quaker prison reformers Thomas Gurney and J. J. Gurney—were quick to criticize. They argued that the large staff and continual recourse to punishment needed to enforce silence would make it both expensive to maintain and socially costly. Hill, Inspector for the Northern District, considered the very ethos of the silent system to be unnatural: 'When prisoners are brought together . . . they should really associate as human beings, and not be doomed as under the silent system, to eternal silence, with their heads and eyes fixed, like statues in one direction.' [25] Henry Mayhew made the impassioned claim that the very aim of preventing all communication was a denigration of 'the highest gift of the Almighty to man—that wondrous faculty of speech.' [26] He insisted that only communication upon 'vicious subjects' was in itself corrupting. He argued that speech was not only harmless but, if properly directed, potentially beneficial. Similar sentiments were expressed by those obliged to implement the system. For example, the Matron of Tothill Fields House of Correction at Westminster, argued that it was contrary to human nature to expect total silence from women of the lowest class. [27] Since women were considered to be naturally sociable, imposing silent association upon them was seen to be both more difficult and, possibly, more damaging than imposing it on men. Other reservations arose from the fact that women were considered to be naturally weaker and, therefore, lacking in the self-discipline necessary to maintain silence. In the long term, she argued, the more excitable state of the female nervous system made women less able to endure the considerable strain that living in silent association entailed.

In fact, as Matthew Davenport Hill, Revd J. Kingsmill, Henry Mayhew, and many others pointed out, it was often impossible to enforce silence amongst prisoners held together in association. Prisoners could, and did, evolve subtle and sophisticated means of secret communication which effectively undermined the entire system. Cynics observed that prisoners

[25] Thirteenth Report of the Inspector of Prisons: Northern District, *PP*, 36 (1847–8), 368.

[26] H. Mayhew and J. Binney, *The Criminal Prisons of London and Scenes of Prison Life* (1862), 107.

[27] GLRO WA/G/13. Minute Book of the Westminster House of Correction (1868–71), Minute of 3 July 1869.

were more likely to acquire a facility for deception than habits of obedience. Moreover, since women were believed to be more impressionable than men, they were thought to suffer particularly from the corrupting effects of association. Relatively innocent women, convicted only of some petty crime, would, it was argued, be irrevocably corrupted when fellow prisoners unashamedly whispered details of their sexual exploits or extolled the advantages of prostitution. The adverse effects of association among women were highlighted by the close attention paid to the individual moral character of female convicts. According to evidence given by Revd G. de Renzi, Chaplain of Millbank Convict Prison, to the Royal Commission into the Penal Servitude Acts 1878–9: 'the injurious effects of the women being placed in association are immediately seen; there is less disposition to attend to the scripture reader, there is not the same desire to see the Chaplain, the evidence of deterioration is palpable.'[28]

Attempting to maintain any semblance of silence involved a level of surveillance that could only be oppressive. The implications of this surveillance, whether over men or women, were incisively criticized again by Frederic Hill. He argued: 'by turning the officers into constant organs of punishment, it must greatly weaken their moral influence among the prisoners, and tend to prevent that feeling of respect and attachment which it is so desirable to create.'[29] The intended agents of moral reform were reduced to mere disciplinary guardians despised by the prisoners for their pettiness. That their rule provoked much frustration is only too evident; for example, amongst the women at Tothill Fields House of Correction there were 'older criminals, who upon being reported for breaches of silence . . . almost invariably pour forth upon the Officers a torrent of obscene and blasphemous abuse'.[30] The extent to which these practical drawbacks threatened to undermine the very operation of the silent system will become apparent when we examine the internal life of Tothill Fields in chapter 5. In England, despite widespread and trenchant criticism, the silent system, or a

[28] Royal Commission into the Working of the Penal Servitude Acts, Minutes of Evidence, *PP*, 37 (1878–9), 540.
[29] Thirteenth Report: Northern District.
[30] GLRO/WA/G/7. Minute Book (1852–5), 6208.

poorly supervised variant of association, was still common well
past mid-century. Significantly, the Middlesex magistrates (led
by magistrate Peter Laurie), whose jurisdiction extended over
many of the London prisons, remained committed to silent
association. Their main motivation was economy. They both
condemned cellular isolation as ineffective and cruel, and
sought to avoid the heavy expense of constructing enough
separate cells to accommodate every prisoner. Association
evidently persisted even up to the end of the century; as late as
the Paris Prison Congress of 1895 delegates were still condemning the damage done to women by associating in prison with
'thieves, brothel-servants, prostitutes.'[31]

THE 'SEPARATE SYSTEM'

The separate system, as its name implies, sought to isolate
prisoners from one another as far as possible by holding them
in individual cells and, when bringing them together was unavoidable, by reducing contact to a minimum. As early as 1791,
Sir George Onesiphorus Paul, High Sheriff of Gloucestershire,
had pioneered the containment of prisoners in separate cells at
a penitentiary in Gloucester.[32] His initial impetus was the
desire to prevent prisoners from corrupting one another. In the
free association allowed in unreformed prisons corruption was
so rife that it seemed that incarceration encouraged rather than
eliminated vice. Even in prisons which attempted to classify
prisoners and to separate the comparatively innocent from the
'hardened offender', the possibilities for contact were considerable. Paul's experiment had a further, more ambitious aim in
that it hoped that isolation would provide the prisoner with an
opportunity for self-examination and reflection. Under the
separate system, prisoners were completely isolated from one
another. Locked in individual cells for most of the day, they
slept, worked, and ate alone, for it was believed that reflection

[31] Madame d'Abbadie d'Arrast, 'Women in Prison', in Howard Association, *The
Paris Prison Congress 1895* (1895), 8–9.
[32] William J. Forsythe suggests that penal reformers such as Jonas Hanway,
Samuel Denne, and John Howard were promoting such ideas as early as the 1770s, see
his *Reform of Prisoners*, 19.

was most effective if the individual was held apart from the distraction of associating with other prisoners.

Arguably even more influential were similar experiments carried out by Quakers in Pennsylvania and visited by many English observers. Here separation was held up as a means of inducing a profound spiritual transformation in prisoners subject to the weight of isolation from all but beneficial influences. Although this use of isolation was reviewed favourably, most English proponents suggested that separation be mitigated by visits from the chaplain, scripture reader, and warder or matron.[33] As a result it was somewhat less severe than the completely solitary confinement advocated in America. From the establishment of convict prisons, separation was adopted as the official means of holding convicts in the first period of confinement or 'probationary period' (women served this initially at Millbank, see Chapter 5).

Separation was effectively championed in relation to the local prison by the two most influential prison inspectors for the important 'Home District', William Crawford and the Reverend Whitworth Russell. Crawford had travelled to America in the 1830s on a government commission to report on penitentiary systems. There he had become convinced of the value of separation, not least because it fitted with the prevailing view of criminals as sinners and provided a plausible means of achieving moral and spiritual reform. Russell shared his conviction as a result of his own experiences as chaplain to the convict prison at Millbank.[34] In their annual reports during this period the inspectors argued that the loneliness and boredom inflicted by separation would wear down even the most recalcitrant of prisoners and make them receptive to the exhortations of the chaplain. However, they also supported separation partly as a covert means of maximizing their personal control over penal discipline. Inspectors could only achieve this by reducing their reliance on prison warders as agents of discipline and reform. Separation was, therefore, a useful means of reducing human agency to a minimum. It

[33] For fuller discussion of the promotion of separation see U. Henriques, 'The Rise and Decline of the Separate System of Prison Discipline', *Past and Present*, 54 (1972).

[34] Ibid. 61–93. See also J. Field, *Prison Discipline* (1846) concerning the operation of separation at the model prison at Pentonville; Second Report of the Surveyor-General, *PP*, 29 (1847).

seems that both Crawford and Russell overlooked the fact that, in doing so, it placed far greater onus on the powers of those with whom the prisoner did come into contact, not least the prison chaplain, whose own aims were not always in accord with the secular purposes of the regime.

In England the prison system was headed by the innovative penal policy maker Joshua Jebb (appointed Surveyor-General of Prisons in 1844).[35] He looked to the separate system as a means of achieving the three main aims—first to prevent corruption, second to secure moral reform, and third, least controversial of all, to deter. Whilst Jebb had reservations about the psychological impact of long periods of separate confinement, he saw it as a useful first stage in convict discipline. The benefits of limiting contact with other prisoners were widely accepted as even being greater in relation to women. Whitworth Russell advised: 'with Women . . . I would have Silence and Separation strictly observed, for Women contaminate one another even more than the Men do.'[36] Yet again the supposed vulnerability of women demanded even greater protection from the adverse influences of other inmates than was deemed necessary for men.

Only when absolutely necessary were prisoners brought together—for prayer or for hard labour which could not be done in the cell. Even then, elaborate devices were employed to maintain isolation. Prisoners walked in single file, chapels were partitioned into hundreds of tiny stalls, and face masks (or, as in Pentonville, long peaked caps) were issued to reduce any possibility of recognition to a minimum. Though every effort was made to prevent corruption under the separate system, in practice the prisoners found new and ingenious ways to communicate, as will become evident below in Chapters 4 and 5.

After preventing corruption, the second aim of separation was moral reform—a goal that was even more difficult to achieve. The theory was that, denied the companionship and approbation of fellow prisoners, inmates would become more receptive to the good influence of the chaplain. Even the most

---

[35] See E. Stockdale, 'The Rise of Joshua Jebb 1837–1850', *British Journal of Criminology*, 16:2.

[36] W. Russell to Select Committee of the House of Lords on Gaols and Houses of Correction, Minutes of Evidence, *PP*, 11 (1835), 124.

hardened offenders, once burdened with solitude would become crushingly aware of their sins. Of the male prisoner, Henry Mayhew remarked: 'he is urged to make his conduct the subject of his own reflections, because it is almost universally found such self-communion is the precursor of moral amendment.'[37] On the common assumption that women were more impressionable than men, some optimistic observers suggested that they must, therefore, be more susceptible to reform. For example, the Chaplain at Stafford County Gaol argued that young female offenders benefited particularly. Such women, he claimed, 'under the old system, must have gone out corrupted and ruined by association with the most depraved and basest of their sex, have under that now in operation (the separate system) been discharged from prison impressed with better principles, and possessed of a real desire to retrieve their characters and to become useful members of society.'[38] The supposed intensity of women's emotions meant that a successful appeal to the heart might win a woman's trust and secure her enduring loyalty to a higher ideal.

The chaplain became a key agent in the process of reform. As Revd John Field, Chaplain of Reading Gaol, stressed in a lengthy discussion of the advantages of the separate system, the role of the chaplain was to exploit the prisoner's potential for remorse and ultimately for penitence.[39] Significantly Revd John Clay at Preston House of Correction insisted that all men should recognize themselves as sinners and that given time, sympathy, and guidance, even criminals could be reformed.[40] Prison chaplains like Clay, Kingsmill, and Burt spent a great deal of time trying to build up good relations with prisoners. They felt that by treating prisoners as fellow men they gave them the chance to recover their self-esteem and to regain hope.[41] It was in this emphasis on regular exposure to good religious influence that the English version of the separate system differed most markedly from its American blueprint.

[37] Mayhew and Binney, *Criminal Prisons*, 102.

[38] Nineteenth Report of the Inspector of Prisons: Southern and Western District, *PP*, 34 (1854); see also Carpenter, *Our Convicts*, i. 42.

[39] Field, *Prison Discipline*.

[40] W. L. Clay, *The Prison Chaplain* (1861).

[41] See J. Kingsmill, *Chapters on Prisons and Prisoners and the Prevention of Crime (1854)*, 114–18.

Gauging to what extent moral reform had been achieved proved to be irresolvably difficult. The evangelical fervour which often characterized the reform process tended to promote hypocrisy. Prisoners found no difficulty in feigning the appearances of penitence in order to gain whatever benefits might then accrue. Whilst men like Clay were highly talented—perceptive and subtle—in their approach, many other chaplains were far less adept. Clumsy reproaches and admonition were liable to make prisoners feel that they were being judged, and pressurize them to 'confess' their guilt. Joshua Jebb recognized the need to be highly circumspect in accepting that prisoners were truly reformed. Significantly he considered women convicts to be especially adept at dissimulation, and advised 'that as the most favourable side of the character is presented to the Chaplain, the impressions that I may receive need the correction or confirmation of the discipline officers, who see the women constantly'.[42] As supposedly reformed characters reappeared for second and third sentences, faith in the efficacy of moral reform was badly shaken. The historian William Forsythe rightly points out that women in particular 'were seen as a problematic group whose conduct was frequently "uninfluenced by the word of God"'.[43] Despite its continuing commitment to moral reform, even the Howard Association recognized that: 'It is well-known that the least hopeful subjects of moral influence are habitual criminals, and most of all, criminal and debased *women.*'[44] By the end of the century, confidence in the feasibility of reform was greatly undermined. This was partly due to the increasing penological interest in the 'habitual offenders' who defied all reformatory efforts, returning to the prison time and again. As we will show in the final chapters, even those penologists who had been most committed to the idea of reform had to admit that this group at least were irredeemable.

The third and perhaps most important facet of separation was deterrence. The deterrence principle relied on the assumption that its subject was a rational man who could estimate the consequences of his actions and who was capable of weighing

[42] J. Jebb, untitled report, Jebb Papers Box 7 (*c.*1861).
[43] Forsythe, *Reform of Prisoners*, 64.
[44] Howard Association, *Annual Report* (1880), 11.

up the pains of imprisonment against the possible benefits of crime. In practice, offenders rarely operated on the basis of such calculations. Women, considered to be emotional and impulsive in all their actions, were thought to be less prone to such rational calculus and less likely, therefore, to be deterred.

Deterrence was less frequently the subject of debate amongst penologists. This may have been because this purpose was so long established as to be uncontroversial. Or possibly the implication of locking a fellow human being in a single cell for months at a time was too terrifying to contemplate. Michael Ignatieff asserts that the greatest achievement of Pentonville prison was that it achieved that fine balance between terror and humanity—the measure of pain that could be accepted as 'just'.[45] And yet the writing of penal reformers and administrators up to mid-century seem to reflect considerable unease about the damage done by imprisonment. They questioned prisoners' capacity to withstand, physically and more especially mentally, the pains of solitude. Callous or obdurate characters might become more hardened still, whilst those of more ordinary human frailty might collapse into melancholy or even madness.[46] Men had such terrible nightmares they screamed in their sleep and even fell prey to daytime hallucinations. Many considered women even less able to cope, whilst young girls especially were said not to have the mental equipment to benefit from solitude at all. Elizabeth Fry was particularly concerned about the adverse effects of separation on 'very nervous' women. She urged that separation should be applied very cautiously, with matrons given full discretionary powers over its application.[47]

Despite the doubts expressed by those actually working in local prisons, policy makers at the national level were surprisingly sanguine about imposing separation on women, even arguing that the impact of close confinement was less harmful to women than men. They considered women to be 'naturally'

---

[45] Ignatieff, *Just Measure*, 77.

[46] Such consequences were widely feared, see Ch. 5. Also see the writings of Joshua Jebb, for example, letter from J. Jebb to H. Waddington, 'On the Treatment of Female Prisoners' (11 April 1855), Jebb Papers Box 7; Clay, *Prison Chaplain*, 218–19; Mayhew and Binney, *Criminal Prisons*, 103–5.

[47] Evidence by E. Fry, Select Com. of the House of Lords on Gaols and Houses of Correction, Minutes of Evidence (1835), 527.

more sedentary and passive in their habits, and therefore better able to withstand this restriction of their mobility.[48] The troubling vision of active men caged like beasts pacing behind bars simply did not apply to women since they were thought to be inherently lethargic and used to the confines of the home. Many of the witnesses called to give evidence to the Commission into the Working of the Penal Servitude Acts (1878–9) concurred that women stood separation better than men, suffering less from 'depression of the spirits' and 'physical deterioration' and, according to some, even preferring solitary confinement to association.[49]

After the appointment of Sir Edmund Du Cane, a former military man, to head the penal system in 1869, reform undoubtedly took second place to the more readily achievable purposes of deterrence and punishment.[50] A Royal Commission appointed in 1863 to investigate the implementation of penal servitude had expressed a widely held feeling that the regime needed to be tightened up. Whilst Du Cane recognized the importance of reform, he was most concerned to increase the deterrent impact of the prison on the wider population.

Any efforts by humane authorities to alleviate the burden of isolation by implementing a careful reformatory programme might achieve success only at the price of deterrence or punishment. Reading Gaol, for example, was derided for teaching inmates to read—a luxury not enjoyed by the mass of the population. Avoiding such disparities with the outside world placed considerable constraints on penal policy makers. The Poor Law Amendment Act 1834 had laid down in relation to paupers that conditions provided in workhouses should be of a lesser standard than those enjoyed by the poorest workers outside. 'Less eligibility', as this injunction was known, had implications for every aspect of penal provision. The 1834 'less eligibility' clause and its underlying philosophy signalled a moral imperative to penal reformers to apply the same principle in prison administration. For, given the example set by the

[48] For example, see evidence given by Prison Inspector J. G. Perry, Select Committee on Prison Discipline, Minutes of Evidence, *PP*, 17 (1850), 153.

[49] See, for example, evidence given by W. Morrish, a Director of Convict Prisons, and by Mrs S. Seale, Lady Superintendent of Fulham Prison, RC on the Working of the Penal Servitude Acts, Minutes of Evidence (1878–9), 264, 381–3.

[50] E. Du Cane, *The Punishment and Prevention of Crime* (1885).

workhouse authorities, they could hardly justify treating the criminal to better conditions than those of the 'honest' poor.[51]

In practice, to implement a 'less eligible' regime posed an absurd dilemma: to provide conditions lower than those of the honest poor would be to impose filth, squalor, and starvation, so endangering the very lives of the prisoners. And yet to provide decent, clean accommodation and food sufficient to ensure good health would be to provide conditions of relative comfort, and the extremely poor with a positive incentive to crime. Basic humanitarian instincts set the level of provision in many cases; for example, a sufficient diet was considered essential for women as it was feared that their reproductive system would be irreparably damaged by prolonged privation of food.[52] Rusche and Kirchheimer claim that, erring on the side of deterrence, standards were kept at an inhumanely low level, causing malnutrition and even death.[53] Yet, as will become evident in Chapter 4, some women undoubtedly regarded the conditions of prison life as preferable to absolute destitution and certainly better than the more punitive level of provision in some workhouses.

As the historian W. J. Forsythe has observed, the success of the separate system is partly attributable to the propitious circumstances of is promotion.[54] In the 1830s successive Whig governments under Grey and Melbourne fostered a rationalistic approach to social problems, particularly in pursuit of the moral reform of the poor (for example, by the 1834 Poor Law Amendment Act). The House of Lords Select Committee on Gaols and Houses of Correction (1835) came out strongly in favour of the separation, not least because the system had, as we have seen, several influential supporters. Of these the prison inspectors Russell and Crawford were undoubtedly best

[51] The development of a 'less eligible' regime was a major concern of the Select Committee on Prison Discipline in 1850, see especially its Report, *PP*, 17 (1850), 5 and 6.

[52] This fear was frequently a subject for discussion in the reports of the Surveyor-General, the Inspectors of Local Prisons and the Directors of Convict Prisons. It also concerned several of those giving evidence to the two Royal Commissions on the Penal Servitude Acts, see especially *PP*, 21 (1863), 457, 463, 684; *PP*, 38 (1878-9), 846-8. On the issue of diet more generally, see V. Johnston, 'Diet in Workhouses and Prisons 1835 to 1895', D.Phil. thesis (1981).

[53] G. Rusche and O. Kirchheimer, *Punishment and Social Structure* (1939), 107-8.

[54] Forsythe, *Reform of Prisoners*, 28-9.

placed to ensure that the notion of separation was translated into architectural reality. Many of the powers given to the Home Secretary under the 1835 Act—to issue Home Office circulars, to certify prison rulebooks, and to approve architectural plans—devolved on to them. Whilst they met opposition and downright hostility from opponents of the system, they did much to ensure its general adoption. By the middle of the century, fifty prisons had been converted or newly built to run on the separate system, providing cellular accommodation for 13,000 prisoners. However, as we shall see in the following chapter, many local prisons remained unreformed and the realities of life inside were very different from the regime envisaged by reformers.

ELIZABETH FRY AND THE LADIES' PRISON ASSOCIATIONS

We have noted briefly how women were seen to fit into the major theories determining the course of nineteenth-century prison reform, and how far these policies were themselves seen to be appropriate for women. At least as important in determining how women in prison were treated was the impetus generated by women reformers who, not unnaturally, were primarily concerned with members of their own sex. Their thinking was heavily influenced by prevailing assumptions about gender and notions of appropriate femininity. These assumptions underpinned the development of theories and policies specific to women's prison treatment.

Whilst there was some doubt about whether women should be imprisoned at all, it was generally accepted that, if imprisoned, they had to be treated differently from men. Whitworth Russell insisted throughout his lengthy examination by the Select Committee on Gaols and Houses of Correction (1835) that his general proposals on penal policy should not be taken to apply to women and that it was necessary to devise 'a very different System of Penal Discipline' for them. He concluded emphatically: 'I see hardly anything in common between the Case of a Male and a Female Convict.' [55]

---

[55] Select Com. of the House of Lords on Gaols and Houses of Correction, Minutes of Evidence (1835), 124.

Though historians of imprisonment in England have largely
ignored the particular problem of the prison treatment of
women, historians of French and American penal history have
demonstrated how in both these countries it was generally
accepted that women must be held separately and treated
differently from men.[56] In America this resulted in distinct and
highly innovative 'domestic reformatories' whose regimes were
based on notions of ideal female behaviour towards which
women inmates were encouraged to aspire.[57] These reformatories were set up in rural surroundings, usually organized
around large houses or a series of cottages designed to maxim-
ize the potential for domesticating their inmates. The impetus
behind this reformatory movement has been well described by
historians like Estelle Freedman and Nicole Hahn Rafter. In
part it can be situated in the wider context of 'female institution
building'.[58] In part it arose from the American social purity
movement, which demanded the segregation of 'immoral
women' from decent society.[59] The assurance with which these
policies for women were developed in America was never
paralleled in England, where most reformers appear to have
been far less confident of the appropriate response to female
offenders. For example, writing in 1864, Mary Carpenter
lamented: 'It is well known to all persons who have the care of
criminals, either old or young, that the treatment of females is
far more perplexing than that of males. It demands, indeed,
peculiar consideration and comprehension of the special difficulties to be grappled with.'[60] That such a leading and innovative figure in prison reform remained so uncertain, well after
mid-century, as to the appropriate treatment of women does
not imply that the issues were new or unconsidered. Since the
pioneering work of Elizabeth Fry had forced the plight of
women prisoners into the nation's conscience, their treatment
had been the subject of continuing debate. Fry, 'the more than

[56] See, for example, Freedman, *Their Sisters' Keepers*, 54 ff., P. O'Brien, *The Promise
of Punishment* (1982), 62–3, Rafter, *Partial Justice*, and C. Lesselier, 'Les Femmes et la
prison 1820–1939' (unpublished paper, no date).
[57] See, in particular, N. Hahn Rafter, 'Chastizing the Unchaste: Social Control
Functions of a Women's Reformatory 1894–1931', in Cohen and Scull (eds.), *Social
Control*.   [58] Freedman, *Their Sisters' Keepers*, 46.
[59] N. Hahn Rafter, 'Prisons for Women 1790–1980', in M. Tonry and N. Morris
(eds.), *Crime and Justice*, v (1983).   [60] Carpenter, *Our Convicts*, i. 31.

female Howard', became widely known throughout Britain and abroad for exposing the appalling conditions of 'riot, licentiousness, and filth' suffered by women in the female wards of Newgate, and for her lauded efforts to improve the condition of women prisoners generally.

Shocked by her first visit to Newgate in 1813, Fry and a growing number of other Quaker women launched a radical experiment to reform both prison conditions and, by personal influence and religious instruction, women prisoners themselves. The high profile given to their efforts and the enthusiastic publicity awarded their claimed successes drew much public attention to the plight of women in prison.[61] This, in turn, led to the setting up, in 1817, of the pioneering Ladies' Association for the Reformation of Female Prisoners in Newgate; soon to be followed by the wider reaching British Ladies' Society for the Reformation of Female Prisoners. Fry's travels throughout England during the following decades promoted the establishment of numerous associated organizations, and her travels led to the founding of Ladies' Committees in Scotland in 1828 and Ireland in 1834.

These organizations put considerable energy into improving standards of accommodation, establishing regimes appropriate to women, and promoting programmes for moral reform. Eulogistic accounts of these female penal reformers, in the best Whig tradition, have presented their work as having achieved an overwhelming transformation, or, at very least, steady progress in implementing humanitarian reform.[62] Although their work was undoubtedly highly influential, their efforts to organize prison visiting were by no means an unqualified success. For example, whereas initially Fry succeeded in commanding considerable public interest in her work, she found it hard to sustain this level of interest. In many of the Ladies' Prison Associations set up around the country the early enthusiasm generated was not sufficient to keep up the time-consuming work involved. Since their work was often greatly

---

[61] See accounts by T. Timpson, *Memoirs of Mrs Elizabeth Fry* (1847), M. Wrench, *Visits to Female Prisoners at Home and Abroad* (1852), and Clay, *Prison Chaplain*.

[62] In addition to Elizabeth Fry, Sarah Martin at Yarmouth Gaol and Susanna Meredith in London were the subject of similarly admiring biographies, e.g. the anonymous *Sarah Martin: The Prisoner's Friend* (n.d.) and M. A. Lloyd, *Susanna Meredith* (1903).

resented by prison officials, it is hardly surprising that many women simply gave up.[63]

Most accounts focus on Fry's personal achievements and, in particular, on the supposedly miraculous reform of women prisoners who came under the care of the Ladies' Prison Associations. Less attention has been paid to Fry's wide-ranging ideas for prison reform and the impact that her proposals had on subsequent penal policy. Probably the most detailed account of Fry's ideas can be found in her *Observations on the Visiting, Superintending, and Government of Female Prisoners* (1825). Unlike many other works on penal reform of the period it provided a minutely detailed account of exactly how prisons should be run. She insisted on the importance of cleanliness, plain decent clothing, and warm, orderly surroundings as vital to the inculcation of womanly self-respect.[64] Concerning the classification of women prisoners, she proposed that they should be grouped not according to offence but with reference to character, past record, and conduct.[65] Writing about the best means to achieve reform, she argued that work was integral to the process: 'We cannot promote the reformation of such persons more effectually than by making them experimentally acquainted with the sweets of industry.'[66] And finally she believed that, in all aspects of prison treatment, religion was of paramount importance. Fry's own position as a wealthy and highly connected member of an influential group of Quakers meant that she was particularly well placed to promote the non-sectarian religious education of women prisoners. And it was in this emphasis on religious instruction, personal influence, and individual attention from voluntary Lady Visitors that Fry's approach differed most markedly from that proposed by Bentham. Whereas Bentham's scheme advocated uniform treatment, formal direction, rigid adherence to rules, and no individual differentiation between prisoners, Fry went so far as to suggest that willing co-operation and cheerful submission to rules by the

[63] See e.g. the criticisms made by Revd Clay in his *Prison Chaplain*, 81–6.

[64] E. Fry, *Observations on the Visiting, Superintending and Government of Female Prisoners*, 2nd edn. (1827), 55–64.

[65] Ibid. 35–40; see also Fry's evidence, Select Com. of the House of Lords on Gaols and Houses of Correction, Minutes of Evidence (1835), 525.

[66] Fry, *Observations*, 51.

women was a prerequisite to their reform.[67] Her insistence on
the need for individualization became widely accepted as the
most distinctive feature of the treatment of women.

Later in the century, the Directors of Convict Prisons accepted this as a crucial difference between prison regimes for
women and men and stressed: 'Male convicts must be treated
in masses rather than according to their individual characters.
Individuality must be more regarded with female convicts.'[68]
This view was generally accepted and reflected the common
belief that women would benefit from close, personal attention. That this must be provided by women was perhaps the
most influential of Fry's tenets. She insisted: 'It is absolutely
essential to the proper order and regulation of every prison,
that the female prisoners should be placed under the super-
intendence of officers of their own sex.'[69] Quite apart from
pointing out the appalling examples of corruption and exploitation which occurred when female prisoners were guarded by,
or even merely accessible to, male warders, Fry emphasized
the positive role a 'respectable' woman might fulfil as 'a consistent example of propriety and virtue'.[70] That women should
be held separately from men, that they should be supervised
only by women officers, and that no male officer should be
allowed to visit the female part of the prison unless accompanied by a female officer were all subsequently made legal
requirements by Peel's Gaol Act of 1823 (4 Geo. IV, c. 64).
Significantly, the responsibilities that this imposed on the
female warder were far more exacting than the mere wording
of the legislation suggested, and certainly more so than those
imposed on her male counterpart. Though male officers were
also expected to be honest and to demonstrate qualities of
integrity and impartiality, in women's prisons the character
and conduct of the warders became the very means of effecting
reform and had, therefore, to be exemplary.

A report by the predominantly Quaker Society for the Improvement of Prison Discipline, published the year after this

---

[67] E. Fry, *Sketch of the Origins and Results of Ladies' Prison Associations* (1827), 7–8.
[68] Reports of the Directors of Convict Prisons, Millbank for 1856, *PP*, 23 (1857),
48. Some went so far as to argue for the 'complete individuality of treatment' for
women—A Prison Matron, *Prison Characters drawn from Life with Suggestions for Prison
Government* (1866), i. 297.   [69] Fry, *Observations*, 27.   [70] Ibid. 31.

Act, indicates the very high expectations this placed on the female warder: 'In the exercise of her duties she is at once the representative and guardian of her sex, and she ought to be a bright example of its purity, disinterestedness, and love.'[71] The aim was to make the female warder a maternal figure who would, by benevolent encouragement, engender in inmates a strong desire to please. There was a marked gap between the informal and subtle process of modifying women prisoners' behaviour and the highly symbolic tariff of punishment that officially applied, and which was indeed generally enforced in male prisons. Over the course of the century female super-intendence in women's prisons became an undisputed necessity, so that by 1881 Louisa Hubbard (reformer and editor of *The Englishwoman's Year Book*) could applaud as generally recognized 'the obvious impossibility of masculine superintendence in feminine institutions'.[72] This view was prompted above all by a vision of the ideal female prison as a supremely feminine institution founded on such attributes as sympathy, compassion, caring, and kindness.

Towards the latter part of the nineteenth century, energetic women philanthropists, seeking to extend their sphere of jurisdiction, insisted not only that female prisons be staffed by women but increasingly on the importance of female government in all types of public institution.[73] Louisa Twining, an eminent philanthropist and Secretary of the Workhouse Visiting Society, argued that women's knowledge and experience of domestic management, household economy, sickness, and nursing made them particularly suited to run institutions such as prisons, workhouses, and asylums. Such demands ensured that the role of women in the staffing and management of the female prison was an important topic of consideration for the Departmental Committee on Prisons set up in 1894. Indeed, several of the Committee's subsequent recommendations

[71] Society for the Improvement of Prison Discipline, *Sixth Report* (1824), 65.

[72] L. M. Hubbard, *The Englishwoman's Year Book for 1881* (1881), 95. In fact this view was in wide circulation some years earlier, see, for example, 'Women's Work in Prisons', in E. Pears (ed.), *Prisons at Home and Abroad: Transactions of the International Penitentiary Congress 1872*; and J. A. Bremner, 'What Improvements are Required in the System of Discipline in County and Borough Prisons?', *Transactions NAPSS 1873* (1874), 282–5.

[73] L. Twining, 'Women as Public Servants', *The Nineteenth Century* (Dec. 1890); also F. K. Prochaska, *Women and Philanthropy in Nineteenth Century England* (1980), *passim*.

related to the need to increase the numbers of female warders in prisons, to create new management posts for women, and to appoint a lady superintendent as a general supervisor of women's prisons.[74] By the end of the century, then, the need for distinctively female government of women's prisons appears to have been officially recognized.

A ROLE FOR 'LADY VISITORS'

Debate concerning the role of women in official posts within the female prison was paralleled by a more controversial debate over voluntary women workers or 'Lady Visitors'. Their work was based on the fundamental assumption discussed above that women had special powers of influence. It was felt that bringing female prisoners into contact with their benevolent 'betters' would provide them with concrete demonstration of qualities to which they might aspire. The British Ladies' Society felt that women prisoners were respectful to Lady Visitors because: 'They appreciate refinement of manners and high motives, while conscious of the sad contrast in themselves.'[75] And the hope that women would seek self-improvement in this way was not necessarily limited only to the co-operative and impressionable; for example, the Howard Association insisted that 'the most degraded females can, in general, only be elevated and reformed through the instrumentality of the best of their own sex.'[76] Clearly here, 'best' referred primarily to social status. In later reports the Association demanded such visitors 'must be real ladies', whilst Florence Davenport-Hill applauded the fact that 'they are *ladies* in the truest sense of the word.'[77] This stress on the social standing of Lady Visitors clearly differentiated their personal qualities, and therefore the high example they set, from the less exalted attributes of the lower-class women actually employed as prison warders. It also suggested

[74] See Report of the Departmental Committee on Prisons, *PP*, 56 (1895).

[75] C. Frazer, 'The Origin and Progress of the British Ladies' Society for Promoting the Reformation of Female Prisoners', *Transactions NAPSS 1862* (1863), 498.

[76] Howard Association, 'County and Borough Prisons', a letter from William Tallack to *The Times*, 6 Dec. 1879 (1880), 8.

[77] Howard Association, *Annual Report 1878*, 9; F. Davenport-Hill, 'Women Prison Visitors', *The Englishwoman's Review* (1885), 543.

that merely by association with such 'pure, high, and holy women', inmates would somehow be spiritually uplifted. And, throughout the literature on prison visiting, the language of 'raising up' intermingled imagery of religious salvation with the more prosaic purpose of 'bettering oneself'.

If, in this respect, the purpose of visiting was as much based on class as on gender, the means by which Lady Visitors were to exert their influence was peculiarly feminine, and it is no doubt significant that there was no male equivalent of the Lady Visitor. Modelled on the devoted, emotionally charged relations common between upper-class Victorian women, the close relation between Lady Visitor and female prisoner was intended to break down the latter's stubborn resistance to external influences which so often obstructed the rather less energetic efforts at reform carried out by overworked or ill-qualified staff. It was argued: 'Official staff may do their duty and achieve much, but it is not possible for them to give the individual sympathy and patient attention to every member of a community . . . needed to win the confidence of persons made suspicious and distrustful by years of guilt, and maybe of ill-usage too.'[78] A philanthropic lady with little call on her time might devote sufficient attention to win confidence and so establish a friendship which would continue throughout the sentence and, ideally, even after discharge. They operated on a one-to-one basis in the belief that 'confidence can only be won through the medium of the sympathy of woman with woman.'[79] And wherever possible they elicited close, if deferential, relations similar to those between a servant and her middle-class mistress. It is perhaps instructive to note that middle-class women were inculcated with a very strong sense of responsibility for the moral welfare of the servants in their household and this was clearly carried over into the relationships they developed with women in prison. Significantly, the Lady Visitors sought to gain a degree of personal influence that would not have been possible within the more rigid, quasi-militaristic organization of male prisons.

Lady Visitors themselves extolled the benefits of their intervention. However, others not only feared that those sheltered

[78] Davenport-Hill, 'Women Prison Visitors', 542.
[79] M. Wrench, *Visits*, 30.

middle-class women who took up visiting would be corrupted by the experience, but expressed doubts about the efficacy of their work. For example, Arthur Griffiths, Deputy Governor of Millbank, dismissed Lady Visitors as no more than 'amateurs'; he claimed their work was 'imperfect and incomplete', that, by over-emphasizing religion, they encouraged lip-service and hypocrisy, and that, by focusing on the worst-behaved women, they tended to neglect 'good' prisoners and so place a premium on misconduct.[80]

How far these criticisms were prompted by the failings of an endeavour which was ill-thought out and which operated on a voluntary and consequently highly variable basis, and how far they were prompted by prison officials' fear of the threat that outside visitors posed to their authority is not clear. Less ambiguous is the direct confrontation between the values on which prison visiting was based and those of the official regime. The growing concern under Du Cane's administration for uniformity, deterrence, and even bureaucratic secrecy was inimical to the individualistic, emotional, and often dilettante activities of the Lady Visitors. As central government control over prisons increased, many women philanthropists were pushed out of the prison and forced to concentrate their activities on providing aid for discharged prisoners. After the nationalization of local prisons in 1877, unofficial intervention of any sort (including prison visiting, academic inquiry, and journalistic investigation) was decried as intrusive and disruptive to the smooth running of the regime. In order to circumvent these new objections, visiting necessarily became more organized, scheduled, and, as a result, more impersonal. This new approach meant that visiting was no longer organized on an *ad hoc*, voluntary basis by philanthropists but became integrated as formal, institutional provision. It led eventually to the setting up of a National Association of Lady Visitors in 1901 under the presidency of Adeline, Duchess of Bedford. A highly organized and efficient body, it successfully gained the approbation of the Chairman of the Prison Commissioners, Evelyn Ruggles-Brise.[81]

[80] A. Griffiths, *Memorials of Millbank and Chapters in Prison History* (1884), 202–5.
[81] See PRO PriCom 7/174 for details.

THE LIMITS TO REFORM

The 'revisionist' prison histories examined earlier in this chapter focus on the birth of the prison and they close at mid-century. They tend to suggest that by this time issues of penal policy were resolved; that once the separate system prevailed in England it became uncontroversial and was therefore rapidly introduced throughout the system. In fact, as Ignatieff later conceded,[82] uncertainty about the system continued well after mid-century. Under the leadership of the Director General of Convict Prisons, Sir Joshua Jebb, attempts were made to mitigate the severities of the regime in convict prisons. But these were ended after his death in 1863 and, in the midst of public concern about the efficacy of penal servitude (sparked by the 'garrotting panic' of 1862),[83] were subsequently reversed. The efforts of local prison administrators to introduce separation continued despite growing disillusionment with its use in convict prisons. Certainly, implementing separation in convict prisons had necessitated a vastly expensive building programme. Once cellular convict prisons had been built the separate system was not so easily alterable, as we can see from the persistence of such prisons even today.[84] However, the effects of these architectural constraints should not be read to indicate general satisfaction with the system. Nor should it be assumed, as Ignatieff seems to imply, that the aura of success surrounding separation in the 1840s meant that all issues of penal policy were subsumed within, and settled by, this single philosophy.

Other issues, such as how to classify offenders, how to impose rewards and punishments, the reformative role of religion, and, not least, the imperatives of economy, engrossed and perplexed penal reformers. Moreover, the implications of containing in the home country, long term, those who had

[82] Ignatieff, 'State, Civil Society'.

[83] J. Davis, 'The London Garrotting Panic of 1862', in V. A. C. Gatrell *et al.* (eds.), *Crime and the Law* (1980).

[84] See M. H. Tomlinson, ' "Prison Palaces": a reappraisal of Early Victorian Prisons 1835–77', *Bulletin of the Institute of Historical Research* (May 1978); also R. Evans, *The Fabrication of Virtue: English Prison Architecture 1750–1840* (1982), 388–91.

formerly been transported were only beginning to impinge. As prisoners showed themselves to be remarkably impervious to 'good influence', the simple need to maintain order became an imperative. Under the leadership of Du Cane, increasingly punitive discipline replaced Jebb's loftier vision of moral reclamation.[85] The reform of local prisons was even slower and less uniform. Since separation was expensive to introduce, given the massive structural alterations it generally required, it was not widely adopted until well after mid-century. Revd Clay, a proponent of separation concerned to secure its extension, freely admitted that by 1856 only about one-third of local prisons had fully implemented the system (see Chapter 4 below).

Debate about the appropriate prison treatment of women was stepped up in the last quarter of the century by the growing number of feminists who not only saw social and welfare issues as a female province but who, like women reformers earlier in the century, saw women prisoners in particular as demanding their special attention. They, too, decried the consistent failure of penal policy makers to take sufficient account of women in prison. But whereas earlier women reformers had demanded only a supervisory role, by the end of the nineteenth century women were seeking to establish their influence over prison administration and in the wider development of penal and social policy. Such claims became intimately bound up with demands for their participation in all aspects of government. The reformer Eliza Orme, for example, lamented that 'our prison administration is entirely in the hands of men, and partly from ignorance of the wants and characteristics of women, and partly from fear of doing more harm than good, the Commissioners turn a deaf ear to suggestions of radical reform.'[86] She called for a greater role for women not only as prison managers but by the creation of national posts for women as prison inspectors and commissioners in order that

[85] As propounded in his annual reports as the first Surveyor-General of Prisons and throughout his writings held in the Jebb Papers at the LSE. For more detailed discussion of changing developments in policy relating specifically to convict prisons, see Ch. 5 below.

[86] E. Orme, 'Our Female Criminals', *Fortnightly Review*, 69 (1898), 790; see also similar sentiments expressed by S. M. Amos, 'The Prison Treatment of Women', *Contemporary Review* (June 1898), 807–9.

they might gain positions of influence. Others saw the problem as more profound, being only one aspect of a fundamental disregard for the rights of women extending throughout society. Sarah Amos argued: 'It must also be remembered that a Parliament which cannot find the right time for considering the openly proclaimed civic grievances of free women citizens cannot be in the mood to give due weight to those cloistered griefs that crush the womanhood of those who have fallen so far below the ideal as to be classed among criminals . . .'[87]

As was shown in Chapter 2, criminal women were seen by many feminists as victims of a social structure that provided few opportunities for all women and was especially inhospitable to those who sought to support themselves. Consequently, much feminist thinking on penal policy opposed prevailing views on the deterrent and primarily punitive role of the prison. Instead, like earlier moral reformers, the feminists argued that the primary purpose of penal treatment should be to bring about some good in its charges. Yet, since they saw criminal women not so much as innately evil but as socially disadvantaged, they placed far less emphasis on moral improvement and more on a practical programme intended to equip women better to lead an honest life once released. They deplored the overriding concern amongst prison administrators to reduce running costs, for example by employing women prisoners in drudge work around the prison, as a false saving. And they called instead for proper industrial training which would fit women for careers in skilled work on release.[88] By thus providing women with the means to self-sufficiency, such a regime would strike at the very source of much crime.

In addition to the economic reasons for female crime, it was increasingly recognized that many women in prison suffered from mental or physical incapacities which left them unable to survive unprotected in society. Such women, it was argued, should be removed from prisons to specialized institutions orientated towards dealing with their particular problems (see Chapters 6 and 7 below). To ensure recognition of that class of women who should not have been in prison at all, feminists

[87] Amos, 'Prison Treatment', 803.
[88] See Orme, 'Our Female Criminals', 793–4, and Amos, 'Prison Treatment', 809.

called for the appointment of female medical officers in
women's prisons and of female medical inspectors at national
level.[89] It was not by chance that the first Lady Inspector of
Prisons, Mary Gordon, appointed after persistent demands
from various women's organizations, was herself a qualified
doctor and a committed feminist.[90]

Perhaps the greatest influence on the treatment of women
prisoners came as a quite indirect result of the suffragette
campaign for votes for women. When suffragettes turned in
desperation to militant and often violent tactics in the pursuit
of their cause, they quickly found themselves in the hands of
the criminal justice system. The repeated imprisonment of
militant suffragettes between 1905 and 1914 gave numbers
of educated and influential women personal experience of
conditions in prison for the first time.[91] Despite the fact that
they insisted that they were not ordinary criminals but political
prisoners, they nonetheless took a keen interest in the criminal
women confined alongside them. The suffragettes gave unprecedented publicity to the plight of women prisoners. They
continually barraged the Home Office with demands for the
differential treatment of women. And they made many specific
recommendations for changes to practices and regulations, not
least those they regarded as unnecessarily humiliating and
degrading to women (such as the requirement that they have
their hair shorn off).[92] Critics as vociferous as these could not
easily be ignored and a frantic, private correspondence between
top Home Office officials gave 'grave consideration' to their

[89] Such demands, remaining largely unmet, continued into the 1920s—see the
mass of letters, proposals, and petitions from leading bodies such as the Consultative
Committee of Women's Organisations and the Women's Freedom League in
PRO HO45/16184.

[90] Though the medical orientation of her work was denied by the Home Office and
her very appointment derided as a mere 'sop to feminism'. E. Troup, 21 December
1919, in PRO HO45 10552/163497.

[91] B. Harrison, 'The Act of Militancy: Violence and the Suffragettes, 1909–1914',
in his *Peaceable Kingdom* (1982); 'The Political Offender', ch. 13 in L. Radzinowicz and
R. Hood, *A History of English Criminal Law and its Administration from 1750*, v (1986).
Papers, petitions, reports, and correspondence relating to the imprisonment of the
suffragettes held at PRO HO144 were kindly made available for consultation by
agreement of HO Notice Office.

[92] See, for example, PRO HO144 1042/183256/20; see also the later account in
M. Gordon, *Penal Discipline* (1922), 37–9.

'weight and importance'. Perhaps surprisingly, the officials admitted that 'the superior training and culture of many of them have induced useful criticisms as regards the treatment of women prisoners.'[93] If some suffragettes made positive proposals, their ally Mary Gordon, on taking office as Lady Inspector, became highly disillusioned with the philanthropic, reformist spirit, which she saw as masking the profoundly damaging impact of imprisonment.[94] She published a damning condemnation of the prison system, insisting: 'When we adopt the puerile view that "reformative influences" of any such kind, or that systematic marks, stages, and rewards, or a kind, mild and even rule, or an evening's pleasure, or any other creation from *our* phantasies will serve our turn with our crushed but untameable man or woman, we only cover up the root causes of our failure.'[95] Gordon went on to develop psychologically informed strategies designed not to subjugate women prisoners but to give them back a sense of control. Earlier penologists had insisted that women lacked the rationality necessary for self-control and self-improvement. Women's 'natural dependency' had dictated a prison regime which sought to exploit this lack of confidence and reliance on others to good effect. Gordon, however, believed that only by promoting independence, albeit circumscribed by the demands of maintaining order, could one foster a sense of social responsibility.

Despite the general silence of penal historians on issues of gender and despite the belief of at least one historian that these issues failed to generate anything that might be identified as gender-specific policy, such policies did exist. As William Forsythe rightly recognizes, 'the almost exclusive male world of prison discipline theorists and analysts was never entirely at

[93] PRO HO144 1042/18325/17. Correspondence between Sir Evelyn Ruggles-Brise, the Director of Prisons, and Sir Edward Troup, the Permanent Under-Secretary. The impact of the suffragettes on the imprisonment of women in the longer term falls beyond the period of my study but is very clearly enunciated in Gordon, *Penal Discipline*. Overlooked by most accounts is the startling fact that Gordon was not only sympathetic and supportive of suffragette militancy but that she secretly contributed to the Women's Social and Political Union Defence Fund. Her involvement was uncovered and vehemently condemned by the Home Office after a surprise raid on the offices of the WSPU in 1914. PRO HO45 10552/163497.

[94] Gordon, *Penal Discipline*, 203.   [95] Ibid. 205.

ease in its attitude to women prisoners.'[96] It was this very unease that led them to adapt, to modify, and often to mitigate the regime developed for women's prisons. The most coherent sources of penal policy for women lay mainly outside official government policy-making circles and arose from highly publicized but largely voluntary endeavours. They were not without influence in promoting a general acceptance that women required separate and different treatment from men—so pressing further demands on a system already fraught with contradiction and controversy.

[96] Forsythe, *Reform of Prisoners*, 131.

4 WOMEN IN LOCAL PRISONS
1850-1877

The histories of penal reform examined in the previous chapter tend to focus on national 'model' institutions for men, failing to take account of the wider translation of theory into policies and practices in the mass of local prisons. However, during the second half of the nineteenth century around 98 per cent of women sentenced by the courts to custody were committed to local prisons. This chapter looks at local prisons after mid-century and up to their nationalization in 1877, and seeks to reveal the extent of their diversity and divergence from the theoretical model of reform. Local prisons were under the juris-diction of local magistrates and managed by boards of Visiting Justices.[1] Their regime, size, and standards of accommodation varied enormously according to local conditions, the attitudes of local magistracy, and the number and nature of the inmates.

Whilst drawing on records, reports, and enquiries pertaining to local prisons in general, this chapter focuses on one prison in particular—the House of Correction, Tothill Fields, West-minster—which catered primarily, and after 1860 exclusively, for women. Whereas the records of many prisons have been lost, its records are remarkably well preserved. As a result it is possible to reconstruct a composite picture more closely approaching the 'realities' of prison life than external reports alone allow. Many recent historians have written about the prison primarily in terms of intentions and goals. Few have attempted to see how penal theory was translated in practice in the national convict prisons, let alone in local prisons (though there are one or two notable exceptions such as Ignatieff's vivid picture of the model prison at Pentonville or Margaret DeLacy's study of local prisons in Lancashire).[2] As DeLacy has pointed out, a major flaw in much of this writing is that

[1] Visiting Justices were first authorized by an Act of 1785; their supervisory activities varied greatly by area from minimal, ritualized inspection to considerable involvement in the management of the prison backed up by frequent, unannounced visits.

[2] M. Ignatieff, *A Just Measure of Pain* (1978); M. DeLacy, *Prison Reform in Lancashire 1700-1850* (1986).

historians 'have tended to rely on the writings of the very men they are so anxious to discredit: middle-class reformers, for their depiction of prison conditions in the mid-nineteenth century. On the whole, however, these reformers neither ran the prisons nor inhabited them.'[3] DeLacy's criticism is a trenchant one, that there is more to understanding prison history than analysing its supposed functions—either the avowed purposes of the penal theorists or even the more pragmatic ones of prison administrators. The significant actors in the life of the prison are not those who merely philosophize about it from afar but those who as inmates and warders are inhabitants within its walls. Only by recreating the daily interaction of staff and inmates, their accommodation to one another, and their evasion of regulations, does the gap between theory and practice become clear. As the American sociologists Georg Rusche and Otto Kirchheimer insisted as early as 1939, 'it is necessary to strip from the social institution of punishment its ideological veils and juristic appearance and to describe its real relationships.'[4]

Belatedly taking up this challenge, this chapter explores the premise that the daily life of the prison was created less by the rulebook than by the interrelations of those obliged to live under its aegis. It is an attempt, therefore to write the history of women in local prisons not only from above but, more ambitiously, from 'within'.

LOCAL PRISONS AT MID-CENTURY

Since the seventeenth century a series of partial provisions and temporary Acts had given justices powers to build, repair, and administer gaols. These powers were consolidated and made permanent in 1719 (6 Geo. I, c. 19.), but the development of prisons under local authority administration remained severely limited by a disinclination to increase the burden on the rates.[5] As a result the development of local prisons up to the nineteenth century remained haphazard and varied greatly depending on

[3] M. DeLacy, 'Grinding Men Good? Lancashire Prisons at Mid-Century', in V. Bailey (ed.), *Policing and Punishment in Nineteenth Century Britain* (1981), 182.

[4] G. Rusche and O. Kirchheimer, *Punishment and Social Structure* (1939), 5.

[5] See C. Harding *et al.*, *Imprisonment in England and Wales* (1985), 97.

the interest, or lack of it, amongst local justices. The passing of the Gaol Act 1823 (4 Geo. IV, c. 64.) was intended to achieve some measure of uniformity amongst local prisons by requiring that they seek to 'improve the morals of the Prisoners' by 'due Classification, Inspection, regular Labour and Employment, and Religious and Moral Instruction'.[6] The Act, however, lacked sanctions and appears to have been widely ignored. Compromise and accommodation to existing structures were more salient features of the efforts to improve local prisons than any coherent programme of radical reform. Given their proximity to the workhouse system, local prisons were far more deeply enmeshed in the trap of 'less eligibility'[7] than national convict prisons. This constraint, combined with an eagle eye for economy, placed severe limitations on any possibility of emulating the ambitious achievements of the model penitentiaries, or even of more limited reform. When separation was introduced into local prisons it was primarily as a deterrent and a means of preventing corruption. The fact that most local prisoners were sentenced to extremely short terms made any effort to launch programmes of moral reform extremely problematic. During sentences of a few days, or at most months, lasting change could scarcely be expected. The most such sentences could hope to achieve was to punish and deter through the severity of their regime. Jebb, though he was committed to the pursuit of moral reform in convict prisons, argued in his evidence to the Carnarvon Committee in 1863, 'no one can hope to reform during very short periods, and if you fail to deter, you miss both objects of a sentence.'[8] Contesting Jebb's conclusion was a determinedly sanguine faith in the possibility of effecting moral reform held by prison chaplains such as Revd Clay at Preston Gaol. Clay saw all imprisonment as a potential means of moral reform and, refusing to admit its impracticability in local prisons, called for longer sentences to allow time for complete rehabilitation.[9]

[6] Quoted ibid. 144.

[7] The requirement that conditions within workhouses be of a lesser standard than that enjoyed by the poorest, honest labourers living outside—this requirement effectively obliged prisons to adhere to the same low levels of provision.

[8] J. Jebb, 'Carnarvon Committee, Minutes of Evidence' (1863), quoted in S. McConville, *A History of English Prison Administration*, i, *1750–1877* (1981), 348.

[9] For a discussion of W. L. Clay's ideas see DeLacy 'Grinding Men Good?', 208.

Around mid-century, debate raged over the possible purposes and conflicting roles of the local prison. Far from representing the zenith of penal reform as many historians have suggested, this period was characterized by greater uncertainty than the early years of reform, not least in the light of actual experiences of its implementation. These experiences revealed how far the various aims of the prison pulled reformers in different directions and were often simply not compatible with one another. As Hepworth Dixon, an authority on prisons in London, guardedly conceded in 1850, 'our prison will be very imperfect until it is both reformatory and deterring. At present we are in an experimental stage.'[10] Attempts to synthesize these twin aims, or more commonly vacillation between the two, reflected an irresolvable conflict in the very purpose of imprisonment which long continued to dog the development of the local prison. With a clarity of perspective acquired only by time, Sidney and Beatrice Webb characterized its history and its continuing heterogeneity, even after nationalization in 1877, as comprehensible only in terms of the ebb and flow of these independent and often conflicting purposes.[11]

Before going on to examine the drama of human relations within prison we must establish the condition of women's local prisons at mid-century: the stage, scenes, script, and props which combined to furnish the dramatic setting of mid-Victorian prison reform. The set, and indeed the plot played out against it, varied from venue to venue, from prison to prison. The general impression gleaned from reports of the Prison Inspectors[12] over the period 1850–77 is of considerable diversity.

Throughout the second half of the nineteenth century, women consistently made up around a quarter of those committed to county and borough prisons (Appendix, Table 6). Details of the length of sentences to which men and women were committed are not available until after the reorganization of the Judicial Statistics in 1893. However, figures for the daily average number of prisoners, which are available for the earlier period, show that women comprised only about one-fifth of

[10] H. Dixon, *The London Prisons* (1850), 18.

[11] S. and B. Webb, *English Prisons Under Local Government* (1922).

[12] Set up in 1835, the Prison Inspectors published annual reports until they were replaced by the Prison Commissioners in 1877.

the local prison population at any one time, suggesting that they were committed for slightly shorter periods than men (Appendix, Table 5). Whether this represented greater leniency on the part of the courts towards women, indicated that women committed less serious offences, or simply reflected a recognition that local prison accommodation for women was, both in quantity and quality, inadequate is debatable. Certainly many observers, not least the editor of 'Judicial Statistics', commented on the courts' greater reluctance 'to convict a woman on evidence on which a man would be sent to prison'.[13] Though this cannot easily be proved, it is clear that around mid-century local prisons had insufficient accommodation for the numbers of women committed to them. In 1857, local prisons were 'constructed to contain' only 4,842 women but the greatest number held at one time rose to 4,962. The resultant shortfall of 120 places does not adequately convey the extent of overcrowding this entailed in what were often already cramped quarters. Although accommodation increased dramatically (most larger prisons had sufficient accommodation, and were systematically organized in rough accordance with official regulations) the larger number of small rural prisons with low prisoner populations were run on a more pragmatic basis with little regard for the dictates of contemporary penal thinking or regulations. In the smallest prisons, which often had very few women, at times only one, or none at all, provision was at best makeshift and women were unhappily housed in inferior rooms, an attic or wing of the male prison. The governor's wife or sister became unofficial matron or schoolmistress, and if the local chaplain held services at all, women were often excluded for lack of space. Sean McConville cites an instance, reported in 1865, of a woman held for five months in solitary confinement in Helston Borough Gaol whilst awaiting trail—'for much of this time she was the only inmate.'[14]

In many smaller gaols and houses of correction, female prisoners were barely separated from the males and could communicate and even meet with them. Inspectorate reports for the early 1850s repeatedly complained that the separation of the sexes was inadequate. When women were housed in a single room within the male prison it was virtually impossible

---

[13] 'Judicial Statistics', *PP*, 108 (1895), 86.   [14] McConville, *History*, 366.

to prevent communication. Many governors frankly admitted their inability to segregate prisoners. One revealed, 'I frequently detect communication going on by notes and otherwise, between the male and female prisoners, and often hear obscene conversation between them.'[15] More disruptive still were the possibilities such proximity gave for outrageous violations of prison discipline, as Revd Clay, Chaplain of Preston Gaol, admitted: 'I have known even females climb over the chevaux-de-frise (a wire fence), which I should have thought utterly impossible, in order to get into the ward of the other sex.'[16]

The ideal solution, to remove all the women to a separate institution, was all too often considered impracticable on grounds of expense as there were too few women to warrant entirely separate provision. Nonetheless, Hepworth Dixon, in praise of a detached building for women at Wakefield, stressed that 'women are found a great deal easier to manage when removed to a distance from the men. The spirit of reckless stubbornness and bravado dies within them when they know that they are out of sight, hearing, and notice of their fellows of the other sex.'[17]

Accommodating women generally remained an afterthought to be fitted in with least effort and expense, with the result that they often suffered much worse conditions than men convicted of similar offences. Provision was generally maintained at a minimum, so that even a small rise in the number of commitments could overburden available accommodation. For example, in the female division of Carmarthen Gaol, in 1857, inspectors found fourteen women and two children sharing only five beds in filthy conditions. Overcrowding in inadequate accommodation probably did more to undermine the intentions of the disciplinary regime than any other single fact, and often resulted in open breaches of the law. Female debtors were often held together with women criminals. Untried, and therefore technically innocent, women shared beds with tried and convicted criminals.[18]

[15] Report of the House of Correction, Spalding, in the Sixteenth Report of the Inspectors of Prisons: Northern and Eastern Districts, *PP*, 29 (1851), 590.

[16] Report of the County House of Correction, Preston, in the Fifteenth Report of the Inspectors of Prisons: Northern and Eastern Districts, *PP*, 28 (1850), 567.

[17] Dixon, *London Prisons*, 304.

[18] Sixteenth Report of the Inspectors of Prisons: Southern and Western Districts, *PP*, 27 (1851), 711–12.

Over the period many prisons were amalgamated or closed, so that the number of those in use fell from 187 in 1850 to 112 in 1877. This undoubtedly reduced such anomalies and made the imposition of uniformity in the prisons remaining somewhat easier. But so long as prisons remained under local government the potential for rationalization remained largely unfulfilled. Taking stock of existing provision in 1877, the newly established Prison Commissioners were clearly horrified by the lack of uniformity. In a few cases they could find no identifiable system at all. Indeed it is difficult to identify the salient traits of prison policy even within a single county. To say, for example, that the Lancashire County Bench's prime consideration was economy [19] or that the Middlesex magistracy was renownedly retrograde [20] is to oversimplify. Policies varied over the period and, perhaps more importantly, were often largely determined by the temperament and opinions of the governors and matrons of individual prisons. A strong-minded or an incompetent keeper could do as much to determine the nature of the regime as any outside influence.

Given the extent of this diversity, it seems barely meaningful to talk of the local prison in general terms at all. I intend to take the example of Tothill Fields, Westminster, though not because it was in any way typical (there being no such thing). Whilst the separate system was increasingly taken up by many local prisons during the 1870s, Tothill Fields continued to allow prisoners to associate. However, its very atypicality as a prison run on the silent associated system, solely for women and children, serves as a convenient point of reference for analysing more common regimes. At the same time it highlights many of those traits less readily observable, but nonetheless important, in the many other local prisons with few female inmates. Most importantly, because its records are unusually complete and deal exclusively with women, they amplify characteristics common to women's prisons, rendering them discernible in a way not found elsewhere.

Tothill Fields was originally established in 1618 as a Bridewell, one of the early houses of correction that served as forerunners of the modern prison in instituting the idea of reform through punishment. It was reopened in new buildings as a

19 As Margaret DeLacy suggests in 'Grinding Men Good?', 184.
20 As Philip Collins claims in *Dickens and Crime* (1964), 70.

local prison in 1834. After 1845 it took convicted prisoners only
and in 1850 a committee was set up to consider how to ration-
alize its administration further still.[21] As a result it was decided
to remove all women from the neighbouring prison at Coldbath
Fields and likewise men over the age of 17 from Tothill Fields
in order that separate supervision might be carried out more
cheaply and efficiently in each. Tothill Fields was redesignated
to receive convicted boys under 17 years and all women sentenced in Middlesex to imprisonment or awaiting removal to
national convict prisons (under sentences of transportation or
penal servitude). Boys were housed in a separate wing, women
before trial were kept separate from those already convicted,
and a nursery was provided for women bringing babies or
having them in prison.

Over the subsequent decade the number of boys fell until, in
1859, the Visiting Justices petitioned the authorities of Coldbath Fields to admit all boys then held at Tothill Fields.[22] In
1860 Tothill Fields became a prison for convicted women only.
(There were a very few other local prisons which took only
women, such as Wymondham in Cambridge and Borough
Compter in Southwark, London.) Because Tothill Fields was
dealing mainly, and latterly solely, with women, debates about
appropriate treatment and the specific provisions, problems,
and solutions pertinent to their sex, were of a length and depth
rare in the records of institutions which contained only a small
minority of women. Despite the possible distortions that this
unusual, exclusive concern with women may have engendered,
the records of Tothill Fields are valuable in that they make
explicit the unique characteristics of the prison treatment of
women.

THE FABRIC OF THE LOCAL PRISON

Although Tothill Fields achieved higher standards of accommodation than were common at mid-century, separate cells

---

[21] GLRO/WA/G/5. Minute Book of Westminster House of Correction, Tothill
Fields—minutes of meetings of Visiting Justices (1849–50), 2868.
[22] GLRO/WA/G/9. Minute Book (1857–60), 7419–20.

were available for only 240 of the daily average of 430 women.[23] To the evident concern of the Visiting Justices the remainder were accommodated 'in Dormitories of a most objectionable construction, wanting in space and ventilation, and so imperfect are the means of Supervision as to render any attempt to keep up discipline during the many hours they pass in them, not only doubtful but absolutely impossible.'[24] Ironically the inspector's evidence to the Select Committee in the same year claimed that, despite the large numbers held in each dormitory, effective supervision was possible.[25] This inconsistency between the evident concern of the Visiting Justices and the complacent assertion of the Prison Inspector indicates that the latter's acquaintance with the prison was somewhat superficial. It underlines the danger in relying exclusively on external accounts of institutions by remote outsiders.

The standard of the separate cells in Tothill Fields was scarcely better. Before 1865 women were locked up for up to twelve hours a day in small, unlit, unheated, and inadequately ventilated cells. Moreover, the walls were so thin it was impossible to prevent communication between them. Although accommodation was repeatedly increased (a whole new wing was added in 1857 providing 131 new, adequately fitted cells), by 1869 the Visiting Justices resolved that the prison was so badly overcrowded that it was simply impossible to maintain discipline. In desperation they applied to the Secretary of State for an order to remove up to a hundred women to Newgate.[26] Despite this, the problem continued well into the 1870s. Although by 1870 the prison had 771 cells, fit at least for sleeping in, the number of women often rose well above this figure, reaching a high, in 1873, of 828. Thus whilst separation spread rapidly through other local prisons, Tothill Fields provided separate cells only for sleeping in at night and, even then, evidently not for all its prisoners.

According to the testimony of one prisoner, Susan Willis Fletcher, conditions remained poor well into the 1880s: cells

[23] H. Mayhew and J. Binney, *The Criminal Prisons of London and Scenes of Prison Life* (1862), 368.

[24] GLRO/WA/G/6. Minute Book 1850 (1850–2), 3001.

[25] Evidence of Captain W. J. Williams, Inspector for the Home District, Minutes of Evidence, Select Committee on Prison Discipline, *PP*, 17 (1850), 68.

[26] GLRO/WA/G/13. Minute Book, minutes of April and May 1869 (1868–71).

were cold, damp, and dark; women were expected to keep their cells clean but not provided with any means of doing so; and bathing facilities were almost non-existent. As an upper-class woman, and thus a wholly untypical prisoner, Fletcher was cogent and scathing in her criticisms of such a regime: 'Cold, darkness, silence, and solitude . . . are not curative or reformatory, or humanizing influences. They disease the body, and depress, stupefy, and debase the mind. Their tendency is to fill it with gloom, hatred, and desperation.'[27]

DISCIPLINARY SETTINGS

Prison Labour

Despite the emphasis on productive labour salient in Benthamism, local prison administrators found it difficult to secure employment that was both productive and punitive. An additional problem was the need to find work that demanded little skill or which could be learned quickly by prisoners undergoing very short sentences. The problem occupied the energies of Visiting Justices and inspectors who were critical of the compromises resorted to by governors and matrons in order to impose sentences of hard labour. Devices such as the treadwheel were resorted to as the only sufficiently punitive method of enforcing hard labour on short sentence prisoners. Writing in 1856, George Laval Chesterton, a prison governor, recorded the banning of its use for women by a Middlesex magistrate, Mr Walesby. Walesby regarded its use as 'unfeminine', 'indelicate', and even 'cruel', but his decision was highly controversial.[28] Chesterton clearly thought such concern was inappropriate in relation to the women under his care: 'it was absurd to argue that a harmless mechanical occupation could degrade nine-tenths of such women as we beheld consigned to it.' In Chesterton's view, women prisoners could not expect the same chivalrous sympathy accorded to their more morally upright sisters.

---

[27] Fletcher was imprisoned for obtaining jewellery and clothing by false pretences. S. W. Fletcher, *Twelve Months in an English Prison* (1884), 404.

[28] G. L. Chesterton, *Revelations* (1856), i. 220-3.

Denied use of the treadwheel, the authorities of Tothill Fields found great difficulty in substituting suitable alternatives. In his report for 1857–8 the Inspector for the Southern District, J. G. Perry, castigated the prison for imposing only 'the lightest and least disagreeable pursuits' upon women supposedly sentenced to hard labour.[29] In subsequent years Perry continued to upbraid the Visiting Justices for allowing a regime so soft it could only fail to inculcate habits of industry. His specific criticisms included: women were allowed twelve hours per day in bed; they achieved only half the output required of comparable workhouse inmates; and the tasks allotted were so small that the women completed them by 3 p.m. each day.[30] The debate between Perry and the Visiting Justices highlights the contradictory nature of the local prison's aims and purposes. To leave labour at present levels provided an obvious incentive for workhouse inmates to commit offences in order to enjoy the 'more eligible' regime of the prison. To lengthen the working day would entail the potential dangers of bringing prisoners together in workrooms on dark evenings; an innovation which the Governor assured the Visiting Justices 'would lead to much mischief, as they would delight in causing confusion and committing damage when they could not easily be detected'.[31] And yet to impose labour upon inmates in the safety of isolation involved the considerable cost of extending gas lighting to all cells. The purely penal aims of deterrence and 'less eligibility' were here in open conflict with the more pragmatic exigencies of security and economy.[32] Rejecting 'repulsive and unfeminine forms of hard labour',[33] the search to find more suitable work for women was further complicated by the need to avoid undercutting poor seamstresses and laundry women in honest labour outside the prison. At Tothill Fields, the women employed in these typically female trades worked only in the service of the county, and so avoided the risk of undercutting ordinary

[29] Twenty-third Report of the Inspector of Prisons: Southern District, *PP*, 29 (1857–8), 133.

[30] GLRO/WA/G/9. Minute Book (1857–60), 7343–4.   [31] Ibid. 7348.

[32] The issue was partially resolved by the allocation of six additional staff but the working day was not lengthened to ten hours until 1864 when 400 of the cells were fitted with gas lights. GLRO/RS/2/75, Reports of Middlesex Prisons (1865), Appendix B: Keeper's Report.

[33] Mayhew and Binney, *Criminal Prisons*, 476.

women workers. Mayhew, on his visit to the prison in 1862, praised the feminine appearance of women sitting in total silence knitting, straw-plaiting, and bonnet-making. Yet the prison's statistics show that over half the women were, in fact, employed at oakum picking, retrieving the tar from old rope. In subsequent years the proportion of women employed in this monotonous, unpleasant, and dirty task rose steadily, reaching nearly three-quarters of inmates by 1877.[34] The realities of prison management clearly confounded the avowed aim of penal reformers to return women to the ideal state of femininity.

The Prison School

The educational achievements possible in local prisons were similarly limited. Evidence as to literacy levels amongst the local prison population is of dubious reliability. There appears to have been no directive from the Home Office to prison chaplains as to how to assess prisoner literacy. Therefore the criteria used very probably varied both over time and from prison to prison, depending on the judgement of the chaplain. Also, in the 'Judicial Statistics', the headings under which levels of literacy were reported were very vague, referring only to 'imperfect' or 'superior' levels of instruction. Although the general literacy of prisoners slowly improved over the period, they continued to represent the least literate sections of the population. Well over 90 per cent of prisoners remained in the bottom two categories: wholly illiterate, or only able to 'read, or read and write imperfectly'. Though this latter category increased, especially toward the end of the century, prisoner literacy did not keep pace with that of the population as a whole (especially since this improved considerably after the 1870 Education Act). Women prisoners were consistently less well educated than men and were more likely to be completely illiterate. The higher proportion of illiterate women prisoners partly reflected the poorer education of women in general but also suggests that women prisoners tended to come from an even more socially deprived, outcast group than men.

Women received at Tothill Fields were marginally better educated than the national average for women in local prisons

[34] GLRO/MA/RS/2/71–90. Reports of Middlesex Prisons (1862–77), *passim.*

but not much. Roughly a third of them could neither read nor write; of the rest, all but a very few were poorly educated. As late as 1877, scarcely more than 1 per cent were said to have a superior education or were able to read and write well.[35] No proper schoolmistress was appointed to improve these standards but instead, under the supervision of the chaplain, three female warders were selected from the rest to act as teachers to the 500 to 800 women (the daily average population fluctuated considerably over the period 1850 to 1877). Although open to all women, classes were in practice attended only by those below twenty-four years, and mostly by only the very young.

Schooling was supposed to impart basic skills in reading, writing, spelling, and catechism, but prisoners sentenced to only a few days or weeks had no time to acquire proficiency in any of these. Proud claims by the Chaplain, in the prison's annual reports, about his exertions in distributing Bibles, prayer books, religious and moral tracts seem ridiculous given the illiteracy of the vast majority of the women. It is scarcely surprising that even those with basic literary skills often lacked the motivation to attempt to read such sober literature. Instead they tore out pages to send notes to their 'pals' or even ripped whole volumes to shreds in outbursts of anger.

The Prison Chapel

In theory the greater spirituality of womankind and woman's supposed susceptibility to external influences should have ensured a key role for religion in the reclamation of women prisoners. In practice the local prison chaplain could hope to achieve very little influence over women who knew that they would be free at the end of that week or the next. At least one prisoner recognized the limited power of the chaplain's reformative powers over 'the constantly changing swarm of bloated, drunken, miserable women sent to prison for short terms'. She observed: 'If the chaplains could make any impression upon them when they got sober, the moment they got outside the

[35] These, and all the following statistics relating to Tothill Fields given in this chapter, are compiled from the Annual Returns of Tothill Fields as bound in volumes in the series GLRO/RS/2/71–90, Reports of Middlesex Prisons (1862–77).

walls, it was drowned in drunken riot.'[36] The Chaplain of Tothill Fields began by claiming modest success for his attempts to promote reform, through 'the influence of kind and Christian treatment'.[37] But significantly his reports became steadily shorter and less varied over subsequent years, suggesting a decline in his influence, or, perhaps, that he slowly became resigned to the hopelessness of his task.

His role was complicated by lengthy and increasingly embittered altercations over the rights of access and influence accorded to Roman Catholic priests. In the late 1860s, successive priests sought to enforce their legal right to visit Catholic prisoners in Tothill Fields. As more than a third of the women professed to be Catholics,[38] the priest's claim had both considerable justification and serious implications for the religious life of the prison. Attempts by Visiting Justices to impede the priest's freedom of access, to restrict facilities to hold mass instruction, or even to distribute religious literature to Catholics in the prison, led the Secretary of State to intervene in 1867 condemning the Justices for their intolerance. However, the Visiting Justices, to say nothing of officials within the prison, remained unsympathetic to the rights of Catholic women.

For more than a decade, the minutes of the Justices' Committee reflected a depth of religious acrimony which confounds the Christian example supposedly set before the prisoners. Only with the dismissal of the Catholic priest Revd A. White for his part in a leak to the *Pall Mall Gazette* (27 August 1869) on the ill-treatment of Catholic women prisoners, and the appointment of a new priest, does the issue finally appear to have been resolved. This dispute clearly struck at the heart of Church of England dominance within the prison. Religion, in this lengthy and often acrimonious debate, appears less as a focus for spiritual concern than as a point of conflict between powers competing for influence. And the women prisoners appear less as potential subjects for conversion than mere pawns in a political battle for their allegiance.

[36] Fletcher, *Twelve Months*, 403.

[37] GLRO/WA/G/6. Minute Book (1850–2), 4222.

[38] This high proportion of Catholics may partly be explained by reference to the 'Judicial Statistics'. These show that, nationally, twice as many women (20–23 per cent) as men (11–12 per cent) coming into local prisons gave their place of birth as Ireland. It is likely that many of these Irish-born women were Roman Catholics.

The Prison Infirmary

The infirmary spanned a similarly ambiguous and difficult range of roles: as a means of escaping the rigours of the regime for malingerers; as the source of unprecedented comforts for the enfeebled destitute; and even as a site of moral reform. Fully aware of the ethical difficulties involved in providing a level of care for the sick not available to many outside the prison, most accounts played down the relative luxuries of infirmary provision. They laid stress, instead, on sickness and suffering as both agents of and companions to reform. The most hardened women, once enfeebled by disease, were seen to become more susceptible to reformative influence. One influential lady prison visitor, Matilda Wrench, actually found proof of redemption in suffering. It was as if the penitent woman paid for her sins through illness and even death. She glorified the penitent decline of one prisoner, Eliza Cooper, thus: 'It was very encouraging to witness the child-like submission with which she endured her often agonizing pain. Her eyes would sparkle with love and gratitude when God's mercies to her were mentioned.'[39] Through sickness the woman was reduced to a weak, passive, and dependent state far more closely akin to the ideal of femininity than her bold, independent ventures in crime. It would be wrong, however, to see illness as a punitive device malevolently welcomed by officials as the ultimate means of subduing the unruly. In fact, the sickly condition of many women received into local prisons caused considerable concern.

The steady increase over this period in the number of women committed for very short sentences of seven or fourteen days for being drunk and disorderly[40] ensured that an increasingly large proportion of prisoners were in varying degrees of poor health. More generally, the Surgeon of Tothill Fields, John Lavies, found the women under his care to be a class of woman 'whose constitutions are less robust, who are more susceptible to disease, and are liable to many ailments peculiar to themselves'.[41] Frailty and ill-health placed severe constraints on the

[39] M. Wrench, *Visits to Female Prisoners at Home and Abroad* (1852), 23–4.

[40] At Tothill Fields the proportion convicted of being drunk and disorderly rose from 21 per cent of those committed in 1860 to 62 per cent in 1877.

[41] GLRO/WA/G/11. Minute Book (1862–5), Report by the Surgeon, 28 Feb. 1863.

requirement that the diet be 'less eligible' in its quality and quantity. It also had to be sufficient to ensure good health. Whilst those women imprisoned for very short sentences might endure a minimal diet without ill effect, the Surgeon reported that the health of those under longer sentences deteriorated alarmingly: 'women under imprisonment for long terms are very apt to lose the healthy performance of functions peculiar to their sex',[42] in other words their menstrual periods stopped.

In common with all closed institutions, food was a focus of great interest amongst the women themselves.[43] They continually applied for alterations or additions to their diet on the basis of ill-health or the greater exertions of their labour. Rejection of such applications was a frequent 'source of injustice', whilst additional allocations of food gave rise to 'bickering and jealousy' amongst other women. So disruptive were these effects that, on the nationalization of prisons in 1877, a special committee set up to investigate local prison dietaries recommended that no differentiation of diet be allowed on grounds of labour alone.[44] The conflicting requirements, that the prison diet be sufficient to ensure good health, adequate to sustain hard labour, and yet 'less eligible', proved insoluble.

The Prison Nursery

One might have expected the prison nursery to play a highly positive part in the process of reform: fostering responsibility and pride in prison mothers, whilst encouraging the remainder of the women to aspire to motherhood. At the very least it might have provided the opportunity of preserving some vestiges of the ideal female role. Instead the prison nursery was decried as 'the most deeply pathetic of all the scenes in the world'.[45] Whereas those areas described so far were to a greater or lesser extent components of reform, the nursery was widely deplored as the most disturbing distortion of the intentions of the regime. Since it was undesirable, unjust, and in

[42] Ibid.

[43] For more on the role of diet in prisons see V. J. Johnston, 'Diets in Workhouses and Prisons 1835 to 1895' (1981).

[44] Report of the Committee into the Dietaries of the Prisons (1878), *Home Office Printed Memoranda*, iv. 427.   [45] Mayhew and Binney, *Criminal Prisons*, 475.

any case impossible to subject babies to penal discipline, their presence necessarily disrupted the regime. As a former deputy governor of Liverpool Prison, reflecting back on the problems of maintaining discipline, remembered: 'A baby disorganizes the entire female prison, from the matron downwards. Everyone wants to play with it, and, as sunshine is followed by storm, so the baby causes jealousy and uncharitableness to divide the oldest friends . . . the only person who seemed never to get a "look in" was the mother.'[46]

In Tothill Fields about 200 infants were admitted each year on the commitment of their mothers. In addition, up to forty births within the prison each year combined to pose the considerable problem of how to accommodate such large numbers of very small children. Housed in a large barn-like nursery, mothers and their babies were subjected to a poor adaptation of the prison regime. Since it would have been inhumane to attempt to enforce the silent system, mothers were allowed to talk to their babies but not to one another. The difficulties of imposing this absurd compromise solution all but relieved prison mothers of the usual rigours of discipline. For those with children under eight months, even the requirements of hard labour were waived (though thereafter they were expected to pick about two-thirds of the usual allocation of oakum). Ironically these compromises placed the prison mother in a far easier position than that of many 'honest Mothers' outside the prison, trying to earn enough to support their offspring. Clearly, the demands of 'less eligibility' were not being met. The Middlesex magistrate, Sir Peter Laurie, castigated Tothill Field officials: 'A Mother in your Prison is practically a Lady.'[47]

Throughout the 1850s and 1860s the Visiting Justices sought to persuade magistrates to send all but the very smallest children to the workhouse on their mother's committal and also to oblige workhouse authorities to receive, from the prison, prisoners' children once they were fully weaned. A sense that the parish in which the mother was committed ought to bear the cost of maintaining her children seems to have motivated the Justices even more than their anxiety over the internal problems caused

---

[46] B. Thomson, *The Criminal* (1925), 33.
[47] For condemnation of this disparity see letter to the Visiting Justices in GLRO/WA/G/11, Minute Book (1862–5), minutes of 6 Feb. 1864.

by retaining them in prison. Economy, then, not discipline, was the overriding consideration in efforts to reduce the nursery population to a minimum.

Efforts to remove children from the prison came not only from prison authorities but also from more energetic Lady Visitors such as Susanna Meredith.[48] She shared the common conviction that a child born or raised in the dismal surroundings of the prison and who remained under the influence of its criminal mother was inevitably doomed to a life of criminality itself. She sought to obtain legal powers to separate such children from their mothers, or other unsuitable guardians, and to confirm the guardianship of a charitable agency instead. The Industrial Schools Act 1866, providing for the removal of children to industrial schools, applied only to the children of women who had been previously convicted twice. In 1871 Meredith and her supporters secured the insertion of a clause into the Prevention of Crimes Act of that year extending the 1866 Act to include women with only a single previous conviction. If their children were under 14 years, with 'no visible means of subsistence, or . . . without proper guardianship', they became liable to be sent by the courts to a certified industrial school.[49] Initial hostility to this concerted deprivation of the most basic of maternal rights—that of a woman's custody over her children—was, by Meredith's account at least, soon overcome. Strikingly, the feelings of the mother were rarely discussed unless she absolutely refused to surrender her children. And even then discussion centred on the inconvenience of having to acquire legal powers of removal. The criminal mother had apparently forfeited all rights and all claims to sympathy.

LOCAL PRISON STAFF

In 1958 Gresham Sykes argued that to understand the prison one must examine 'the wide range of social interaction between

---

[48] She had set up a refuge opposite the gate of Tothill Fields to offer aid to women on their discharge and was a regular visitor to women within the prison. For further details of her life's work see M. A. Lloyd, *Susanna Meredith* (1903).

[49] W. Crofton, 'Female Criminals: Their Children's Fate', *Good Words* (1873), 172.

inmate and inmate and guard and inmate . . . (for) . . . in this interaction we can begin to see the realities of the prison social system emerge.'[50] To do this one needs to identify the principal actors, not least the prison warder, whom Sykes identified as 'the pivotal figure on which the custodial bureaucracy turns'. Although we are circumscribed by the uneven survival of records, such details as are available provide at least a shadowy portrait of the local prison matron and her staff.

By 1877 the staff at Tothill Fields comprised 56 women and 13 men. Since the passing of Peel's Gaol Act in 1823, male staff's access to female prisoners had been severely curtailed so that even the chaplain could visit only in the company of female staff. Effectively the matron, 55 female warders and subwarders managed the daily life of Tothill Fields prison. Their background was often barely socially separated from those whom they sought to control. Of the warders in office in 1877, 18 had no previous occupation.[51] Most of the rest had been domestic servants, nurses, or held similar work in other prisons or lunatic asylums. The remainder included a sewing machinist, a dressmaker, a milliner, and shirt and bonnet makers. Only one was considered to have received a 'very good education', the rest ranging from good to merely indifferent.

Those giving evidence to official enquiries frequently lamented the difficulty of attracting any but the most hard-pressed women into prison employment. In 1835 Fry had deplored the 'very poor Sort of Matrons' engaged, stressing the difficulties that their lack of calibre entailed for maintaining discipline: 'some of the Women are superior to themselves in point of Power and Talent, so that they have scarcely any Influence over them.'[52] The testimony of the Matron of York County Prison to the Prison Inspectorate in 1849 reveals only too clearly the sheer inadequacy of a woman in charge of an entire female wing: 'I know what a journal is; it is some one higher than myself. I keep a girl, but no one else. The rules are hung up in the day-rooms, and those who can read, read the

---

[50] G. Sykes, *The Society of Captives* (1958), 6.

[51] These and all the statistical details on staff given below are taken from GLRO/MF/487, Particulars of Officers at House of Correction, Westminster (1877), *passim*.

[52] First Report of the Select Committee of the House of Lords on Gaols and Houses of Correction, Minutes of Evidence, *PP*, 11 (1835), 520.

rules to the others, but I do not read the rules to them, as I am not a good scholar.'[53] Similarly, the disciplinary records of Tothill Fields indicate that many warders were incapable of fulfilling their duties. Reprimands for negligence were frequent: warders left doors unlocked, left prisoners unsupervised or allowed them to wander unaccounted for around the prison, and even failed to intervene when women under their control fought one another.[54] Others were not just naturally incompetent but even appeared on duty 'in a disgraceful state of intoxication'.[55] One subwarder was so repeatedly drunk on duty that the Matron took the unusual step of seeking evidence from prisoners as to her conduct. She reported 'that they have had to call Mrs Hoare every hour to answer the Watchman and that on Monday night she was so much under the influence of Drink that she fell against the Stove and but for the Guard would have been in the fire'.[56] Mrs Hoare was immediately dismissed but the very details of the case seem more reminiscent of what is supposed to have happened in eighteenth-century gaols than the usual historical portrayal of the reformed penitentiary of the nineteenth century.

Such dismissals were relatively rare. In contrast to the very short stay of many prisoners, the warders endured extremely long 'sentences'. Length of service returns for 1877 show that 24 women had served for more than ten years and 14 for six to ten years. Since 1861 only single or widowed women had been engaged, so that, short of marriage (upon which event they were obliged to leave the service), they were likely to remain for the rest of their working lives. The typical warder was, then, of relatively lowly social class, generally without a family of her own, and forced to seek subsistence and security in prison work which, once entered, effectively engulfed her life.

The extent to which the female warder was subject to official regulation is revealed by the mass of prison rules which minutely defined and delimited every aspect of her life—outside working hours as well as within. Whilst the conduct of all prison staff, male and female, was subject to minute regulation, female

[53] Fourteenth Report of the Inspector of Prisons: Northern District, *PP*, 26 (1849), 210.

[54] Evidence taken from GLRO/WA/G Minute Books, *passim*.

[55] GLRO/WA/G/15. Minute Book (1875–8). Subwarder Sarah Neave was subsequently dismissed from service for this offence.

[56] GLRO/WA/G/7. Minute Book (1852–5), 4547.

officers were burdened with additional requirements concerning their demeanour and temperament. Included in the rules
respecting the duties and responsibilities of the Matron at
Tothill Fields was the prescription that 'she shall require that
humanity and good temper be shown towards the prisoners by
her several subordinate officers.'[57] As we have already observed,
it was officially recognized that close relations in women's
prisons could have a beneficial impact. The following comment
synthesizes many of the assumptions behind the warder's role:

successful treatment of female criminals . . . largely depends on the
tone and disposition of female prison officers. Harshness and im-
patience—the frequent characteristics of an ill-trained officer—are
only calculated to aggravate violence and insubordination. Despair
and hopelessness fill the minds of most criminal women; the antidote
to which is the feeling of hope—the first necessary amendment of life.
Were all female prison officers really imbued with sympathy for the
fallen creatures under their charge, and were they trained, as a part of
their profession, to endeavour to assume a hopeful manner in their
general treatment, much of the scandal would be spared of criminal
women who, in desperate violence, set at defiance the whole gaol
authorities.[58]

That many prison staff were not, in practice, capable of
assuming this desired role of model, feminine, 'maternal'
compassion fuelled demands for the voluntary involvement of
Visiting Ladies. Only by persuading capable, intelligent ladies
that the care of their 'fallen' sisters was an essential feminine
vocation could they compensate for the inadequacy of the
official paid staff.[59]

THE PRISONERS

Rates of Turnover

Perhaps the most marked difference between life in local as
compared to convict prisons was the very high turnover in the

[57] GLRO/MA/G/GEN 1252. Middlesex House of Correction at Westminster
amended proposed Regulations and Rules (1865), 16. Rule 81.

[58] J. A. Bremner, 'What Improvements are Required in the System of Discipline in
County and Borough Prisons?', *Transactions NAPSS 1873* (1874), 282.

[59] See the transactions of the International Penitentiary Congress in E. Pears (ed.),
Prisons at Home and Abroad (1872), and also L. M. Hubbard, *The Englishwoman's Year
Book* (1881), *passim*, for discussion of the duty of middle-class women to their less
fortunate sisters.

prison population. Examination of the length of sentences in local prisons tends to undermine the notion of the prison as a closed institution cut off from wider society. It also belies much of the contemporary rhetoric of moral and spiritual reform. The majority of sentences to Tothill Fields were extremely short and the proportion of these short sentences (mainly for drunkenness, disorderly conduct, and petty theft) increased over the period. By the time prisons were nationalized in 1877 just under three-quarters of women entering Tothill Fields were sentenced to terms of less than one month. Even more strikingly, over half of the women were sentenced to less than fourteen days. Correspondingly few were sentenced to longer terms—well below 10 per cent were sentenced to more than six months. Although details of sentence lengths are not given in the Judicial Statistics, it does appear that sentences were becoming shorter nationally, for although the numbers of women committed to local prisons increased from 34,586 in 1857 to 46,538 in 1880, the daily average female population remained steady at just under 4,000 (Appendix, Tables 5 and 6).

These figures indicate how very different a place the local prison was to the convict prison. In the latter, sentences of five, six, and seven years effectively cut inmates off from the outside world and endowed the convict prison with many of the attributes of a so-called 'total institution'. In contrast, local prison inmates, knowing that they would be free at the end of the week or month, could usually maintain their previous life and, above all, keep up contacts with friends and family (if any) outside. Their attitudes, and those of the prison authorities towards them, were necessarily very different. Many inmates were, arguably, less institutionalized 'prisoners' than 'women in prison'.

Administratively this level of turnover caused considerable problems. One observer, Dr Nichols, reported: 'The prison-vans bring them, fifty or sixty a day, from all the police-courts of the metropolis, as well as from the criminal courts and sessions. So many come in every day: so many discharged, mostly to come again. What a work for the chaplains! What a work for reformers!'[60] Reception facilities at Tothill Fields

[60] Dr Nichols's article in the *Herald of Health* (May 1882), quoted in Fletcher, *Twelve Months*, 461.

were extremely limited and, despite the fact that many women were filthy and verminous, Nichols noted that they all had to wash their hair in the same water. That 10 per cent of the prison population entered or were set free every day not only taxed reception facilities but was enormously disruptive more generally. However, the level of discontinuity in the inmate population was not quite so high as these statistics seem to suggest. For many of the women returned time and again to Tothill Fields: 'Women are often released on Thursday and come back to their old cells again on Saturday.'[61] Such 'regulars' might return to the same prison half a dozen times or more in a year, each time for terms of no more than a few weeks but staying in all for several months a year. These women did become highly institutionalized, so much so that the prison authorities began to doubt their ability to survive outside at all.

Statistics available for Tothill Fields for the period 1860 to 1877 indicate the types of women who found their way into prison. There is such variety in the characteristics of the women, whether committed at any one time or over this period, that it is not possible to identify a single, typical, female prisoner as some historians have attempted to do. Priestley, for example, perhaps rashly asserts that 'the prostitute was the typical woman prisoner of the later nineteenth century.'[62] It is with grave reservations as to the usefulness of trying to identify common characteristics of such a heterogeneous group that I present some of their more salient traits.

Previous Occupation

It is difficult to say anything with certainty about the lives of the prisoners at Tothill Fields prior to conviction. Prison records relied entirely on the evidence given by women on committal. Even if all the information were honestly given, it would be problematic in so far as the figures given indicate only the woman's supposed occupation on commitment. Her past occupations, however many and various, were not considered. To build up a more coherent picture of the types of women who became prisoners in local prisons one would need better

[61] Fletcher, *Twelve Months*, 326.
[62] P. Priestley, *Victorian Prison Lives* (1985), 72.

evidence of their previous life histories, at work and whilst unemployed.

The temptation for women to give false details about their lives in order to hide past delinquencies, to assert their previous respectability, or most probably to elicit sympathy must have been great. Consequently, well over half the women coming into Tothill Fields claimed to have had no previous occupation at all. Nationally, the percentages of women with no previous employment coming into prison fell from over half of those committed to prison in 1857 to around 28 per cent in the 1870s (Appendix, Table 8).[63] Many may have made such a claim simply to exonerate themselves from crimes they could then attribute to the destitution or demoralization resulting from unemployment. However, most were probably housewives and mothers.

Similarly, a suspiciously small and, over the period, declining proportion entering Tothill Fields admitted to being prostitutes. Although just over 14 per cent admitted to being prostitutes in 1862, this fell steadily to below 4 per cent by 1877. Women were no doubt keen to evade condemnation by prison officials and it seems likely that a far greater proportion of women prisoners, if they had not been full-time professionals, had at some time resorted to prostitution. The rest of the women admitted to Tothill Fields claimed to have been in some sort of respectable employment; most were labourers, charwomen, or needlewomen. This picture may be distorted, however, by the incentive to lie about one's trade in order to secure more palatable prison employment. For example, many women claimed to be laundresses simply to secure the most popular of positions—in the prison laundry.

The fact that never much above 10 per cent (both nationally and in Tothill Fields) described themselves as servants calls into question the common assertion that most criminal women were drawn from the ranks of domestic service. The small number of servants partly reflects the fact that domestic service was a relatively respectable area of employment and, more importantly, one which largely sheltered women from circumstances liable to lead to crime. As many social commentators

[63] This compared strikingly to the relatively small proportion of previously unemployed male prisoners—11 per cent in 1857 falling to 6 per cent in the 1870s.

observed, it was only once a domestic servant lost her post or had been dismissed for improper conduct that, lacking the 'character' (reference) necessary to secure another post, she fell prey to temptations to crime. Possibly some of those listed under the heading 'no previous occupation' had at one time been in domestic service.

Overall, both men and women prisoners, if employed at all, came from the lower end of the job market, which was prey to seasonal slumps and often paid barely subsistence wages. Women tended to come from the lowest, least skilled areas of employment. This, combined with the greater proportion who had had no occupation, suggests that, in parallel to respectable society, criminal women occupied more poorly paid and less secure jobs than criminal men.

Offences

Local prisons were primarily a means of punishing less serious offences (more serious crimes being dealt with by penal servitude in convict prisons) and by far the majority of men and women going into local prisons did so on summary conviction—61 per cent in 1857, rising to 77 per cent by 1880 (Appendix, Table 6).[64] A rather higher proportion of women than men were committed on summary conviction (71 per cent of women as opposed to 58 per cent of men in 1857, rising to 87 per cent of women as opposed to 75 per cent of men in 1880). This disparity was produced mainly by the lesser numbers of women committed for debt or on civil process, and, of course, the complete absence of women committed under the Mutiny Act (for military and naval offences).

For what offences were women committed to Tothill Fields?[65] By far the largest number of women were committed for being

[64] The Summary Jurisdiction Act 1877 led to an increase in summary convictions for cases that previously had been tried on indictment. For example, the number of larcenies tried summarily for both males and females rose from 42,011 in 1879 to 51,125 in 1880. See V. A. C. Gattrell and T. B. Hadden, 'Criminal Statistics and their Interpretation', in E. A. Wrigley (ed.), *Nineteenth Century Society* (1972), 356. See also the note on the Judicial Statistics in the Appendix below.

[65] The subsequent figures about prisoners at Tothill Fields are taken from the Annual Returns in series GLRO/RS/2/71–90, Reports of Middlesex Prisons, giving total numbers committed to Tothill Fields for the previous year.

drunk and disorderly. This category grew from 21 per cent of commitments in 1860 to 62 per cent in 1877. Notably, commitments under the Vagrancy Act (the second largest single category) declined over the period from 31 per cent in 1860 to less than 7 per cent by 1877. Given the likely overlap between the two types of offence, these two opposing trends may have been due to changes in prosecution policy rather than absolute changes in types of crime.

Thus, over the period 1850 to 1877, on average more than half the commitments were for these largely moral, and most strikingly victimless, offences. By 1875 this proportion had risen to nearly three-quarters of all commitments. Women sentenced for violent offences, including common and aggravated assault and assaults upon the police, never made up more than 16 per cent of those committed to Tothill Fields. And commitments for felony to imprisonment with hard labour never rose above 9 per cent.

Ages

The average age at which both sexes were committed to local prisons rose considerably over the period. This apparent ageing of the prison population is supported by the increasing number of previous convictions recorded against those committed to local prisons. Taking these two trends together it seems that the local prison population was increasingly drawn from a narrowing group of confirmed criminals, repeatedly offending and repeatedly imprisoned, even into old age.[66] Women committed to local prisons tended to be older than men. They appear to have been less likely than men to have begun their criminal careers in their teens and so became embroiled in the criminal justice system later.

Similar trends appear in the figures for Tothill Fields, indicating an ageing in the prison's population between 1864 and 1877 (the only period for which comparable figures are available). In 1866 (the high point), just over 29 per cent of the women were less than 21 years old, but by 1877 this proportion

---

[66] This apparent 'localization' of crime was widely applauded by penal administrators; see, for example, E. Du Cane, 'Address on Repression of Crime', *Transactions NAPSS 1875* (1876).

had dropped to less than 10 per cent. The main adult population of women between 21 and 45 rose from 59 per cent in
1866 to 73 per cent in 1877; the proportion of older women,
over 45, rose from 12 per cent in 1866 to 18 per cent in 1877.

This gradual shift towards a more elderly population may
have resulted, in part, from the development after mid-century
of alternative provision for juvenile criminals (up to age 16)
in industrial and reformatory schools. Increasingly, women
came to prison only after they had defied the efforts of the
reformatories.

Numbers of Previous Commitments

Establishing the numbers of prisoners' previous commitments
is hampered by the inefficiency of record-keeping and the
obvious temptation for offenders to use aliases in order to
conceal their past criminal careers. In so far as these data can
be relied on, they reveal far higher rates of previous commitments for women than men. This seems to suggest that women
were far more likely than men to reoffend once they had been
in prison. However, such figures may reflect the fact that
women were far less likely than men to be committed to prison
for a first offence. Consequently, most women were committed
to prison only if they were persistent reoffenders.

For both sexes the portion of those committed to local prisons
who had one or more previous commitments increased fairly
steadily over the period, from about 30 per cent in 1857 to
39 per cent by 1880 (Appendix, Table 7). Significantly a far
larger proportion of women than men had been previously
imprisoned; so that whilst the proportion of men who had been
previously imprisoned rose from 26 per cent in 1857 to 34 per
cent in 1880, the comparable proportions of women were
40 per cent in 1857 rising to 53 per cent by 1880.

Either women had a much higher rate of recidivism, or
possibly the courts were unwilling to send women to prison for
first or second offences but more ready to recommit them once
they were designated criminal. Whatever the cause, the difference between the sexes becomes clear when we examine the
numbers of those who had more than ten previous commitments to prison. In 1857, 5 per cent of women, as opposed to

less than 1 per cent of men, had been in prison before, more
than ten times. By 1880, this had risen to 15 per cent of women
as opposed to only 3 per cent of men. This disparity not only
reflected a marked difference in the relative proportions of each
sex who could be regarded as confirmed recidivists but the
much more remarkable fact that women actually outnumbered
men in this category. For much of the period, well over two-
thirds of those imprisoned more than ten times were women.
Since women accounted for little more than a quarter of all
those committed, their extremely high showing in this category
indicates how much more likely they were to be recommitted
than men. Although it is not possible to infer the reasons for this
disparity from these statistics alone, likely causes include the
greater hindrances faced by discharged women prisoners trying
to reform, and the tendency of public, police, and the courts to
assume that a woman, once corrupted, was irredeemable.

Figures for previous commitments at Tothill Fields are not
exactly comparable with national statistics but do reveal
similarly high rates of recommitment amongst women. The
proportion of those women recorded as having one or more
previous conviction never dropped below half and for much
of the period was much higher. The category of 'habitual
offenders', drawn by the prison's authorities at those with four
or more previous convictions, consistently made up around a
third of all commitments. However, officials at Tothill Fields
recognized only too clearly the difficulty of eliciting any certain
information about past criminal careers. When, in 1855, the
Visiting Justices suggested introducing a basic form of classification, their proposal provoked a scornful rejoinder from the
Governor. He pointed out the impossibility of knowing how
many times a prisoner had really been convicted. A woman
currently convicted for some minor misdemeanour might well
have had previous convictions for felony or as a habitual rogue
or vagrant.[67] The only practicable solution seemed to be a
crude separation of first offenders from those considered to be
'confirmed profligates' and the suggestion that all those known
to have been committed three or more times be made to wear a

[67] GLRO/WA/G/7. Minute Book (1852–5), minutes of 13 and 27 Oct. 1855
incorporating the Governor's Report, 6207.

letter 'R' on their sleeve denoting, with pitiful vagueness, 'Repeated'.[68]

STAFF-INMATE RELATIONS

As we have seen, in theory the prison regime sought to reduce prisoner–warder contact to a minimum, to limit the extent to which human frailties might impede the process of reform. In addition, the prison rulebook sought to minimize possible interactions which might mitigate the supposedly awesome impact of confinement. At Tothill Fields, Number 47 of the Prison Rules dictated that:

Subordinate Officers are strictly forbidden to hold familiar conversa-tion with the prisoners, or to communicate with them on any subject whatever unconnected with their duties . . . it is expected that they will carefully abstain from forming intimacy or acquaintanceship with discharged prisoners of any class or description, and that they will maintain their respectability by avoiding the company of all such persons.[69]

Thus the prison authorities sought to establish what Erving Goffman has described as 'the staging of a grim social distance . . . between two constructed categories of persons'.[70] Many historians have accepted the claims of nineteenth-century penal theorists that prison regimes were indeed based on a rigid segregation not only of inmates from one another but of warders from inmates. In practice, the divide was continually eroded at the many points where warders and prisoners met and talked. Even the strictest hierarchical barrier could not completely divide two groups of women who found themselves obliged to coexist under the peculiar, punitive constraints of the prison environment. Living and working in the narrow world of a closed institution, officers could not easily maintain a view of their prisoners as anonymous captives. Nor could prisoners remain indifferent to the temperament—the weak-nesses and the kindnesses—of their captors.

[68] Ibid. 6209.
[69] GLRO/MA/GEN 1252. Amended proposed Regulations and Rules (1865), 9.
[70] E. Goffman, *Asylums* (1970), 103-4.

It is not possible, therefore, to understand relations within the local prison as those of coercion alone. Despite the massive imbalance of formal power in favour of the custodians, order was not dictated solely by them or by the rules in which their power was invested. By their responses to, and resistance to, rule from above, inmates could ensure stability or create disorder. In practice, the relationship between warders and prisoners was a symbiotic one, of voluntary compliance as much as of coercion. The punishment records and the matrons' reports reveal a range of activities and interactions all too often overlooked or simply dismissed as failures of the official regime. Officially both warders and inmates were required to submit to a whole complex of norms and rules. Beneath this requirement lurked a life which, though strictly illicit, was in fact crucial to the maintenance of a more general order. Forbidden activities and behaviour that was strictly speaking deviant, by both prisoners and warders, combined to mitigate the severity of the regime. Only massive breakdowns of order, manifested in open riot, reflected a real failure of the regime. The mass of punishments appear, therefore, less as failures than the outer delimiting boundaries of accepted behaviour.

By complex relations of exchange and by 'informal patterns of social accommodation',[71] the warders sought to elicit the 'voluntary' compliance of the prisoners. For example, they tolerated minor illegalities in return for more general acquiescence to their rule and even a degree of loyalty. In part at least, they were successful. The records of Tothill Fields include numerous actions by prisoners which cannot be accounted for other than in terms of close, and perhaps even congenial, relations between inmate and warder. For example, throughout the minutes of the Justices' meetings there are references to incidents in which the warder was protected or even rescued by prisoners from assault by another prisoner. A lengthy investigation into a fearsome fight between two prisoners in the summer of 1856 provides several interesting insights into staff–inmate relations. One subwarder reported that her attempt to intercede was blocked when 'one of the prisoners took hold of me, and told me not to go, as I was a New Officer, I should be

[71] R. A. Cloward, 'Social Control in the Prison', in R. A. Cloward *et al.* (eds.), *Theoretical Studies in the Social Organisation of the Prison* (1960), 36.

ill-treated . . .'[72] That this intervention was not simply a rare act of pity for a novice warder is evidenced in similar accounts by the other officers. Even the Chief Warder reported, 'as I was going up to the room some of the Prisoners called out to me not to go, as I should be killed.'[73] Some of the women may have been motivated less by genuine concern than by a desire to prove themselves to be 'model' prisoners; but the frequency with which individual prisoners interceded on behalf of warders, or simply took their side in almost daily altercations, suggests some degree of solicitude.

Occasionally the authorities discovered evidence of closer relations than was thought consistent with the general order. Evidence of sexual intimacy between warders and prisoners, heterosexual or lesbian, was treated very seriously. The existence of heterosexual relations was possible only in those prisons where discipline was so lax as to allow male warders private access to the female side. And yet well into the century occasional cases came to light. In 1855, for example, the Prison Inspector for Scotland was horrified to discover that ex-prisoner Elizabeth White had given birth to a child which, if not actually conceived in Berwick Gaol (there was some confusion about the exact timing of conception), was certainly the product of 'an improper intimacy' which had developed between White and the Turnkey whilst she was under his jurisdiction there.[74]

More common were relations of close emotional and even sexual intimacy between female warders and prisoners. Despite the formal injunction that warders should keep conversation with prisoners to a minimum, most accepted that a female warder could exert good moral influence by her personal interest and advice. Officially the rulebook forbade undue familiarity. However, in practice the common assumption that women were more susceptible than men to reform through personal influence justified a greater intimacy than was ever countenanced in male prisons. There was, it seems, a carefully placed dividing line between emotional attachment and lesbianism. Cases of 'tampering' with prisoners (the coy and conveniently ambiguous euphemism commonly used) appear

---

[72] GLRO/WA/G/8. Minute Book (1855–7), 6391.   [73] Ibid. 6393.
[74] Twentieth Report of the Inspectors of Prisons: Scotland (including the Northern Counties of England), *PP*, 26 (1854–5), 100.

with surprising frequency. Penalties ranged from simple reprimand to immediate dismissal. The divide between emotional and sexual intimacy is not always easy to discern, especially given the common use of sentimental expressions by Victorian women to one another. What, for example, is one to make of the effusive letters by a subwarder to former prisoner Susan W. Fletcher, one of which began:

MY DEAR DARLING BABY,—If I may still call you so,—and I think you will let me, for indeed you are very dear to me,—you don't know how miserable and unhappy I feel, now you are gone. It is not like the same place. It was very bad, but now it is much worse. As I am passing that old cell, I look in. It is empty—no one there. Then I don't know what to do with myself. Oh, *do* forgive me! I ought not to remind you of this dreadful place, but I do miss you so much![75]

The warder was clearly so besotted that she valued the possibility of continuing their association above the security of her job (which she subsequently lost). The letter indicates how in the monotonous, limited life of a warder the prisoners might become an important source of emotional sustenance. Incidently, it also reveals how much the warder remained a captive of the institution well after her prisoner had left. That such intimate relations developed at all reflects the degree to which both warders and prisoners, cut off from human contact with the outside world, found sanctuary in one another from the cold hostility of prison life.

Other warder–prisoner relations were perceived as a more serious threat to prison discipline. Occasionally a warder was discovered to have close links with the criminal fraternity outside. Chief Warder Frances Pigrum was dismissed in February 1862 after suspicions were aroused that she was 'holding intercourse' with prisoners and their friends outside the prison. The Justices directed that she be watched when off-duty and they soon discovered 'that she was in the habit of frequenting a notorious House well known to the Police as being visited by thieves and other bad characters'.[76] This incident again points to the narrowness of the social gap between many warders and prisoners, so that they might share the same friends and social life outside the artificial hierarchy of the prison.

[75] Fletcher, *Twelve Months*, 409.
[76] GLRO/WA/G/10. Minute Book (1860–2), minute of 15 Feb. 1862.

Other equally illicit warder–prisoner relations were more materially instrumental. The disciplinary records of Tothill Fields reveal wholly illegal arrangements by which prisoners obviated some of the hardships of prison life and warders eked out their meagre wages. Such activities ranged from simple bartering of services for petty privileges and goods to sophisticated systems for trafficking 'luxuries' into the prison. For example, in September 1860, four female prisoners 'confessed they had done plain and embroidery work for these Officers in return for which they received bread, butter, cheese, tea, fruit, and cake'.[77] Subwarders whose lowly status and pay weakened their loyalty to the regime were most likely to yield. They were frequently reprimanded for ignoring minor misdemeanours, for applying regulations laxly, or even allowing a favourite prisoner special privileges or providing her with simple treats. However, by mitigating the intended severities of penal discipline in this way, warders made it easier for the women to comply with their rule.

Access to a warder prepared to bring in food and drink on a regular basis did much to alleviate the hardships of the regime. But organized trafficking, whereby a prisoner's friends outside the prison paid warders to take in highly illicit goods such as meat or even alcohol, constituted a far more overt and damaging attack on the official order. Substantial sums were offered as inducements. One warder, Charlotte Howe, posing as a potential trafficker, was offered 10 shillings and a gold watch by one woman to take in 'some Grub, Meat, and a Little Wine' to her nephew's wife who was in Tothill Fields. In the prison's disciplinary records, reports of trafficking are so frequent, and the subsequent punishments imposed so mild (warders were generally reprimanded or at most fined but rarely dismissed), that one can only assume that the activity was very widespread. One prisoner, Emma Steiner, shopped subwarder Ann Bailey for supplying goods to a number of her peers, though, significantly, not to Steiner herself: 'The Prisoners say subwarder Bailey receives so much a week to bring parcels into them: it is talked about outside, and when a prisoner gets her sentence they say never mind she has a good screw inside.'[78] That this particular offence was reported by a

[77] Ibid. minute of 1 Sept. 1860.
[78] GLRO/WA/G/8. Minute Book (1855-7), 6536.

prisoner points to a crucial facet of the warder–prisoner relationship—that is, the delicate balance of power between them. The warder obviously occupied a position of superiority and could largely dictate the terms of their relationship. She determined whether it was of the coldest formality or approaching the various degrees of intimacy indicated above. But any relationship which strayed beyond that strictly allowed immediately involved a whole range of illegalities. As a result the warder effectively abrogated much of her formal authority. She risked being reported by the very recipient of her largesse, by other prisoners jealous of her favouritism, or, most commonly, by other more scrupulous warders. The Justices had difficulty in eliciting the truth in a constant stream of accusation and counter-denial. On investigation many accusations led to the punishment of those found to be engaged in illicit activities, but others were so totally unfounded they were dismissed out of hand. Significantly, the Justices showed themselves very ready to disbelieve all the more horrifying accusations. When, for example, one woman accused a male warder of giving her beer and food in return for 'improper intercourse' in her cell they recoiled in automatic disbelief and refused to take any action.[79]

Prisoners also stored up contraventions of rules and regulations by warders as ammunition against them when threatened with punishment themselves. Warnings of disciplinary action would be countered by threats to reveal awkward or incriminating information. Blackmail, bartering of favours, and implicit agreements to avoid revealing one another's misdemeanours maintained a precarious balance of apparent order. Occasionally it toppled completely, leaving the underlife of the prison wholly exposed to the horrified gaze of the prison authorities. In covert recognition that many prisoners' complaints against warders sought to undermine this 'negotiated peace', the Justices sometimes chose to ignore a charge that was blatantly retributive. When one prisoner, Johanna O'Brien, was punished by three days confinement in the dark cells on the report of a subwarder, Ann Manning, she attempted to retaliate by claiming that Manning had 'tampered' with her two months previously.[80] Ignoring the possible veracity of the charge, the

---

[79] GLRO/WA/G/12. Minute Book (1865–8), minute of 15 Dec. 1866.
[80] GLRO/WA/G/8. Minute Book (1855–7), minute of 15 Nov. 1856, at 6502–3.

Justices simply dismissed it on the grounds that she had not reported it earlier. Clearly they recognized that to take such charges seriously was to endow prisoners with an unacceptable degree of power. To do so could only undermine the ability of warders to maintain their authority.

On the other hand, the prison authorities responded very seriously to warders found to be unduly severe or cruel. Warder Charlotte Howe was reported for abusing her command by committing 'an Act amounting to Cruelty in reporting a prisoner . . . for impertinence and disobedience of Orders, the Warder having aggravated the Prisoner to commit the offence'.[81] Warders were found guilty of 'winding-up', or provoking misbehaviour by taunts or abuse, and their cruelty ranged from 'harsh language' to outright violence. If discovered, they were heavily penalized by reprimands, fines, and even dismissal. Given the potential for abusing their considerable power over prisoners, it was evidently thought necessary to employ potent deterrents. Even when prisoners deliberately behaved outrageously to provoke a warder beyond self-control, the latter was severely castigated if she retaliated. When subwarder Mary Ann Hershall boxed the ears of a woman who had spat in her face, the Justices warned her that 'under no circumstances whatever is she justified in striking a prisoner'.[82]

INMATE SUBCULTURE

The illicit underlife of the local prison was not confined to relations between warders and prisoners but also flourished amongst the inmates. Inmate subculture is far more difficult to discern, largely because, for the most part, it remained unrecorded. Reconstructing inmate relations relies on those occasions when their misbehaviour was so blatant as to attract the attention of the prison authorities. Inmates' own accounts of their life in prison are so rare as to call into question their typicality. For example, the only surviving memoir by an inmate of Tothill Fields is Susan Willis Fletcher's *Twelve Months in an English Prison* (1884). Although it is an eloquent

---

[81] GLRO/WA/G/6. Minute Book (1850–2), Matron's Journal reported at meeting of 8 May 1852, a 4360.

[82] GLRO/WA/G/15. Minute Book (1875–8), minute of 29 July 1876.

and informative account, Fletcher's own status as an upper-class American spiritualist convicted of obtaining jewels and clothing 'of great value' by false pretences or 'undue influence' makes her a very untypical prisoner of Tothill Fields.

Most official accounts of inmate interaction are to be found used as evidence in tirades against the dangers of corruption under the associated system. The partisan purpose of these accounts (to promote the separate system) make them a source of dubious reliability, not least because there is a suspicious similarity in the scenes, incidents, and examples given throughout these texts. Commonly they depict a relatively innocent young woman exposed to the corrupting influence of degraded recidivists. For example, in 1849 Inspector D. O'Brien took up the case of a respectable women committed in Coventry for a debt of 40 shillings: 'contrary to law, common decency, to common humanity, to common sense, and justice, she was condemned to be the hearer of the foul language of the stews, the unchecked blasphemy and ribaldry of the lowest prostitutes, and the outpourings of thieves.' [83] The lasting damage done by such exposure was repeatedly condemned. The following scenario painted by the Earl of Cathcart, a member of the Select Committee of the House of Lords on the State of Discipline in Gaols (1863), is typical: 'in one prison there was a prostitute associated with a young woman who had been tolerably decent in her conduct, and . . . arrangements were made by which the prostitute induced this girl who was formerly comparatively innocent, to come to her house and become afterwards a prostitute.' [84] Such tales played on wider fears of innocent young women being lured into the 'white slave trade' and corrupted irrevocably.

As we saw in the previous chapter, the evils of association between prisoners were thought to be especially harmful to women, for 'the power exercised by the abandoned prisoners of that sex in corrupting their comparatively innocent fellow-prisoners has always been acknowledged to be greater than that possessed by male prisoners.' [85] Such claims about women's

[83] Fourteenth Report of the Inspector of Prisons: Midland and Eastern District, *PP*, 26 (1849), 9.   [84] Minutes of Evidence, *PP*, 9 (1863), 57.

[85] This common assumption is here expressed by the Inspector of Prisons for the Southern District. Report on the House of Correction, Tothill Fields, Westminster, in Twenty-Third Report of the Inspectors of Prisons, *PP*, 29 (1857–8), 146.

susceptibility to corruption probably tell us as much about official attitudes and assumptions as about the prison experiences of the women themselves. However, they also provide evidence of channels of information and influence amongst prisoners under the supposedly 'silent system' at Tothill Fields. The regulations of the prison delineate the broad range of those illicit activities known to the Justices. One single regulation (Reg. 224) outlawed breaches of silence by singing and whistling 'and any attempt to barter or exchange provisions, any marking, defacing or injuring the doors, walls, forms, tables, clothes, bedding, books, or utensils, whatsoever of the prison; any attempt at communication by signs, writing, or stratagem of any sort; any unnecessary looking round or about . . . any secreting of money, tobacco or forbidden articles or any willful disobedience.'[86] Yet the women, with remarkable creativity, continued to devise means to confound these rules, for example, by signalling along ventilating flues when confined alone in their cells, passing notes on scraps of paper, and evolving elaborate ruses to visit friends in other parts of the building. For example, in 1867, Dr Lavies, the prison surgeon, attributed the very high numbers of applications to see him not to genuine illness or even malingering but to the fact that his surgery provided a meeting place for women from different parts of the prison. By such strategies the women proved themselves capable of exploiting any weakness in the regime of silence and solitude.

Certain gross defects allowed prisoners to evade silence with ridiculous ease. The following example confounds all claims for the disciplined severity or the reformatory influence of the regime. As late as 1866, A, B, and C sections at Tothill Fields were supervised at night by a single male warder whose access and means of control over the women in his charge was, by law, extremely limited. The prisoners were quick to exploit his inability to intervene, as the Matron, Maria S. Billiter, lamented:

It is also most desirable to prevent, if possible, Prisoners holding improper communication, by shouting to each other at the break

---

[86] GLRO/MA/G/GEN/1244. Rules and Regulations for the House of Correction at Westminster (1861), 56.

of dawn, when they know there is no female warder about, and not only detail the news of the day, but fabricate all sorts of falsehoods about Institutions to prevent Prisoners taking the advice of the chaplain, when he has offered them a home at the expiation of their imprisonment.[87]

In such circumstances, then, the women were clearly only too ready to take the opportunity to confound the very purposes of the prison.

The best documented area of inmate resistance was that overt, often violent level of reaction which could not fail to attract the authorities' attention. Certain women were seen to repudiate every attribute of femininity, both in their appearance and conduct. Described as Amazonian in stature and strength, and possessed of vicious tempers, they were not only prone to individual acts of destruction but also incited other inmates to riot. For example, at Coldbath Fields Prison, the Governor G. L. Chesterton encountered a 17-year-old 'untamable vixen' who urged other 'young furies' to rip off their clothes. Clad only in *'chemises de nuit'* they threatened the male officers called in to restrain them by brandishing their chamber pots![88] This was a particularly outrageous contravention of the norms of female decorum and passivity. But lesser acts of violence, both verbal and physical, were common. In the records of Tothill Fields there are innumerable incidents of 'obscene and disgusting language', 'filthy assault', and even 'riot and tumult'. These records seem to indicate defiance, anger, frustration, and lack of self-control. That these feelings were so forcefully expressed belies the image of the subdued, contrite female prisoner so prevalent in the rhetoric of reform. Even those who did not 'break-out' so blatantly, contravened the rules by singing, shouting, and occasionally even laughing.

DISCIPLINE AND PUNISHMENT

The means of punishing women prisoners were debated lengthily, not least because notions of femininity ruled out many of

---

[87] GLRO/WA/G/12. Minute Book (1865–8), minute of 2 June 1866.
[88] Chesterton, *Revelations* (1856), i. 186–8.

the sanctions commonly applied to men. Women were never subjected to whipping and rarely to irons or handcuffs, though, if punished, they were more likely than men to suffer stoppage of diet or to be sent to solitary confinement or the dark cells. Some prison officials clearly regretted that powers of corporal punishment did not extend to women. According to Chesterton: 'The female attendants, in the extremity of their disgust and horror, used to exclaim "What a blessing it would be, if we could employ some stout-armed woman to give them the rod!"'[89] Throughout their reports, the Prison Inspectors deplored the women's indifference to the limited punishments that could be imposed upon them. Solitary cells were simply derided or even welcomed as a change of scene and an opportunity to create a disturbance by banging and singing all night. At Tothill Fields they were never used for more than a tiny proportion of punishments (around 5 per cent). More commonly, stoppage of food or punishment diets were imposed; if they eventually broke down a woman's resistance, they did so only at the cost of her health.

Warders, struck with the futility of prescribed punishments, resorted to more subtle stratagems: ignoring refractory women until they tired of trying to gain attention, or shaming or even cajoling women into better behaviour. Psychological pressure, manipulative actions, and sophisticated attempts to outwit prisoners were probably more important means of overcoming resistance than isolation or deprivation of food.

That the prison was often simply inadequate to its disciplinary task was starkly impressed upon central government in 1870 by the discovery that Leeds Prison still employed stocks as a means of restraining refractory women. Doggedly resisting an order made in May 1870 to throw out these archaic devices, the Visiting Justices argued that 'some mechanical means of this nature are indispensable to secure proper discipline and good government in the Gaol to control most disobedient and violent women.'[90] This incident, and comments on the punishment of women made more generally, reveal how far accounts by both Whig and revisionist historians fail to capture

[89] Ibid. ii. 133.
[90] PRO HO45 9685/A48397. Letter from the Clerk to the Justices of Leeds Prison to the Secretary of State, 14 June 1870.

the continuing weaknesses, anomalies, and failures of the prison. Discipline was maintained more by the personal conduct of warders towards the women than by recourse to the rulebook. And when these manipulative attempts to modify behaviour by influence or argument failed, it was to the rod and the stocks that defeated prison officials turned.

THE LOCAL PRISON AS REFUGE

One final question necessarily modifes our understanding of the prison as a penal institution in quite a different way. Did women regard the local prison as the powerful deterrent and place of punishment that it was intended to be?

Perhaps surprisingly, the answer was very often no. Observers were repeatedly flummoxed by this confounding of the very purposes of the prison. Susan Fletcher, a prisoner at Tothill Fields, was shocked to discover that amongst her fellow inmates 'some spend more time in prison than out of it, and seem to prefer the freedom from care and the more orderly life.'[91] Many women in desperate straits of destitution, sickness, or pregnancy looked upon it as a place of asylum, providing comparative comfort and care. Sick women used 'the prison as an hospital, and came in with the express object of being cured of their disorders'.[92] 'Low street-walkers would deliberately demolish a pane of glass, whenever she (*sic*) coveted a refit, or desired her linen to be washed and got up.'[93] Pregnant women saw the prison 'as a comfortable asylum or lying-in hospital, where they are properly cared for during their confinement, or while nursing their little ones'.[94] And in 1853 the Justices of Tothill Fields, investigating a sharp rise in commitments of women, were shocked to discover 'that this prison enjoys an undesirable popularity amongst Prisoners generally', as it was known to be less severe than neighbouring

[91] Fletcher, *Twelve Months*, 338.
[92] Report on Borough Prison, Leeds, Thirteenth Report of the Inspectors of Prisons: Northern District, *PP*, 36 (1847–8), 428.
[93] Chesterton, *Revelations*, i. 170.
[94] Report on Dundee Prison, in Twentieth Report of the Inspectors of Prisons: Scotland, *PP*, 26 (1854–5), 145.

prisons.[95] As those who propounded the principle of 'less eligibility' had feared, the condition of the very poor was so low as to endow the meanest prison regime with material advantages. Far from being deterred from crime, many women deliberately sought access to their local prison as a welfare agency, preferable even to the workhouse. This continued to be true well into the twentieth century. For example, the first Lady Inspector of Prisons, Mary Gordon, found that women prisoners referred to sentences of a few days' imprisonment as 'a wash and brush up'; that they would deliberately seek sentences long enough to last the winter—a four-month winter sentence was for one woman 'a bit of luck . . . takes me right through the cold weather'; and that they considered the prison to be infinitely more comfortable than the workhouse—'Here I can have a room to myself, and what with three meals a day, and the doctor whenever I want him, I'm better off here.'[96]

Brought up to consider themselves naturally dependent, women lacked the resources to maintain themselves when a benefactor, however heavily disguised, appeared to be at hand. At certain periods of their lives, most obviously during pregnancy, women were reduced to abject dependency. With remarkable expediency they found in the prison the family and home they lacked outside. A tale recounted by Matilda Wrench, a Lady Visitor, epitomizes this particularly feminine and highly subversive attitude: 'A poor country girl came up to London in the hope of finding a situation; not succeeding, she gradually fell into deep destitution; friendless and penniless, to avoid worse sin, she threatened to break a window, and was consequently imprisoned. Three successive times she adopted the same plan of obtaining food and shelter.'[97]

In what ways is our understanding of the prison modified by these 'revelations' of women's experiences in Tothill Fields and elsewhere? Prison reform had attempted to embody coherent principles in comprehensive regulations designed to fix the character and purpose of the prison; but in local prisons these were often evaded and obstructed by the very people intended

---

[95] Committals of women by the Central Criminal Court to Tothill Fields rose from 260 in 1849 to 701 in 1853. GLRO/WA/G/7 at 4792-3.

[96] M. Gordon, *Penal Discipline* (1922), 27 ff.

[97] M. Wrench, *Visits to Female Prisoners* (1852), 132.

to enforce and to suffer its rule. The world they created was a fluid one capable of dramatic changes inconceivable in the immutable vision of the ideal penitentiary. In practice, surveillance and control were eroded by the continual movement of women in and out of the prison, by the inadequacies of the staff, and by the remarkable resilience of the inmates. Order was maintained not merely by authoritarian suppression of their misbehaviour but by continuous negotiation conducted at a personal, even intimate, level. By official directive, informal adaptation, and outright resistance, staff and prisoners evolved a regime that was in many respects distinctly feminine.

5 FEMALE CONVICT PRISONS
1852–1898

As we noted at the beginning of the previous chapter, almost all those sentenced to custody in the second half of the nineteenth century went to local prisons. Only those found guilty of more serious crimes were subjected to penal servitude (long-term confinement with hard labour) in convict prisons. These long-term prisoners, or 'convicts' as they were generally called, made up only 2 per cent of those committed to prison. Short sentences in local gaols or houses of correction were, therefore, the 'typical' prison experience of female criminals. However, it was the far more contentious problem of how to provide for serious offenders sentenced to long terms that perplexed and engrossed contemporary policy makers and administrators. Since terms of penal servitude were set at a minimum of five years (later reduced to three), the proportion of women held in convict prisons at any one time was greater than figures for commitments alone would suggest. The implications of holding prisoners for five or even ten years necessarily meant that convict prisons were very different from local prisons (where as we have seen most prisoners were sentenced only to a few weeks or months). The basic components looked alike: both had their cells, workrooms, chapel, school, and nursery, both confined and regulated the lives of their inmates. However, the very purposes and priorities, as well as the experiences of staff and inmates, of the two tiers of the prison system were otherwise very different.[1]

Female convict prisons were hastily established to replace transportation by enabling sentences of penal servitude to be served in England. Their official regime was adapted from that existing in the model prison for men at Pentonville. Superficially, it is easy to dismiss female convict prisons as mere

[1] Although this differentiation ought to be immediately obvious, it is largely overlooked by a number of recent studies, for example P. Priestley, *Victorian Prison Lives* (1985); R. P. Dobash *et al.*, *The Imprisonment of Women* (1986)—both works discuss convict and local prisons as if they were interchangeable or, more confusingly still, talk simply of 'the prison'.

imitators of their male counterparts, and yet in a society where gender differences were acute, where idealizations of masculinity and femininity were located at extreme poles, the female convict prison necessarily evolved a wholly distinct ideology, regime, and character. However much the convict prison sought to divorce itself from wider society, it inevitably embodied that society's views of women, and of criminal women in particular. This chapter will consider how perceptions of the female criminal and notions of femininity differentiated the nature and purposes of penal servitude for women. It will examine how the theories discussed in Chapter 3 were embodied in the various component parts of the convict prison regime. And, finally, it will show how these were translated into practice by the governors, chaplain, warders, schoolmistresses, and lady scripture readers who staffed the prisons.

Before examining this relationship between ideology, policy, and practice, let us consider why long-term imprisonment came to be adopted for serious female offenders.

HISTORICAL BACKGROUND: TRANSPORTATION

In the eighteenth century, transportation was a convenient means of punishing and simultaneously disposing of criminals convicted of serious crimes. Up to the 1820s those sentenced to transportation were held, sometimes for considerable periods of time, in local prisons or on the hulks (large ships moored on the Thames and used to contain up to 500 convicts each) before being assembled for the long journey in convict ships. In this early period only elderly women were exempt, and there was no reprieve from transportation for the many women who had children.[2] These mothers were obliged to reconcile themselves to never seeing their children again. On board the convict ships conditions were appalling—convicts were chained or confined in cramped holds where they suffered successively from cold and damp, rough seas and, in the tropics, from unbearable heat. Food rations were meagre, medical provision minimal,

---

[2] On the social background, age, marital status, former occupation, and offences for which women were transported, see L. L. Robson, *The Convict Settlers of Australia* (1965), ch. 4, 'The Female Convicts'.

and segregation of the sick virtually impossible. Consequently, disease was widespread and mortality rates high. Women suffered particularly because they were commonly assumed to be prostitutes and were, therefore, liable to sexual abuse by sailors and fellow convicts alike. Such conditions on convict ships continued more or less unchecked until the 1820s, when Elizabeth Fry instigated a major campaign to improve provisions on board. As a result some important changes were made; for example, the use of irons on women was outlawed, children under 7 years were allowed to accompany their mothers, and nursing mothers were no longer to be transported until their baby was weaned.[3]

For those who did survive the journey, conditions in the colonies were similarly hard.[4] Originally women were sent to New South Wales, where, up to the 1820s at least, virtually no special provision was made for them. As philanthropists gradually recognized the vulnerability of women abandoned in a country where men outnumbered them three to one, they began to campaign for some sort of government provision. As a result a 'Factory' was opened at Parramatta in 1821 to provide work and shelter for those women who had failed to secure jobs in service. Unhappily, it quickly came to be used as a quasi-brothel and marriage mart by the male settlers and convicts. If selected, the woman had no choice but to accept her 'assignment' and move in with her new 'master'. In domestic service women were often isolated and unprotected from the attentions of the male members of the household. Obliged to consent to sex if they were to keep their jobs, they were only too quickly sent back to the Factory if they became pregnant, labelled depraved and immoral. Though some women found protection by linking themselves to one man, not necessarily by marriage, others continually sought to escape. Parramatta, along with the two Factories subsequently set up in Van Dieman's Land, came to be looked upon by many women as a place of protection rather than punishment. In order to escape unpleasant or

[3] For more detailed discussion of the conditions women suffered on board, see A. Smith, *Women in Prison* (1962), 113–17; D. Beddoe, *Welsh Convict Women* (1979), *passim*; R. Hughes, *The Fatal Shore* (1987), ch. 8, 'Bunters, Mollies and Sable Brethren'.

[4] Hughes, *Fatal Shore*; Robson, *Convict Settlers*, ch. 6, 'The Female Convicts in Australia'.

violent masters they would deliberately commit crimes with a
view to being returned to the relative sanctuary of the Factory.
Their chances of respectable re-employment were limited since
there was little female work apart from domestic service. Most
importantly, whereas male convicts could earn their passage
home as crew on ships, convict women could do the same only
by prostitution.

Historians such as Deirdre Beddoe have argued that, in
seeking to correct the huge imbalance of the sexes in Australia,
the government transported women much more readily than
men guilty of the same crimes.[5] The question of a woman's
culpability for an offence was less important than her youth
and good health in ensuring she would be transported.[6] Certainly the colonies were in dire need of healthy young women,
not so much to work, for there was little suitable work for
women, but as wives and, bluntly, as sources of sexual grati-
fication.[7] In practice, the types of women sent fell well below
colonists' expectations. Australian officials had hoped that the
women would soon marry and act as a moralizing influence
over the rough, masculine society of settlers and ex-convicts.
Instead, the common view that women employed as domestic
servants could simply be used as prostitutes or even held as
virtual slaves to their masters' whims gave men little incentive
to marry them. One disillusioned official argued that efforts to
even the sex imbalance were pointless since 'the influence of
female convicts is wholly valueless upon male convicts; women
of depraved character do them no good whatsoever.'[8] They
failed to recognize that the huge sex imbalance made it extremely difficult for even the best intentioned women to retain
their good character. Nor did they see how often female
'depravity' was caused by the brutalizing nature of life in
colonial society. They were deeply concerned, however, about
the demoralizing effect the influx of such women appeared to

[5] Beddoe, *Welsh Convict Women*, 20.

[6] See D. Beddoe, *Carmarthenshire's Convict Women in Nineteenth Century Van Diemen's
Land*, repr. from the *Carmarthenshire Antiquary* (1979).

[7] M. Sturma, 'Eye of the Beholder: The Stereotype of Women Convicts
1788–1852', *Labour History*, 34 (May 1978), 3–10; also A. Summers, *Damned Whores
and God's Police: The Colonization of Women in Australia* (1976).

[8] Evidence of Thomas Frederick Elliot, Assistant Under-Secretary of State in the
Colonial Department, Select Committee on Transportation, Minutes of Evidence, *PP*,
17 (1856), 38.

be having on their society. Western Australia refused from the
start to take female convicts; according to one official, this was
because: 'The accounts from the other colonies . . . have given
them the idea that all female convicts are alike; that they are all
prostitutes and drunkards, and all the rest of it.'[9] After New
South Wales closed down as a penal colony in 1840, all female
convicts were sent to Van Diemen's Land, but again not without vehement opposition. In 1844 Captain Matthew Forster,
Controller General of Convicts in Van Diemen's Land, fought
off the establishment of a female convict prison in his colony,
declaring that 'women cannot with advantage be made the
subjects of lengthened Prison Discipline.'[10] Ironically, less
than ten years later it was the very refusal of his colony to
take any more convicts that forced the British government to
substitute long-term imprisonment at home in England as the
pre-eminent penal solution.[11]

ESTABLISHING THE CONVICT PRISON SYSTEM

The decision of Van Diemen's Land in 1852 to refuse further
admission to female convicts abruptly forced the government to
provide comparable punishment for them in Britain by instituting a system of long-term incarceration. Joshua Jebb, then
Chairman of the Directors of Convict Prisons, calculated that
1,000 places, in addition to those already at Millbank Prison
for convicts awaiting transportation, would be needed to
accommodate all those women who would formerly have been
transported.[12] In 1852 the government bought the House of

[9] Evidence of Captain E. Y. W. Henderson, Superintendent of Convicts in
Western Australia, Select C. on Transportation, Minutes of Evidence (1856), 656. For
more detailed discussion of division within Western Australia over whether or not to
admit female convicts, see the evidence of T. F. Elliot, Select C. on Transportation,
Minutes of Evidence, *PP*, 13 (1861), 540–1. For Home Office response to this decision
see letter from H. Waddington to J. Jebb (3 Feb. 1855), PRO HO22/9.

[10] PRO HO45/959. Letter from M. Forster to His Excellency Sir Eardley Wilmot
Bart., 17 Sept. 1844. PRO HO12/19/5448 and PRO HO12/17/4826 both contain
series of letters concerning the problems of disposing of female convicts in
Van Diemen's Land without placing them at risk and so, in turn, rendering them
sources of demoralization.

[11] PRO HO12/19/5448. Letter from N. Merivale to H. Waddington, 23 Sept.
1853.

[12] Report of the Surveyor-General for 1852, *PP* (1852–3), Irish University Press
repr., vol. li (1970), 36–7.

Correction at Brixton from Surrey for £13,000 to provide an extra 700 to 800 places. The urgent need to provide for a system of penal servitude at home ensured that the first stage of its conversion for use as a convict prison was completed within a year.[13]

In August 1853 the Penal Servitude Act was passed, extending all rules and regulations relating to male prisoners to the places of confinement to be established in England for women.[14] Although the male system of discipline was annexed to the female prison as a simple matter of contingency, its implementation was far more problematic. Apart from the obvious imperative to provide secure accommodation, the manner in which transportation was to be translated into long-term penal servitude for women was not at all clear. Jebb repeatedly admitted that 'the plan to be pursued became a matter of anxious consideration.'[15] Only too aware that there was no existing model according to which Brixton should be organized, Jebb argued that it was necessary to evolve a completely new method of penal discipline appropriate to women. His writings toy with alternative schemes for separate confinement and for association, coming down finally on a compromise solution of several months' separate confinement to be followed by 'modified association'. He also hoped to develop a scheme for the free emigration of selected convict women (not least to reduce the cost of their maintenance at home). The difficulty of deciding how to control female prisoners at home occupied prison officials just as much as national policy makers. The Medical Officer of Brixton, J. D. Rendle, reflected in his annual report that:

the present mode of treating female convicts, and the collecting of so large a number of female prisoners in a separate prison expressly prepared for women, are circumstances altogether new in this country. The result of a system of management which, irrespective of age, of length of sentence, and of health, admits as a general principle

[13] Report of the Surveyor-General for 1853, *PP* (1854), IUP repr., vol. xxxiii (1970), 61–2.

[14] PRO Pri Com 7. *187 Victoriae Reginae Cap CXXI* (20 Aug. 1853).

[15] Report of the Surveyor-General for 1854–5, *PP* (1856), IUP repr., vol. xxv (1970), 32.

of but one mode of dealing with *all* the female convicts of the country was almost unknown.[16]

The system mimicked that set up for men in establishing separate stages through which convicts moved during the period of their sentence, depending on their conduct. Some penologists feared that women lacked the rationality on which the success of the stage system depended and that their intense emotionalism would prohibit them from moving up through the stages in quiet order. More importantly, the very core of the male stage system—the 'public works stage' (hard physical labour on some outdoor project, for example, quarrying, road or dock building)—could not be paralleled for women. Consequently the stages formulated for the female system were rather more contrived, the conditions and privileges of each one being carefully constructed to differentiate it from that before. Denied the change of environment and routine provided for male convicts by public works, it was all the more necessary to hold out 'a degree of hope and encouragement' to women by the prospect of better conditions in some new place of confinement.[17] Initially it was intended that all female convicts should be moved from Millbank and contained at Brixton. However, the latter soon became overcrowded and in 1855 a pentagon at Millbank Prison was reallocated for women. In the following year a new prison—the Fulham Refuge—was opened, so providing three quite separate places of confinement and making it possible to implement what was known as the Progressive Stage System (a system whereby convicts moved from class to class throughout their stay in prison).

The Probationary Class, set originally at four months for women (instead of nine for men), was served in separate, solitary confinement at Millbank. Depending on their behaviour the women then progressed to the less punitive regime of the so-called Third Class. They remained in this class, initially at least, after their transfer to the more liberal regime of Brixton. Whilst movement between stages, represented by this movement between prisons, was supposed to depend on the

[16] Report of the Directors of Convict Prisons (RDCP), Brixton for 1856, *PP*, 23 (1857), 338.
[17] Extract of a letter from Colonel J. Jebb to H. Waddington, Date April 11/55, On Treatment of Female Prisoners (1855), copy in Jebb Papers Box 7.

woman's conduct, in practice it more often accorded simply with the 'exigencies of the service'. This abandonment of principle to convenience was fiercely condemned by Mary Carpenter, who argued that it undermined women's confidence in the regime, giving them nothing on which they could rely to encourage their progress.[18] When Brixton became overcrowded, promotion out of Millbank was delayed until a place became available. Under these conditions the system of incentives to good behaviour largely broke down, and those obliged to remain at Millbank became a major disciplinary problem for staff there.

Brixton provided a significant improvement in the quality of the convict prisoner's life. It was said to be 'more relaxed . . . they get better food and have pleasanter work to do, and they are brought more into association.'[19] If well behaved, women were moved to the Second Class and allowed to exercise in pairs around the prison yard, conversing quietly on 'proper subjects'.[20] They were allowed to have tea three evenings a week instead of gruel, and could wear a different dress indicating their class. Although these may seem petty privileges, such allowances made considerable impact on the monotonous routine of prison life. Simply to be allowed to talk to fellow prisoners after months of rarely interrupted silence was a strong incentive to behave well and even the change of dress appears to have been coveted. For transfer to a higher class to be delayed on the grounds that it was not administratively convenient—for example, if the appropriate wing for a given class was already fully occupied—caused intense frustration.

Those women who failed to improve further remained in the Second Class at Brixton until release. However, if they were young, well behaved, industrious, and repentant, and considered, therefore, to be reformable, they became eligible for the First Class and were generally transferred to Fulham Refuge for the last twelve months of their sentence. It was in this last stage that the female convict system differed most markedly from the male. Although Fulham was set up to

[18] M. Carpenter, *Our Convicts* (1864), ii. 212.

[19] Captain D. O'Brien, a Director of Convict Prisons, in Royal Commission on the Operation of the Penal Servitude Acts, Minutes of Evidence, *PP*, 21 (1863), 173.

[20] Report of the Surveyor-General for 1854–5, 273, though it is nowhere specified what these might have been.

parallel the 'special service' class at the male convict prisons of Dartmoor and Portland, its character was distinctly feminine and its purpose directly related to views about women criminals. In part, Fulham was intended to provide a period of lighter discipline that would lessen the severity of long, uninterrupted incarceration, an incarceration which Jebb feared women might not be able to withstand. In part, he hoped that a lengthy period in a so-called 'refuge', reminiscent of benign philanthropic institutions for distressed gentlewomen, might erase the considerable stigma of being recognized as a female ex-convict. Jebb recognized that the distinctions drawn between good and bad women made the prospects of women making their way as ex-convicts far harder than for men. He explained: 'The difficulties in the way of a woman of the character of the majority of these prisoners returning to respectability are too notorious to require description. They beset her in every direction the moment she is discharged.' [21] Perhaps intuitively, Jebb recognized that the prevailing dichotomy of good and bad women would prohibit them from regaining a place in society until such time as they had thrown off the stigma of being an 'ex-con'. Significantly Jebb had hoped to persuade one of the existing voluntary refuges for discharged prisoners, the Refuge for the Destitute at Dalston, to set up a pre-discharge refuge on the government's behalf. [22] Apparently he feared that an institution solely under government control would be seen as a prison and so continue to brand women discharged from it as 'convicted felon'. [23] Though these plans did not work out and Fulham was eventually set up by the government, the aim of ridding its inmates of the stigma of penal servitude remained paramount. The first chaplain of Fulham, whilst recognizing that 'the great danger is the suspicion with which the public will regard them', hoped that 'many persons who would not

[21] Report of the Surveyor-General for 1853, 64. In the word 'character' the full weight of Victorian notions of propriety and impropriety and, at the same time, the 'character', a reference or testimony (particularly vital to securing posts in domestic service), are both clearly implied.

[22] See PRO HO22/9, Correspondence and Instructions relating to Prisons 1854–5, H. Waddington to J. Jebb (5 June 1855).

[23] Letter from J. Jebb to J. Gurney Hoare, 22 Nov. 1853, quoted in Report of the Surveyor-General for 1854–5, 270.

like to receive servants *from a prison*, will . . . be induced to take them from a benevolent institution' such as a refuge.[24]

Fulham was also intended to serve the more practical purpose of preparing women for release. 'Industrial training' in household work, cleaning, and cooking prepared women for jobs in domestic service and in laundering. Discipline was 'relaxed as far as possible', women associated together, and conversation was allowed at all but specified times with the aim of encouraging responsibility and restoring self-respect. It was in these aspects that Fulham differed most from male convict prisons. Given the smaller numbers of women convicts, such experiments could be introduced without risking the large-scale costs that these might entail if attempted within the massive organization of male convict prisons. Whereas the latter continued to force men into silent submission to an unbending authority, Fulham employed 'softening and civilising, and enlightening influences' intended 'to raise the women up in the social scale, as respects personal character and aspirations. In proportion as they acquire and cherish *self-respect*, will they be respected by others, and helped forward in future endeavours to lead an honest and steady life.'[25]

Up to the 1860s, the female convict system was made up of these three institutions—Millbank, Brixton, and the Fulham Refuge. But after 1863, the system underwent continual reorganization. During 1863–4 Parkhurst, previously a prison for juveniles, was adapted for Roman Catholic women, but its use for women ceased on the opening of Woking in 1869 as a female convict prison.[26] In the same year Brixton was closed to women to become a light labour prison for men, and Fulham abandoned the sanguine purposes of its early years to become an ordinary female convict prison. This continual reorganization of female convict prisons was partly a response to fluctuating numbers of male and female convicts. More fundamentally, it suggests dissatisfaction with each attempt to provide a system appropriate to the requirements of the long-term imprisonment of women. Whilst reform was an important feature of the male regime, the narrow, rigidly defined ideal of femininity towards

[24] Letter from Revd J. H. Moran, ibid. 281.

[25] Chaplain's Report, RDCP, Fulham Refuge for 1860, *PP*, 30 (1861), 606.

[26] For a prisoner's view of the regime at Woking, see F. E. Maybrick, *My Fifteen Lost Years* (1905), 50–126.

which women were to be moved placed far greater demands on
the female convict system.

SUBJECTING WOMEN TO PENAL SERVITUDE
AT HOME

As we saw in Chapter 3, the setting up of female convict
prisons was bedevilled by controversy over whether or not the
'weaker sex' would be able to endure long-term imprisonment.
Would women, being 'naturally' more sedentary, suffer less
from lengthy confinement? Or would the supposedly weaker
female constitution be unable to stand the ill effects of such
constraint? Jebb's own view of women was that 'they have not
the same physical and mental powers which enable them to
bear up against the depressing influence of prolonged imprisonment.'[27] The most damaging aspect of female confinement as
compared to male was, as I have suggested, that there was no
female equivalent to the public works. Jebb recognized this and
argued that 'as females must of necessity be employed chiefly
indoors, and will have neither the varied work, nor the
complete change afforded to Male Convicts, by removal to
public works it would be inexpedient to extend the term of
Imprisonment to the same period as may have been determined in the case of males.'[28] This view quickly predominated
and is reflected, for example, in the decision to award women a
more generous sliding scale of remission of their sentence for
good behaviour under the marks system than men (a possible
maximum of one-third remission as opposed to one-quarter
for men).[29]

[27] Letter from J. Jebb to H. Waddington, 24 Mar. 1855, Jebb Papers Box 7. See
also testimony to the same effect by the Superintendent, Medical Officer and Chaplain
of Brixton in Report of the Surveyor-General for 1854–5, 275–7.

[28] Letter from J. Jebb to H. Waddington, 24 Mar. 1855. Jebb's view was based on
medical evidence of a Dr Baly, who warned of 'the deteriorating effect of . . . the
monotonous, and sedentary character of the occupation in which women prisoners are
almost of necessity employed'. Letter from Baly to Jebb (3 Nov. 1853) quoted by Jebb.

[29] On the ending of female transportation, remission had been abolished on the
basis that sentences of penal servitude were much less hard than sentences of
transportation (see Notice to Female Convicts, quoted in the First Report of the Select
C. on Transportation, *PP*, 17 (1856)). However, it soon came to be seen as a necessary
inducement to good conduct (see Report of the Surveyor-General for 1854–5, 8) and
was reintroduced in 1857 (see RDCP, Brixton for 1859, *PP*, 35 (1860)).

In the early years of penal servitude, anxiety about how women would respond to the prospect of long-term imprisonment was borne out. The impact of the dramatic change in the nature of their sentence was clearly appreciated by Jebb, who explained: 'Until very lately female convicts were taught to regard expatriation as the inevitable consequence of their sentence, and when detained in Millbank . . . they were reconciled to the discipline, however strict, by the knowledge that it would soon cease, and that it was only a necessary step to all but absolute freedom in a colony.'[30] Unsurprisingly, the first women detained as convicts in England were indeed extremely refractory; despite the known hardships of transportation, they regretted losing the chance of a 'new life' in the colonies and bitterly resented the prospect of long-term confinement. They stimulated fits, refused to have their hair shorn, refused food, and even smashed up their cells.[31] Their indiscipline was attributed in part to the influence of 'a few very bad women', in part to overcrowding in the female pentagon at Millbank, but primarily to: 'Extreme disappointment at finding that they were not to be sent abroad; and an anxious and feverish desire on the part of many of them to ascertain what was likely to become of them, which prevented their settling down to their regular employments and submitting to the routine of discipline with the more quiet resignation that ordinarily characterises male convicts'.[32]

This view that the prison treatment of women was more problematic than that of men cannot be dismissed merely as early teething troubles. Female convicts continued to be seen as temperamentally unstable, less likely to submit to prison discipline, and more prone to riotous behaviour which only confirmed suspicions of their moral degeneration. Whether these views simply reflected pre-existing assumptions about female criminals, or whether, by way of self-fulfilling prophecy, the women acted as they were expected to do, is unclear. The difficult character of the female convict became axiomatic. Throughout the 1850s, annual reports by convict prison directors, lady superintendents, chaplains, and medical officers reiterate again and again the greater problems of controlling

[30] RDCP, Brixton for 1853, *PP*, 33 (1854), 307.
[31] A. Griffiths, *Memorials of Millbank and Chapters in Prison History* (1884), 59 ff.
[32] RDCP, Millbank for 1853, *PP*, 33 (1854), 87.

women. By 1865 the prison chaplain Revd Walter Clay could declare as a truism: 'it is well known that women are far worse to manage, and resist what is for their good far more vehemently than men.'[33] This resistance will be considered at length later in the chapter, for it is vital to any understanding of the internal life of the female convict prison.

PURPOSES OF THE PRISON REGIME

As we saw in the first two chapters, Victorian perceptions of criminal women were quite distinct from views of criminal men. These differences were clearly reflected in the regime of the female convict prison. Nearly every aspect of this differentiation can be traced back to views about the character of the female criminal or forward to the goal of idealized femininity which the women were to strive to achieve. The fact that the notion of the 'lady', by definition, presumed a social class most convicts could never hope to attain and set standards remote from the realities of the life they faced outside did not prevent prison reformers from holding it up as the ultimate goal.

Although discipline and punishment actually characterized much of female convict life this was less an aim in itself than the inevitable result of trying to impose an alien regime on unwilling subjects. The punitive element was legitimated as an unfortunate but necessary precondition to moral treatment intended to restore criminal women to honesty, propriety, and 'womanliness'. Whereas male convict prisons were primarily concerned with inculcating discipline, orientated as they were around arduous labour at public works, the predominant role of the female prison was the psycho-medical treatment of women, aimed at effecting moral regeneration. A visitor to the female convict prison at Mountjoy, Dublin, in 1862 described her journey as 'passing through the progressive convalescent wards of this moral hospital'.[34] In doing so she encapsulated both the purpose and character of female convict prisons generally.

---

[33] W. L. Clay, 'On Recent Improvements in our System for the Punishment and Reformation of Adult Criminals', *Transactions NAPSS 1865* (1866), 192.

[34] A. Jellicoe, 'A Visit to the Female Convict Prison at Mountjoy, Dublin', *Transactions NAPSS 1862* (1863), 439.

The primary aim was to provide the woman with the opportunity and means to reform. Cut off from wider society much more effectively than in local prisons, her past history, socio-economic background, even the very crime which had led to her imprisonment could all be regarded as peripheral to the real focus of attention, the woman herself: 'A woman on entering a convict prison should feel that however vicious her past life has been, she is come to a place where she has a character to regain and support.'[35] This attitude partly reflects a simple recognition that the convict prison could have no real influence on the wider causes of her crime (degrading housing conditions, the lack of full-time female employment, etc.). More importantly, it arose from the belief that female crime was the product of emotional, intellectual, and behavioural inadequacy over and above economic need or social circumstance. Women were described as being both incapable of moral judgement and yet at the same time as morally degraded, as being shameless and yet desperate for self-respect. These contradictions were resolved by a programme aimed at revealing to the woman her moral degradation and teaching her to differentiate between 'vice and virtue'. Only once she was convinced that she occupied the moral depths could she begin to reform. Any concrete demonstration of effort to amend was rewarded. Good conduct badges were used as a means of discipline and, after the introduction of remission in 1857, could be accrued towards remission of sentence. The Lady Superintendent of Brixton reported: 'The badges continue to be great incentives to good conduct, and form the subject of much conversation amongst them and often of great anxiety . . . the restraining influence of the badge . . . acts more powerfully than deprivation of food or any amount of other punishment.'[36]

The Royal Commission into the Penal Servitude Acts (1863), finding that penal servitude was not feared as much as had been hoped, sought to make it more deterrent. They recommended that badges be replaced by marks, also accruable towards remission. Denying the concept of positively good behaviour within the confines of the prison, they recommended that these be awarded not merely for good conduct but for

[35] Anon., 'The Petting and Fretting of Female Convicts', *Meliora*, 6 (1864), 47.
[36] Quoted in Report of the Surveyor-General for 1856–7, *PP* (1857–8), IUP repr., vol. xxix (1970), 51.

diligence and productivity. Although marks were to be awarded on the same scale for both sexes, women, unlike men, could begin to earn marks during the probationary period whilst still in separate confinement. In practice, whereas men were now awarded marks mainly in respect of their industry at the public works, women continued to earn them for good behaviour and moral improvement. The prison authorities extolled the importance of seeking moral regeneration. The convict prison was to bring about the process of reform by religious education, by generating self-respect, and by inculcating habits of industry, honesty, and cleanliness. This stress on the process of reform was mainly produced by the difficulties encountered in assessing its end result. Whilst the prison authorities could control the means of reform, they felt far less confident of their ability to judge the female prisoner's true 'religious and moral condition'.[37]

Officially the regime sought to eradicate and replace the filth, chaos, and corruption of the eighteenth-century prison. The reformed convict prison combined pervasive supervision with a minutely timetabled regime designed to minimize corruption whilst giving free rein to those influences thought to be reformative.[38] In practice, as we shall see, these aims were not quite so readily achieved as penal experts hoped. Moreover it was commonly accepted, even officially reported, that the treatment of women could not profitably be directed by uniform adherence to regulations in the same way that men were governed. Mrs Emma Martin, Superintendent of Brixton Prison, admitted that discipline in her prison, 'although it is in strict accordance with the rules laid down, yet it varies in some degree according to the disposition and habits of the prisoner. Indeed, without this individual treatment the attempt to reform them would be most superficial.'[39] Thus the very conditions of the prison allowed staff to observe and to get to know the peculiarities of each woman's personality. Rules were bent or even ignored to allow discipline to be adapted to individual temperaments. For example, although female convicts were

[37] RDCP, Brixton for 1856, *PP*, 23 (1857), 322.
[38] For details of the daily regime at both Millbank and Brixton, see the RC on the Operation of the Penal Servitude Acts, *PP*, 21 (1863), 194 and 195 respectively.
[39] Quoted in Report of the Surveyor-General for 1856–7, 50.

supposed to wear uniform outfits, their efforts to alter their dress and to improve their appearance were tolerated as hopeful indications of recovering femininity. The militaristic precision with which regulations were enforced in male prisons gave way to more responsive, even manipulative, treatment in female prisons.

THE CONVICT PRISON SETTING

The 'treatment' process can only be understood in its architectural setting.[40] The female convict prisons were housed mainly in buildings built for men and only latterly adapted for women. They thus inherited buildings designed to isolate prisoners from one another whilst holding them exposed to total and continuous supervision by their captors. However, by the mid-nineteenth century, total separation was becoming less popular.[41] Isolation, if it did not brutalize the prisoner irrevocably, had a worrying tendency to cause insanity. Constrained by existing architecture and nervous of the contaminating effects of association, the regime in female convict prisons operated an uneasy mix of separation in cells and associated work and exercise. This lack of fit between changing regime and immutable architecture is well illustrated by an engraving of Brixton Prison that shows women sitting working silently and in evident discomfort on the narrow landings outside their cells.[42] Attempts were made to lessen the severity of the architecture. At Brixton the women were 'allowed' to whitewash the interior of the prison, and grass plots and flower beds were dotted around the airing or exercise yards in a bid to feminize the dour, masculine environment. Yet even when new buildings were erected, the legacy of separate confinement endured. Although new wings at Brixton were built on the cellular principle, their original purpose, to prevent prisoners communicating at night, was corrupted by the use of corrugated

[40] L. L. Goldfarb and L. R. Singer, *After Conviction* (1977), 38–9. See also R. Evans, *The Fabrication of Virtue: English Prison Architecture 1750–1840* (1982).

[41] See U. Henriques, 'The Rise and Decline of the Separate System of Prison Discipline', *Past and Present* (Feb. 1974).

[42] H. Mayhew and J. Binney, *The Criminal Prisons of London and Scenes of Prison Life* (1862), 196 facing.

iron partitions through which the women could easily communicate. Although Mayhew considered Brixton had 'nothing
of the ordinary prison character or gloomy look',[43] one cannot
help but be struck by the dichotomy between the supposedly
humane and individualized regime and the austere, uniform
environment within which it operated.

One aspect of the environment, at least, accorded with the
intentions of the regime, namely the extreme cleanliness of the
prison. As soon as a woman entered prison she was made to
strip (her clothes were either burned or fumigated), bathe, and
have her hair cut short. This last practice was often criticized as
an unnecessarily brutal attack on the remaining vestiges of
femininity. Yet, quite apart from its obvious hygienic purpose,
it served the symbolic function of divesting the woman of the
tainted character of her former life. In his classic work *Asylums*
(1970), the sociologist Erving Goffman describes this process
as one of 'mortification of self', by which the individual is
dispossessed of her personal identity and is physically and
psychologically stripped, and so laid open to the process of
rehabilitation. Cleanliness, thought to be inimical to the
temper of criminal women, was as much an element of discipline as it was of reform. Whilst men escaped daily to the filth of
the public works, female labour revolved around an endless
routine of scrubbing and cleaning. Significantly the laundry
was the most important and most frequently lauded form of
employment at Brixton, for it exemplified the process of physical and, above all, spiritual purification (quite apart from
providing the woman with a livelihood on release). The extreme
cleanliness of her surroundings was designed to encourage a
parallel self-purification.[44]

Women coming into prison were not only filthy but all too
often also debilitated and diseased—traits that were assumed to
be the wages of her profligate lifestyle. The medical officer of
the female convict prison at Parkhurst commented that 'the
medical records of a convict prison reveal for the most part the
results of a life of vice and crime. A life of criminal excitement

---

[43] Ibid. 179.

[44] Similarly Patricia O'Brien describes French prisons as 'these sanitary
laboratories aimed at creating a therapeutic sterility out of which institutional beings
would emerge to return to civil society.' P. O'Brien, *The Promise of Punishment*
(1982), 19.

and debauchery cannot but leave its traces on the constitution of its victim.'[45] Illness was often construed as retributive, and yet, as we observed in local prisons, it could also feature in the reform process. The idea of salvation through sickness and suffering is clear in this report from Brixton: 'in some the chastening hand of God, in severe bodily affliction, has been laid, and they have passed into the infirmary; where the order and calmness, which ever reign in that department, have had a beneficial influence in subduing the unruly will . . .'[46] The comparative health of convicts versus the rest of the population was lengthily debated. Whether the quality of prison food, sanitation, and ventilation outweighed the supposedly debilitating effects of confinement was unclear. Reflecting the wider view that women, as the 'weaker sex', were delicate and prone to disease, the annual reports of the Directors of Convict Prisons seem preoccupied with women's frailty and their ability to withstand long-term confinement.[47] Although many observers claimed that women convicts were far more inclined to ill health than men, official statistics indicate that women were no more likely than men to suffer either slight indisposition or serious illness and some medical records actually suggest that women spent less time in the infirmary and were generally healthier than male convicts.[48] Nonetheless employment, diet, and discipline were all modified to accord with preconceptions about female frailty. For example, in its very first year, Fulham Refuge found it necessary to introduce a more varied diet and to replace needlework with more active occupations in order to minimize the deteriorating effects of confinement on the 'pale and unhealthy' women held there.[49]

[45] RDCP, Parkhurst for 1868, *PP*, 30 (1868–9), 366.

[46] RDCP, Brixton for 1855, *PP*, 35 (1856), 279.

[47] See the reports of medical officers of the female convict prisons appended to the annual RDCP. Many of the witnesses to the Royal Commission on the Working of the Penal Servitude Acts, *PP*, 37 and 38 (1878–9), were questioned about the health of female convicts as compared to men and about the effects on female health of confinement, separation, labour, and dietary punishments, see esp. evidence given by R. M. Gover, Medical Officer, Millbank Prison; Mrs S. Seale, Lady Superintendent, Fulham Prison; Mrs S. Gibson, Lady Superintendent, Woking Female Prison; and Dr H. W. Hoffman, Medical Officer, Fulham Prison.

[48] W. A. Guy, 'On some Results of a Recent Census of the Population of the Convict Prisons in England; and Especially on the Rate of Mortality at present Prevailing among Convicts', *Transactions NAPSS 1862* (1863), 563.

[49] RDCP, Fulham Refuge for 1856, *PP*, 23 (1857), 363–4.

Perhaps the most incongruous element of the female convict prison was the prison nursery. In the early years of penal servitude, women convicts could take their babies to prison or, if born within the prison, keep them there until the end of their sentence. Henry Mayhew, on visiting Brixton in 1862, found the convict nursery 'the most touching portion of the female convict prison, and what distinguishes it essentially from all the penal institutions appropriated to male prisoners'.[50] The very presence of children demanded that rules be waived and that their mothers escape the rigours of the penal regime. Also, whereas a child could be held in a local prison for a few weeks without doing it any great harm, the convict prison could not hope to provide an acceptable long-term environment for child development. Mayhew was clearly touched by the utter terror that infants, who had had no contact with men other than the prison chaplain, showed on meeting him. He deplored the narrow, artificial existence in which they grew up and noted: 'one little thing had been kept so long incarcerated, that on going out of the prison it called a horse a cat.'[51]

ELEMENTS OF THE TREATMENT PROCESS

Labour

In recent revisionist histories of the prison, labour and, through it, work discipline have been identified as the most crucial facet of penal servitude. The stone-breaking yard, the prison workshop, knitting room, and laundry are seen as the focal sites of moral reform. The prison, the revisionists argue, was not merely similar in form and regime to the new industrial factory but actually became a factory in its own right.[52] Certainly 'hard labour' was specified as part of penal servitude and in male prisons the avowed aim was to create useful, hardworking men inured to discipline and raised out of their perceived physical degeneracy by hard labour. But in female

[50] Mayhew and Binney, *Criminal Prisons*, 189.
[51] Ibid. 190–1.
[52] This view is most exhaustively propounded in D. Melossi and M. Pavarini, *The Prison and the Factory* (1981) but is also clearly enunciated in the works of Michel Foucault, Michael Ignatieff and Patricia O'Brien.

convict prisons work played a lesser, rather more ambiguous role.

One pragmatic reason was quite simply the difficulty prison authorities encountered in obtaining suitable work for women. Whereas male convicts were often employed in isolated areas in extremely arduous labour (quarrying, building breakwaters, etc.), obtaining work for women all too often meant going into direct competition with female seamstresses and washerwomen outside the prison.

Even if work was found, the sedentary, monotonous occupations available for women compared badly with 'the bracing and invigorating effects of out-door labour which is the privilege of male convicts'.[53] Horatio Waddington, Permanent Under-Secretary at the Home Office, explored the possibility of obtaining 'some active and laborious occupation in the open air' for women and suggested setting up a prison with sufficient ground to allow for outdoor work.[54] The suggestion seems to have been prompted by widespread concern about the physical and psychological damage done by sedentary indoor labour. Yet the search for work was constrained by notions of what was thought to be suitable feminine employment. After the first two months of purely punitive work, teasing the fibres from tarred rope or 'picking coir' as it was known, women were employed in needlework. Once promoted from the probationary class and so released from solitary confinement, they worked in association on rough sewing, bag-making, knitting, cooking, baking, and, of course, cleaning and laundering. At Brixton the laundry employed a hundred women in a large-scale enterprise, washing for convicts at Millbank and Pentonville Prisons as well as for themselves. Considered to have failed in their 'natural' role as wife and mother, the women were given a

[53] Mrs Martin, Superintendent, Brixton, quoted in RDCP, Brixton for 1855, 272. Mary Carpenter also commented on the 'dreadful monotonous routine' of female labour as compared to male convicts' experience of 'varied labour in the open air, calculated to exercise their muscles, to occupy their minds, to give them the healthful influences of nature'. *Our Convicts*, ii. 251–2.

[54] PRO HO22/10. Letter from H. Waddington to J. Jebb (14 Nov. 1856). In the same collection see also the letter from H. Waddington to J. Jebb (27 Jan. 1857) in which he quotes evidence in support of his arguments by Dr Daly. The latter advocates that 'some active employment . . . of such a nature as to be the best substitute for out of door employment which circumstances will admit of' be provided, particularly for young women.

second opportunity to develop 'feminine' attributes in the exercise of domestic service.

That the types of work conformed to the stereotype of 'feminine occupation' largely at the expense of more remunerative employment is not surprising. Stress was laid less on acquiring skills which might provide the women with employment on release, and, more, on employment as a means of moral regeneration. Work was presented almost as a privilege, as a relief from the tedium of imprisonment. In turn women were assumed to be grateful for employment and distressed when none was available: 'Unless actively employed they become restless and desponding (*sic*), and brood over the wretchedness their crimes have entailed on husbands and children.'[55] The primary role of labour in relation to female convicts was, then, psychological. It was intended to subdue the supposedly passionate and reckless temperament of the female criminal. Mary Carpenter clearly shared this view, arguing that 'to provide an abundance of active, useful work is absolutely necessary. The restless, excitable nature of these women requires a vent in something, they should have full employment, of a kind which will exercise their muscles and fully occupy their minds, so as to calm their spirits and satisfy them with the feeling of having accomplished something.'[56] The contrast between the labour discipline on the public works for men and the purportedly therapeutic role of female employment is marked.

Schooling

As an agent of penal discipline, education occupied an even more ambivalent position. It was widely seen as a potentially dangerous tool with which to equip the poor, and even more dangerous in the case of the criminal poor. Some feared that education would provide criminals with better skills which would only be used to invent more sophisticated depredations. Moreover, in the case of women, education was considered by many simply to be an irrelevance. Proponents of education stressed instead its efficacy both as a means of preventing crime

[55] RDCP, Brixton for 1862, *PP*, 24 (1863), 257.
[56] Carpenter, *Our Convicts*, ii. 210.

and of bringing about moral reform. Education was a source of occupation during times of idleness and a counter to other temptations; it enabled the prisoner to read improving literature and so to develop moral judgement. In a regime with limited resources, based upon self-improvement and seeking to instil self-discipline, education could play a vital role.

Observers found the apparent correlation between ignorance and crime crucial in explaining women's descent into crime. Quite apart from those women who were actually below normal intelligence, female prisoners had invariably received even less education than their male counterparts. Their lack of learning, especially concerning the Scriptures, was regularly deplored. The women of Brixton Prison were denounced as follows: 'Women more ignorant and stupid than these prisoners it is impossible to conceive; teaching them becomes a hopeless task—the little progress made one week is entirely forgotten the next.'[57] Despite this view, education played a more central role in the long-term treatment of women than men. In male convict prisons little time was allowed for schooling, which was generally relegated to the evening, but female convicts were subjected daily to the attentions of the chaplain, schoolmistresses, and lady scripture readers. Basic literacy skills were taught primarily as a means of access to the Scriptures. Acquiring other useful skills or reading didactic or scientific books were all thought inappropriate for women, whose intellect and power of concentration were supposedly weaker than men. The Chaplain of Millbank stressed that books for women must be carefully selected: 'The reading lessons for the women require to be of a more light and entertaining character than those for men.'[58] Education then took on a particular form in the female convict prison, limited in scope but integral to the process of moral and spiritual regeneration.

Religion

If the schoolroom played an ambivalent role in reform, the chapel was little better. Penal historians focusing on the spiritual quality of prison discipline in this period have cited the

[57] A Prison Matron, *Female Life in Prison* (1862), i. 71.
[58] Report of Chaplain, RDCP, Millbank for 1860, *PP*, 30 (1861), 357.

chaplain as second only to the governor in status and influence. They have discerned many parallels between the cloistered world of the prison and the isolation, reflection, and prayer of the monastery.[59] Early prison reformers, motivated by evangelical zeal, emphasized the importance of spiritual regeneration through religious education, prayer, and contemplation. Penal literature, at least up to mid-century, is imbued with pious religiosity and scattered with grandiose claims for religious teaching as a means of moral reform.

In female convict prisons religion played a central role, with the chapel pivotal to the wider process of reform and the chaplain arguably the only male officer of any importance. The daily routine of silence, reflection, and prayer certainly evokes strong parallels with the nunnery. Compulsory attendance at daily services, endless unsolicited visits by the chaplain, lady scripture readers, and pious lady visitors demanded that the women adopt an outwardly pious demeanour and at least profess to religious faith. Prisoners were said to respond to religious instruction 'with grateful delight', to express rapturous joy at being given the opportunity to convert, and to leave the prison burdened with a sense of their innate sinfulness.[60] More sceptical observers, however, warned of the risks of attempting to elicit instantaneous conversion from women prisoners as:

a doctrine at all times dangerous, but peculiarly so when put before the minds of these poor women. In the female mind it is at all times more likely to find favour. Where the nervous system is more tender, a doctrine that has so much to do with animal as well as mental feeling can more easily be brought to bear. If it be pressed upon frail creatures, when they are just waking up to fearful consciousness of their sins, its application may work the greatest mischief.[61]

Reading the claims of many prison chaplains and lady visitors one can only be struck by the superficiality of their evidence and their glib acceptance that professions of penitence neces-

[59] See N. McLachlan, 'Penal reform and Penal History', in L. Blom-Cooper (ed.), *Progress in Penal Reform* (1974), 9; W. J. Forsythe, *The Reform of Prisoners 1830–1900* (1987).
[60] See, for example, M. Wrench, *Visits to Female Prisoners* (1852), *passim*.
[61] J. Armstrong, 'Female Penitentiaries', *Quarterly Review* (Sept. 1848), 373.

sarily indicated a profound change of heart.[62] It seems only too likely that many women simply feigned interest and piety, whilst remaining singularly unmoved beneath this thin surface hypocrisy.

Many contemporaries believed that the disordered and profligate life led by the female criminal was caused by her lack of that self-regulating quality—shame.[63] She could only hope to be reclaimed, therefore, by being helped to acquire a profound sense of personal sin. The Chaplain of Fulham Refuge clearly supported this stress on atonement; he maintained that 'one way in which this better state of feeling shows itself is in the wider prevalence of a spirit of self-accusation and self-abasement for past sins, and that too apart from definite acts of crime, which is all the more satisfactory as an index of character.'[64] The process of reform was inseparable from the religious model of salvation. It was not merely, as Michael Ignatieff suggests, couched in the religious vernacular but actually demanded full participation in rituals of expiation and amendment in order to achieve salvation. Moral improvement was commonly judged by observing women's attitude to religious education. Again the Chaplain of Fulham Refuge averred, 'improvement in character is to be traced in seriousness of behaviour in chapel, and especially in the attention paid to the explanation of Scripture at the daily services.'[65] The naivety of assuming that moral health could be assessed simply by observing conduct was scornfully criticized by those who stressed women's supposed powers of deception and their obvious incentive to adopt a facade of piety. Despite the difficulty of distinguishing the hypocrite from the truly penitent, the female prison could only judge its immediate success by precarious assessments of moral character and religious faith. Consequently prison reports are scattered with observations on women's conduct, expressions of repentance, and reform. By 1878, this art of character assessment had attained the status of a science whereby the women could be neatly categorized

[62] See the Chaplains' Reports appended to RDCPs; also the writing of Lady Visitors such as Matilda Wrench, Susanna Meredith, and Francis Scougal (pseud.) listed in Bibliography.

[63] See Mayhew and Binney, *Criminal Prisons*, 456, for views on the importance of shame in restraining female behaviour.

[64] RDCP, Fulham Refuge for 1859, *PP*, 35 (1860), 718–19.

[65] RDCP, Fulham Refuge for 1861, *PP*, 24 (1862), 642.

under headings which ranged from 'very hopeful and encouraging' through several intermediate categories to 'hardened and unimpressed'.[66]

The extraordinary confidence bestowed on this process of moral regeneration seems at first to suggest an absurd severance of the prison regime from the environmental causes and the very crimes which had first brought the women to prison. Yet only by defining female crime as the product of personal aberration could the prison find itself a plausible role. Isolated in the extreme, it could have little effect on wider society. By redefining crime solely in relation to the individual over whom it had considerable power, it became possible to seek change within the institutional setting.

Moral regeneration of female convicts played a more pragmatic role in relation to their likely fate on discharge from prison. Whereas male ex-convicts might hope to find work in labour gangs under employers none too fussy about past credentials, female convicts were most likely to seek employment in domestic service. The degree of trust such work entailed required that future employers must be wholly satisfied that the women they took into their homes were irreversibly reformed. The cynic might conclude that the process of moral regeneration was less the outcome of any real faith in reform than an astute exercise in public relations. As Georg Rusche and Otto Kirchheimer have shown, penological theory is often an *ex post facto* justification of penal method, itself determined by economic and social structure.[67]

Under the regime of Edmund Du Cane (the Chairman of Directors of Convict Prisons from 1869 to 1895), religion seems to have declined in importance. Religiously motivated voluntary efforts were subsumed under official provision, and scepticism concerning the durability of reform grew. Henry Mayhew became increasingly scathing in his attacks on a system which relied on appeals to spiritual motives alone and blind faith in the apparent miracles of conversion, insisting instead on the need for more practical incentives to reform. In this increasingly hard-headed, secular climate the status and power of the chaplain waned.

---

[66] RDCP, Woking for 1878, *PP*, 35 (1878–9), 539.
[67] G. Rusche and O. Kirchheimer, *Punishment and Social Structure* (1939).

CONVICT PRISON STAFF

So far we have considered the supposed aims of female penal
servitude and some of the main elements of the programme by
which they were to be achieved. To conclude here would be to
accept Home Office rhetoric at face value, and to assume that
the means were adequate to achieve their avowed end. To get
at the 'reality' of prison life it is necessary to consider how these
aims were actually implemented and by whom.

Below the level of the directorate and with the notable exception of the chaplain, female convict prisons were staffed almost
entirely by women.[68] The unique nature of so large and
complex an institution run entirely by women led Mayhew to
observe that 'one of the main peculiarities of Brixton Prison is
that the great body of officials there belong to the softer sex, so
that the discipline and order maintained at that institution
become the more interesting as being the work of those whom
the world generally considers to be ill-adapted for government.'[69] He admitted himself so much 'a creature of prejudice'
that he found 'it almost ludicrous at first to hear Miss So-and-so
spoken of as an experienced officer, or Mrs Such-a-one described
as having been many years in the service'. Mayhew's incredu-
lity notwithstanding, the responsibility devolved on these
women for the successful running of the prison was huge.
Whereas male convict prisons operated on quasi-military lines,
with the officers in strict hierarchy controlling the convicts as a
body according to closely followed rules, the individualization
considered vital to the treatment of women conferred much
greater independence on female warders. As in local prisons,
the demands made on them were considerable. Yet the majority
entered the service out of necessity rather than vocation and
many were quite unequal to the role demanded of them.[70] A
typical female warder had formerly been in domestic service

[68] Peel's Gaol Act (4 Geo. IV, c. 64) 1823 required that matrons were to be
appointed to every prison in which women were contained, that women prisoners were
to be supervised only by female staff, and that no male officer or keeper was to enter
any part of a women's prison unless accompanied by a female officer.

[69] Mayhew and Binney, *Criminal Prisons*, 178.

[70] See evidence given in Edwin Pears (ed.), *Prisons at Home and Abroad: Transactions of
the International Penitentiary Congress, 1872* (1912), during a discussion on 'Women's
Work in Prisons', at 544.

and had joined the prison service for want of other employment. The fact that she was barely above her charges in social
standing cannot have eased her attempts to assert the authority
she relied on to maintain control.

Once recruited, warders were subject to a lengthy series of
rules and regulations. It is hardly surprising that female
warders were often described as inmates. As in local prisons,
they were expected to exemplify the ideal of womanhood. For
example, the Directors of Millbank insisted that the treatment
of the obstreperous and often violent women of the penal class
required 'patience, a disposition to discover, and give credit
for, the least evidence of improvement, and a sympathy which
can understand and feel for the trials and difficulties even of the
outcast'.[71] By consistently demonstrating these feminine virtues
they were to provide a role model whose attributes, it was
hoped, the inmates would observe, internalize, and eventually
adopt.

The difficulties of maintaining this exemplary façade must
have been enormous.[72] Despite the supposed stress on individualization, the staff–inmate ratio was low; in 1863 Mary
Carpenter estimated it at 1 : 15.[73] Female warders worked a
day of twelve or even fifteen hours, supervising convicts from
6 a.m. to 5 p.m., and often remaining on duty until 8 p.m.[74]
Concern about women's supposed frailty provoked fears that
they could not endure such arduous work without serious
damage. One of the Directors of Convict Prisons, Captain D.
O'Brien, informed Jebb, 'I have the assurance of the Medical
Officer at Brixton prison, that more relaxation must be given to
enable women of ordinary strength to continue in the service
without early loss of health.'[75] The burden was so great that
the exemplary behaviour they were supposed to maintain
remained more an ideal than a reality. The idealized portrait of
zealous women patiently encouraging the penitent to reform is
brought into question by the following observation by Captain

[71] RDCP, Millbank for 1860, *PP*, 30 (1861), 359.
[72] For sympathetic discussion of the stresses placed on women staff see Maybrick
My Fifteen Lost Years, 201–2.   [73] Carpenter, *Our Convicts*, ii. 214.
[74] J. Jebb to the RC on the Operation of the Penal Servitude Acts, Minutes of
Evidence (1863), 355.
[75] PRO Pri Com 2/164. Memo from Captain O'Brien to Joshua Jebb, 1 Nov.
1862.

O'Brien, who noted that at Millbank, 'towards the close of the day . . . some of the officers get irritable and extremely cross with the prisoners, and that other officers get so tired out, they really do not much care whether the prisoners about them conduct themselves well or ill.'[76]

Some warders were patently never the models of virtue they were set up to be. Some failed to complete the probationary period of employment, others resigned of their own accord or were dismissed as being incapable of performing the duties required. The problem of high staff turnover dogged the convict prison service throughout its early years. Others passed through probation but proved unequal to the trust placed in them. In male prisons brutal treatment of prisoners and outright violence were a continual cause for concern. In female prisons abuse was less flagrant, less evident to the observer, and consequently less well documented.[77]

STAFF–INMATE RELATIONS

In male prisons where the main requirement was to supervise and enforce labour, relations between warders and gangs of prisoners were terse and formal. In female convict prisons, as in local prisons, individualized treatment could be realized only through a close relationship fostered between the warder and the inmate. Significantly, in an influential fictional work— *Memoirs of Jane Cameron: Female Convict*[78]—the heroine is reformed by the influence of one warder who takes a special interest and makes allowances for her bad behaviour. The grateful Cameron responds by trying to live up to the kindly expectations of her captor. The story is idealized, its happy ending barely plausible, and yet it embodies views which were widely held. Mary Carpenter maintained that 'an essential part of the work of reforming such women . . . is the healthy development of their affections. These are peculiarly strong in the female sex, and may be made the means of calling out the

[76] Quoted in Carpenter, *Our Convicts*, ii. 215.
[77] Though female prisoners frequently attributed their own misconduct to provocation by the matrons. See evidence given by Mrs Sarah Gibson, Superintendent of Woking Prison to the RC on the Working of the Penal Servitude Acts, Minutes of Evidence (1878–9), 522.   [78] By 'A Prison Matron', see Bibliography.

highest virtues, the most genuine self-devotion.'[79] This view contrasts strongly with the extreme hostility expressed towards the development of similar relationships in male prisons (see, for example, the rules of Pentonville Prison 1842).[80] Even in female prisons it was recognized that such relationships were precariously founded. For example, at Brixton, a single prisoner, E.L., wrested and manipulated power, to the consternation of the prison authorities: 'how she contrives to obtain influence and opportunity over inexperienced officers is remarkable; but certain it is that in every prison she has acquired such an influence.'[81] By the machinations of just one woman, control was undermined, power relations were upset, and the stability of the whole prison threatened. It is impossible to assess how pervasive such corruption was, for only its most blatant and damaging effects are documented. Since there are no surviving day books or other internal records, the underlife of convict prisons is all but inaccessible. To maintain stability, a façade of order had to be upheld and the paucity of surviving records maintains that façade today. However, it is highly likely that, in reality, convict prison life was far less calm, less secure than this impression suggests.

A more overt and therefore more commonly documented destruction of the staff–inmate relationship was outright insurgence by the prisoners. The threat of violence was always present and provided the justification for high security within the convict prison. The penal class at Millbank contained women who were so violent that they had been rejected as unmanageable within the more relaxed regimes of the other female prisons. Renowned for destroying prison property, these women occasionally launched violent attacks on the warders. The frequency of lesser attacks is attested to by the fact that a common plea by inmates seeking remission is that the individuals concerned had intervened to protect warders from assault.

The complexity of the staff–inmate relationship, the precariousness of the balance of power, and the division of loyalties even amongst the inmates evidently combined to make female

[79] Carpenter, *Our Convicts*, ii. 211.
[80] PRO HO20/13. Rules for the Government of Pentonville (1842).
[81] RDCP, Brixton for 1865, *PP*, 38 (1866), 258.

prisons tense and uncertain places. It is, therefore, patently absurd to assume that prison authorities always held sufficient control to implement official policies. Looking beyond the rhetoric of reform we find communities of women living together in close quarters who survived primarily by compromise and bargaining, backed up on one side by warders' institutional power, and on the other by a continual round of verbal and even physical threats. Boredom, frustration, and the anxiety of long-term confinement, suffered by staff as well as inmates, describe their lives more accurately than any penological theory or list of regulations.[82]

THE INMATES

Before going on to attempt to reconstruct the inmates' experience of life in prison it would be useful to determine who these women were. Unfortunately details of their social and occupational background or even of their lives before coming into the convict system are far more scarce than for local prisoners. We can, however, establish the proportion of women in the convict prison population, the offences for which they were sentenced, and their numbers of previous commitments.

Commitments

Over the second half of the nineteenth century, women made up a varying but small proportion of those committed to convict prisons. Whilst they made up 16 per cent in 1860, their proportion fell away again towards the close of the century (Appendix, Table 10). Figures are not available giving length of sentence by sex over the whole period. However, the Commission into the Working of the Penal Servitude Acts (1878–9) provided figures for length of sentences awarded from 1869 to 1876 inclusive.[83] These show that, for this period at least, both men and women were sentenced to an average of between seven and eight years. It seems that this parity of sentencing was main-

---

[82] See the evidence by W. A. Guy, Medical Superintendent at Millbank, and by Revd J. H. Moran, Chaplain at Brixton, to RC on the Operation of the Penal Servitude Acts (1863), 534, 684.   [83] Commission Appendix (1878–9), 1150.

tained over the period, since figures for the daily average number held in convict prisons mirror almost exactly those given above for commitments. The declining number of women in the convict system is striking—whilst just over a thousand women a year went into convict prisons in the years after mid-century, by 1900 the figure was a mere thirty-four. The consequent fall in the population of women convicts was dramatic. It is important to note that over the same period numbers of women committed to local prisons rose steadily, so that women convicts made up a declining proportion of the female prison population as a whole (falling from just below a quarter of the total female prison population in the 1860s to just 4 per cent by the end of the century).

Offences Leading to Penal Servitude

The Judicial Statistics provide limited details of the offences for which men and women were sentenced to penal servitude (Appendix, Table 9).[84] We can establish only the most general idea of the types of crime for which women were most likely to be sentenced to penal servitude. Women made up 4–9 per cent of those sentenced for malicious offences against property, 4–6 per cent of those sentenced for offences against the person, and only 2–5 per cent of those sentenced for offences against property with violence. However, they comprised up to 15 per cent of those sentenced for offences against the currency and up to 24 per cent of those sentenced for offences against property without violence. Figures provided by the Commission into the Working of the Penal Servitude Acts (1878–9) give a rather more informative snapshot picture. They show the offences for which 8,983 men and 1,226 women held in convict prisons on 6 May 1878 had been committed.[85] One should bear in mind that this is a snapshot only and cannot be taken as necessarily

---

[84] Details of the offences for which men and women were sentenced were not presented in raw numbers for each offence but as a percentage of the proportion of men and women committed under five broad classes of offences (together with one 'other' category of offences not included in the other five). This method of presentation gives no indication of trends or the frequency with which different classes of offence led to penal servitude over the period; nor does it indicate the size of each class of offence in relation to the total number of offences committed by each sex.

[85] Commission Appendix (1878–9), 1171–2.

representative of the whole period. Male convicts were sentenced for a far wider range of offences than women and their
offences were spread more widely amongst the various headings. Women were, by sex or socialization, absent from many
types of serious crime and concentrated on a smaller range of
offences. For both sexes, larceny was by far the largest single
type of offence leading to penal servitude. A far higher proportion of women (65 per cent of all women) than men (31 per cent
of all men) were sentenced for larceny. The other major categories under which women were sentenced in 1878 were, in
order, felony (14 per cent), murder or attempted murder (5 per
cent), coinage offences (2 per cent) and feloniously receiving
(also 2 per cent).

Previous Commitments

The annual Judicial Statistics fail to provide details of past
convictions or sentences of imprisonment or penal servitude for
those sentenced to convict prisons. However, the snapshot survey of the Commission into the Working of the Penal Servitude
Acts (1878–9) again provides some indication of the criminal
histories of those coming into convict prisons (Appendix,
Table 11).[86]

Both for male and female convicts, just over half had never
previously been sentenced to penal servitude. Strikingly,
23 per cent of men but only 10 per cent of women had no
previous conviction of any kind. This perhaps reflects the
reluctance of the courts to impose such severe punishment on
women who had not previously been officially designated
criminal. However, once a woman was known to be criminal
she seemed to attract greater stigma than her male counterpart.
This stigmatization is reflected in the more frequent failure of
ex-convict women to recover their 'character' and to reform,
and the frequency, therefore, with which women found themselves back in prison. This applied as much to the more severe,
lengthy sentences of penal servitude as it did to short-term
imprisonment in local prisons. Thus only 25 per cent of male
convicts but over 38 per cent of female convicts in the 1878
sample had one or more previous sentences of penal servitude.

[86] Ibid. 1170.

INMATE CHARACTER AND CONDUCT

As female convicts in national prisons left few records of their own we are largely obliged to seek out their prison lives in the writings of outside observers and in official records.[87] This method is obviously problematic, not least because we are forced to view them through the eyes of their captors.

The snapshot picture from the Commission into the Working of the Penal Servitude Acts (1878–9) of 8,983 males and 1,226 females gives details of their perceived 'prison character', that is, the nature of their conduct and their compliance with the prison regime (Appendix, Table 12).[88] Though contemporaries commonly asserted that women convicts were less compliant and far more rebellious than men, this return shows both sexes arrayed in much the same proportions over the range of conduct categories, from 'exemplary' to 'very bad'. If anything, women seem to have been rather better behaved—roughly a third were classed as exemplary as compared to only a fifth of men—though this disparity was somewhat levelled out by the better showing of men in the 'very good' class. Through the remainder of the scale—good, fair, etc.—the proportions of men and women falling into each category are fairly similar. If we could accept this sample with any certainty as typical, we would need to revise the conclusion arrived at by historians like Michael Ignatieff that female convicts caused prison authorities far more problems through their continued 'violent resistance'.[89] In fact, these statistics tell us more about the perceptions of the Commissioners than about the actual conduct of women within the prison. The latter can be got at only by scouring the prisons' own records for elusive and often obtuse references to prisoner conduct away from the eyes of visiting officials. In this way it is possible to glean something of the ways in which convict women, by secret as well as by flagrant violation, adapted to prison life.

In order to develop any contacts, let alone relationships, inmates needed to secure means to communicate. Historian

[87] Those who did leave records were, by definition, not typical inmates; see, for example, the memoirs of Mrs Florence Maybrick, *My Fifteen Lost Years*.

[88] Commission Appendix (1878–9), 1171–2.

[89] M. Ignatieff, *A Just Measure of Pain* (1978), 203.

Patricia O'Brien has had the good fortune to uncover a volume of letters which had been passed between female inmates in French prisons but intercepted by the authorities and published as evidence of their immorality.[90] These letters expressing affection, loyalty, jealousy, and rejection have all the qualities of love letters. Anecdotal evidence is available for the existence of similarly illicit communication in English prisons.[91] Despite the officers' vigilance, women passed letters, known in the prison jargon as 'stiffs', scribbled on the papers supplied for lighting the gaslights in their cells. And whenever possible they communicated by sign language, lip-reading, and snatched conversation.

Prison authorities, although apparently concerned about the prevalence of 'unnatural practices' in male prisons, claimed that female relationships were strictly affectionate. However, an underlying sense of suspicion and concern is clear. The Medical Officer at Brixton recorded his objection 'to the attachments, which after a few months, are frequently formed between different prisoners. I believe that on all occasions friendships thus formed should invariably be broken.'[92] The tempestuous jealousies that these relationships engendered were evidently a serious threat to prison order. For the women, 'palling-in' (forming close, though often short-lived, relationships) was a means of mitigating the inhumanity of the regime. Whereas local prisoners could hope to maintain relationships outside, convicts were cut off for so long that they were obliged to turn to one another. Modern accounts of women's prisons testify to the centrality of these relationships in inmate life and document the development of complex networks of 'families', primarily as a source of emotional security in a hostile environment.[93]

Inmate interaction, limited though it was by lengthy isolation, separation, and silence, was sufficiently sophisticated to enable women to pass important information and share views. Prison authorities certainly feared the effects of association,

---

[90] O'Brien, *Promise of Punishment*.

[91] For example, in the works of the 'Prison Matron'.

[92] RDCP, Brixton for 1856, *PP*, 23 (1857), 339.

[93] R. Giallombardo, *Society of Women* (1966); E. Heffernan, *Making it in Prison* (1972); P. Carlen, *Women's Imprisonment* (1983).

because it both corrupted the comparatively innocent and gave rise to an inmate counter-culture. Older women were believed to 'exert a secret and powerful influence over the younger prisoners, urging them to commit serious breaches of discipline and to defy authority'.[94] Inmates maintained communal pools of information that were often antipathetic to the official regime. For example, at a meeting of the Social Science Association, a Lady Visitor to Brixton prison deplored the 'sisterhood' that existed among the women.[95] She cited evidence of women sharing information on means of effecting infanticide. Such conversations, overheard by chance, no doubt represented only a minute proportion of those that actually went on. We can only begin to estimate the extent to which the culture of female prisoners operated in opposition to official ideology, denied its destructive vision of them as social failures, and negated the intended effect of its teachings. Far from being weak and isolated outcasts, many of the women worked within a highly disciplined and well-organized 'freemasonry' of thieves and prostitutes. The inmates cemented the foundations of this counter-culture within the very prison walls.

Female convicts were forced to comply, for most of their time, with minutely defined regulations. In doing so they gave their consent, albeit unwillingly, to the validity of the prison rules. However, they were quick to object to conditions or actions which they considered to be outside the prevailing rules or an unjust innovation. Resistance took various forms, from simple non-cooperation to organized protest. Evidence given to the Commission into the Working of the Penal Servitude Acts (1878–9) describes one such incident when a woman 'called upon her fellow prisoners to stand out for their rights'. The punishment of this woman, by isolation in separate cells, provoked other prisoners into 'refusing to comply with the rules, and making a disturbance, and calling for the woman who had previously been the ringleader to attend chapel'.[96] On such occasions at least, the women apparently rejected the model of silent passivity to which they were supposed to aspire.

[94] RDCP, Brixton for 1856, 341.
[95] E. Lankester *et al.*, papers on 'Can Infanticide be Diminished by Legislative Enactment?', *Transactions NAPSS 1869* (1870), 214–15.
[96] Mrs S. Seale, Lady Superintendent of Fulham Prison, RC on the Working of the Penal Servitude Acts, Minutes of Evidence (1878–9), 388.

Instead they seem active and vociferous, with a clear sense of grievance to be righted. In the eyes of those who considered that 'the female character never appears so lovely as when it wears the ornament of meek and quiet spirit,'[97] such behaviour must simply have confirmed their most pessimistic views of female convicts.

A small number of convict women so persistently and violently contravened this female norm that they came to be labelled 'incorrigible'. These women, confined in the penal class at Millbank and later at Parkhurst, were described by Mayhew as 'the very worst women in existence. I don't fancy their equal could be found anywhere.'[98] Totally unrepentant and determinedly defiant, they shocked and frustrated those in authority. One Director, Captain O'Brien, in a memo on the discipline and management of ill-conducted female convicts, exclaimed that 'the language of the penal class women is fouler than anyone who had not heard it could possibly imagine. Nothing but the gag can restrain the abominations they utter.'[99] These women provoked outrage above all because they exhibited a 'deadness to all feminine decency'. As such they repudiated the very precepts of the treatment regime.

The greatest single challenge to the authorities was a phenomenon peculiar to the female convict prison, that is, riotous behaviour 'amounting almost to a frenzy, smashing their windows, tearing up their clothes, destroying every useful article within their reach, generally yelling, shouting or singing as if they were maniacs'.[100] 'Breaking-out', as it was commonly known, was generally of such violence and intensity that it came at once to the attention of the authorities. Since it caused them grave concern it is extensively documented. As I noted earlier in the chapter, it accompanied long-term female imprisonment from the very start. Devastated to find themselves subject to lengthy incarceration at home and uncertain as to their ultimate fate, the first women convicts 'broke-out' with terrifying frequency. Overcrowding in the pentagons at Millbank created a catalyst in which the violence of one woman

[97] Anon., *The School Home of Thornvale* (1859), 47.
[98] Mayhew and Binney, *Criminal Prisons*, 271.
[99] PRO Pri Com 2/164. Memo by O'Brien 14 Jan. 1857.
[100] RDCP, Millbank for 1853, *PP*, 33 (1854), 87.

sparked similar frenzy in many others. Throughout the 1850s and 1860s extreme frustration with the monotony of incarceration continued to manifest itself in violent outbursts of anger and destruction. In 1859 there were 154 cases of breaking-out in Millbank alone.[101] Such outbreaks were always considered a peculiarly female response to long-term imprisonment. They were explained as the product of hysterical tendencies in women incapable of controlling their unnaturally violent passions. Dismissively labelled as irrational, such outbreaks were, none the less, a terrifying spectacle. More importantly they were an inescapable indictment of the prison regime.

DISCIPLINE AND PUNISHMENT

Persistent defiance by the incorrigible few and rebellion, in the form of breaking-out, by a larger proportion of female convicts posed serious problems for maintaining order. As we saw in the previous chapter, corporal punishment could not be imposed on women. Even bread and water was thought to do irrevocable harm and could not be administered for more than three days consecutively.[102] The authorities were left with limited sanctions, all of which they considered unequal to their task. The most common form of punishment, short-term confinement in 'the dark' or isolation cell, was evidently derided; as Dr Guy of Millbank Prison observed, 'if they are put into a dark cell, they shout and sing and make merry. They know that there are prisoners not very far off them who can hear their noise, and they like to go on in that strange way.'[103] Guy was probably not alone in regretting his inability to impose more punitive sanctions on women who appeared to be impervious to existing punishments. He admitted, 'I think that if it were possible to inflict upon them some bodily pain, it would be more advantageous.' Desperate to identify more effective means of restraining violent women, Captain O'Brien visited the lunatic asylums

[101] RDCP, Millbank for 1859, *PP*, 35 (1860), 470.

[102] See evidence by Captain O'Brien, RC on the Operation of the Penal Servitude Acts (1863), 457–8; also RC on the Working of the Penal Servitude Acts (1878–9), 384.

[103] W. A. Guy, Physician and Medical Superintendent at Millbank, to RC on the Operation of the Penal Servitude Acts, Minutes of Evidence (1863), 532.

at Colney Hatch and Hanwell. As a result the canvas dress (a form of strait-jacket) and bedding fortified against attempts to tear it up were adopted. The links made in the minds of the authorities between convict women and the insane are all too obvious.

Eminent authorities such as Jebb and O'Brien shared the view put by the long-standing Deputy Governor of Millbank, Arthur Griffiths, that it was 'a well-established fact in prison logistics that the women are far worse than the men. When given to misconduct they are far more persistent in their evil ways, more outrageously violent, less amenable to reason or reproof.' [104] However, after the unrest which troubled the early years of penal servitude, it is not clear that women convicts actually were so much worse behaved than men. Although women tended to receive a higher rate of punishment than men, this seems to have been due to differing patterns of control rather than to greater misconduct. Since women could not be flogged or half-starved into submission like their male counterparts, prison officials felt that they had to take every possible opportunity to assert their authority. The Lady Superintendent of Brixton, for example, demanded that her officers reported women for every trivial offence as the only 'effectual way of divesting convicts of that daring and independent bearing they endeavour so often to assume, and it is a constant reminder to them that they are undergoing the penalty for crime, and that the slightest alleviation of their punishment must be acquired by themselves, by their becoming orderly, industrious, peaceable with each other, obedient and respectful to their officers.' [105] Whereas male convicts were expected only to work hard and refrain from breaking prison rules, a much higher level of surveillance over women patrolled every aspect of their demeanour and behaviour for possible fault. Such a policy inevitably raised the punishment rate and at the same time heightened levels of tension inside female convict prisons, so making outbreaks of misbehaviour more likely. Mary Carpenter criticized the disproportionately high level of punishment in female convict prisons; finding no evidence that women were any worse behaved than men, she argued: 'This remark-

[104] Griffiths, *Memorials of Millbank*, 198.
[105] RDCP, Brixton for 1863, *PP*, 26 (1864), 426.

able disparity in the punishments inflicted on the two sexes . . . leads to the conclusion that such punishments are in themselves quite inefficacious, and that some different system ought to be adopted.'[106]

Under Joshua Jebb as Chairman of Directors of Convict Prisons (1847–64), the temptation to resort to strict discipline and punishment had been resisted, officially at least, in favour of an apparently unfounded optimism in the possibility of reform. On his death, the 1863 Royal Commission of Inquiry into the Penal Servitude Acts voiced a profound disillusionment with the existing administration, and, in particular, declared the regime to be too lenient. It inaugurated a period marked by increasingly rigid regimentation and severe punishment, introduced by Colonel Edmund Henderson (in post 1863–9) and enthusiastically promoted by his successor as chairman, Captain Edmund Du Cane (in post 1869–95). Du Cane, like his predecessors, had a military background but, unlike Jebb at least, was determined to bring a military style of leadership and discipline to his new work. His autocratic and uncompromising personality, combined with his commitment to uniformity, ensured the implementation of sweeping change throughout the convict system.

Female convict prisons, always on the sidelines of Home Office policy, escaped the worst effects of this change from reform to discipline and deterrence. Nonetheless, the annual reports of the female convict prisons underwent a notable shift in use of vocabulary. We hear increasingly that 'strict discipline has been maintained', orders have been 'rigidly carried out', and offences 'invariably' reported.[107] By 1878 the stress on kindness and affection of earlier years had all but disappeared. The atmosphere of moral regeneration and religious awakening was stifled as the very elements of the reform process were transformed into mere slots on a custodial timetable. The Medical Officer of Millbank noted that, apart from those in the hospital ward, association of female convicts had been ended virtually everywhere. Of this development he concluded, 'it may be one of the first fruits of a salutary measure that the

[106] Carpenter, *Our Convicts*, vol. 2, 222–3.
[107] RDCP, Fulham Refuge for 1868, *PP*, 30 (1868–9), 393–403. See also E. Du Cane, *The Punishment and Prevention of Crime* (1885), 166.

incorrigible virago, who so often in former days defied authority, and was never weary of doing wrong is now rarely met with.' [108]

Earlier efforts to respond to women individually were also subsumed by that obsessive concern for uniformity which had long characterized the regime in male convict establishments. [109] Once the regime no longer held reform amongst its primary aims, the very purpose of differentiated treatment was lost. The only group still held to have any potential for reform were women with no previous convictions 'and whose character and antecedents were such as to show that they do not belong to the criminal class'. [110] For these women a special 'Star Class' was established at Millbank in September 1881. But even this last remaining vestige of optimism was eroded when the prison authorities could find only six women who might be deemed eligible to enter it. The majority, even if they had not previously been committed to penal servitude, had served numerous sentences in local prisons for petty theft or drunkenness.

This reassessment of most women convicts as incorrigible offenders, as the *habitués* of the prison system on whom reformative treatment might have little effect, became widely accepted during the Du Cane regime (1865–98). In part, it was a product of changes in the convict population. As we noted earlier, over the second half of the nineteenth century the annual number of women committed to penal servitude steadily declined from a high of 1,050 in 1860 (1,249 daily average population) to a mere 95 in 1890 (302 daily average population), [111] leading to the removal of women from Millbank and Fulham 'Prison' (as it was significantly now called) in 1886 and 1888 respectively. [112] As a result Fulham Prison was closed down altogether. Those women who remained were held solely at Woking, and were seen to be the hard core of prison incorrigibles. Certainly many

[108] RDCP, Millbank for 1878, *PP*, 35 (1878–9), 247.

[109] From their appointment the Prison Commissioners were determined to standardize provision and remedy the existing lack of uniformity in convict prisons. See Reports of the Commissioners of Prisons and Directors of Convict Prisons, First Annual Report, *PP*, 17 (1878).

[110] RDCP, Millbank for 1881–2, *PP*, 34 (1882), 247.

[111] By 1920 the daily average number of women in convict prisons had fallen to below 100—source, E. Ruggles-Brise, *The English Prison System* (1921).

[112] For discussion of the reasons for this decline in the female convict population see Anon., 'Women Convicts', *The Englishwoman's Review* (Oct. 1887), 473; M. F. Johnston, 'The Life of a Woman Convict', *Fortnightly Review*, 75 (1901).

had been imprisoned forty or fifty times before.[113] The earlier regime of moral reform had supposed that it had as its subjects only impressionable and amenable women. Finding, instead, a number of 'incorrigible virago(s)', penal administrators now held punishment and deterrence to be higher priorities. Moreover, discipline sufficient to control the dangerous few inevitably subjected the whole female convict population to increasingly rigid regulation.[114]

RELEASE

Although a sentence of penal servitude may have appeared endless to women so confined, it did eventually come to an end. Provision for release and disposal of convicts altered considerably over the period as a result of changes in penal policy. Although the refusal of Van Diemen's Land in 1852 and of Western Australia in 1855 to continue to accept women convicts speedily ended transportation for women, male convicts continued to be transported until 1867. Up to 1867 around 10 per cent of those removed from convict prisons each year were disposed of in this way.

Figures for removals between prisons and for release are, therefore, only comparable after 1868, when both men and women came under the ticket-of-leave system (release on licence) in Britain (Appendix: Table 13a), albeit on differentiated sliding scales of remission—the terms of remission for women were more favourable than those applied to men. Thereafter over three-quarters of both men and women were released on ticket-of-leave (approximately the same proportions of men and women failed to earn remission through good behaviour and served right to the term of their sentence). They were, of course, liable to have their licence revoked for failure to abide by its terms. Although figures for revocations were not given in the annual returns of the Judicial Statistics, the Commission into the Penal Servitude Acts (1878–9) provides them

[113] See evidence given by Dr Brayn, the Governor and Medical Officer of Woking Female Convict Prison, to Departmental Committee on Prisons, Minutes of Evidence, *PP*, 56 (1895), 223.

[114] A source of continual frustration to one impeccably behaved prisoner—Mrs Florence Elizabeth Maybrick, *My Fifteen Lost Years*.

for the period 1855 to 1876 (Appendix, Table 13*b*).[115] If we take only those after 1868 (on the basis that before this they are not comparable by sex) we find that, whilst the percentages of men and women released on licence were almost identical, proportionately twice as many women as men subsequently had their licences revoked. Under the terms of licence, men were required to report to the police at monthly intervals whereas women had to report only on release. Presumably, therefore, they were less intensely supervised than men. Given these facts, their higher failure rate provides striking evidence, yet again, of the difficulty women prisoners found in trying to reform. That stigma was a massive obstacle for women seeking employment straight from prison had been recognized in the attempt to establish Fulham as a 'refuge'. However, its re-designation in 1869 as a mere 'prison' was a direct result of its obvious failure to convince the public of this contrived differentiation in status. The continuing problem of how to enable convict women to regain their 'character' was recognized by the government in the setting up of the Carlisle Memorial Refuge (for Protestant women), the Winchester Memorial Refuge, and the Eagle House Refuge (for Catholic women).[116]

A far larger effort was undertaken by the many voluntary agencies, known as Discharged Prisoners Aid Societies, which set up and managed refuges and shelters almost exclusively for women.[117] Their proliferation during the 1860s and 1870s allowed Du Cane to introduce a scheme whereby women 'whose good conduct and character justifies the hope of complete amendment'[118] were released into such institutions nine months before the end of their sentence. Release into a refuge was permitted on licence so that if, by persistent misbehaviour

[115] Commission Appendix (1878–9), 1151.

[116] See the annual reports of these refuges in the RDCPs.

[117] Such organizations generated a mass of literature on the absolute necessity of continuing aid after release: for example, W. B. Ranken, 'On Adults after their Release from Convict Prisons', *Transactions NAPSS 1860* (1861); W. B. Ranken, 'The Origin and Progress of the Discharged Prisoners' Aid Society', *Transactions NAPSS 1858* (1859); papers on aid to discharged prisoners in E. Pears (ed.), *Prisons at Home and Abroad*; A. J. S. Maddison, *Hints on Aid to Discharged Prisoners* (1888).

[118] Du Cane, *Punishment*, 169. For women who did not thus qualify, the only other hope was to be helped to emigrate on release at the end of their sentence. See G. P. Merrick, 'Emigration for Female Prisoners', *The Englishwoman's Review*, NS (Nov. 1877).

or by trying to escape, a woman forfeited her licence she was immediately returned to prison, where she would remain until the very end of her term. Although the receiving institution was guaranteed government financial aid on the proviso that an equal amount was raised by private subscription, there was no other government intervention or even right of inspection.[119] Discipline was more relaxed, and the women were allowed to associate with one another. The bars and bolts of the prison were replaced by a single locked front door and cells by individual bedrooms. Indeed, Du Cane insisted: 'These "refuges" are not prisons either in appearance or in discipline—they are *homes*, and are intended to afford the advantages of a treatment approaching in its characteristics to that of home influence.'[120] Clearly he hoped that the refuges could remain recognizably distinct from the convict system and so provide their inmates with a more favourable opportunity to reform by allowing them to throw off the stigma of conviction. Given Du Cane's renowned concern for discipline and deterrence, the fact that this system allowed women to pass up to a quarter of their sentence of penal servitude in the far more lenient regime offered by the refuge suggests just how important he considered this scheme, or, rather, how pessimistically he otherwise viewed women's likely fate on release.

Although the Departmental Committee on Prisons (1895) acclaimed these arrangements for the aftercare of women convicts as one of the great successes of the Du Cane era,[121] those more closely involved were markedly more sceptical. For example, the Duchess of Bedford, head of the Lady Visitors to the new female convict prison at Aylesbury,[122] found that the women were deeply suspicious of refuges, and were disinclined to go to them. Even if they were persuaded to go, the arrangements within the refuge were not nearly so conducive to reform as the policy makers assumed: young women were often associated indiscriminately with old and hardened criminals, many

[119] Details given by E. Du Cane, RC on the Working of the Penal Servitude Acts, Minutes of Evidence (1878–9), 106–8.
[120] Du Cane, *Punishment*, 170.
[121] Report of the Departmental Committee on Prisons, *PP*, 56 (1895), 37.
[122] Opened in 1896. For two very different impressions of Aylesbury, see Maybrick, *My Fifteen Lost Years*, 127–39, and X.Y.Z., 'England's Convict Prison for Women', *The Englishwoman* (Apr.–June 1912).

of whom were quick to try to subvert them. Perhaps the most serious indictment was the difficulty Lady Visitors found in following women up after they were finally released, so that it was not even possible to assess the relative success of those leaving refuges as compared to the subsequent careers of women released directly from the prison.[123]

Du Cane's leadership of the prison system ended in 1898. By this time his regime of severe discipline and deterrence was no longer generally regarded as necessarily the most appropriate penal response. Indeed, many began to question whether the prison had succeeded in being the panacea for crime that earlier reformers had promised it would be. Around the turn of the century, therefore, penological attention moved away from the prison itself. It is to these debates, and the development of specialist provision on the very borders of the penal sphere, that we must look for continuing innovations in responses to female crime.

[123] See the writings of Adeline, Duchess of Bedford 'Fifteen Years' Work in a Female Convict Prison', *The Nineteenth Century and After* (Oct. 1910), and 'Treatment of Women in Prisons', *International Congress of Women* (1899); also her report in Reports of the Commissioners of Prisons, *PP*, 40 (1897), 129–30; and PRO Pri Com 7/173, Aylesbury Prison, 1895–9 incl. Reports of Lady Visitors, esp. the Report of the Lady Visitors for 1898.

PART III

REMOVING 'INCORRIGIBLE' WOMEN FROM THE PENAL SPHERE

6 HABITUAL DRUNKENNESS AND THE REFORMATORY EXPERIMENT, 1898–1914

Before the central control of local prisons in 1877, the failure of imprisonment to deter or to reform could readily be attributed to the lack of uniformity throughout the system, to the inefficiencies of unreformed prisons, or to the inadequacies of staff, buildings, or regime. However, after nationalization and the consequent development of a unified network of tightly ruled, efficiently run institutions, the continuing failures of prisons were less easy to explain or accept. Towards the end of the nineteenth century there was a serious loss of faith in prisons at both local and convict levels. They were condemned as inflexible, unduly repressive, and, most damningly of all, they were seen to engender, even nurture, a hardened class of recidivists. As we saw in previous chapters, the most hopeless class of these confirmed prison habitués were women. Many were not serious or dangerous criminals so much as petty offenders, social inadequates, and outcasts incapable of surviving in outside society. To continue to commit them to prison, time and time again, was clearly both inappropriate and unproductive of any but the most short-term good. As a result, their very presence in prison was increasingly seen to indict the whole regime. These prison failures attracted the attention both of specialists, interested mainly from a medical perspective, and increasingly of the wider public, who saw them as a major social threat.

This final part will examine the two most important groups of these prison habitués: confirmed drunkards, and the 'feeble-minded' (mentally deficient). It will focus particularly on the question of how they were constructed as 'major social problems', how they were explained, how their cost to society was assessed, and why women predominated in both groups. Most importantly, it will examine how their identification gave rise to specialist policies designed to remove many of the most recidivist women from the prison altogether.

HABITUAL DRUNKENNESS AS A SOCIAL PROBLEM

As I have already indicated in previous chapters, drunkenness formed a considerable proportion of summary offences throughout the nineteenth century (see Appendix, Table 3*b*). Drunkenness or drunk and disorderly behaviour made up 22 per cent of female summary convictions in 1857, rising to 37 per cent by 1890. Drink was also blamed both for causing crime directly and for creating the poverty which led to crime. Over the second half of the nineteenth century, striking changes occurred in the perception of drunkenness. Whether drunkenness was provoked through misery, driven on by dipsomania, or whether it was the product of innate moral depravity became vital points of dissension. It was now seen not only as a cause of wider social ills but as a pathological condition inflicting physical and moral degeneracy. In turn these new concerns provoked heated debate about the status of drunkenness as a social ill, and particularly of that chronic form of alcoholism known as 'habitual inebriety'. As early as 1859, one speaker at the annual meeting of the Social Science Association argued that to get drunk was in itself a crime, irrespective of whether or not the drunkard went on to commit further offences. Foreshadowing later demands, he called for far more stringent measures against 'inebriety', not least the incarceration of 'confirmed inebriates' in special asylums for long enough to effect a cure.[1] Such suggestions opened up debate about its character, etiology, and possible treatment. Habitual drunkenness became a specific form of criminality and, as we shall see, the habitual inebriate a particular deviant type.[2]

To explain these shifts, we need to identify the intellectual and cultural milieu in which drunkenness came to be seen as such a social danger. The relationship between this development and the Temperance Movement is not quite so direct as one might first assume. Certainly the Temperance Movement had identified drinking in general as the cause of numerous

[1] W. Arnot, 'The Criminality of Drunkenness', *Transactions NAPSS 1859* (1860), 456 ff.

[2] See, for example, D. Gibb, 'The Relative Increase in Wages, of Drunkenness and of Crime', *Transactions NAPSS 1874* (1875), 334–6; F. W. Farrar, 'Drink and Crime', *Fortnightly Review* (June 1893), 787–8; W. C. Sullivan, *Alcoholism* (1906), 154–5.

social ills—crime, prostitution, and poverty. It promoted widespread abstention from the use of alcohol, concentrating on cultivating the will power of less hardened drinkers and on promoting teetotalism.[3] However, this policy tended to ignore chronic alcoholics, for they were the least likely to be converted. In so far as Temperance campaigners addressed the problem of habitual inebriety at all, they argued quite simply that if drink was 'locked up' then there would be no drunkards to deal with. Only by restricting the availability of alcohol to all could the greater danger of habitual inebriety be avoided.

Whereas the impetus for the Temperance Movement came mainly from the Church, and particularly from Nonconformists, those concerned with the problem of habitual inebriety were mainly scientific and medical men. Many inebriate reformers thought that legal constraints should be limited to habitual drunkards and saw no need to interfere with moderate social drinking (although some thought that certain people had a pathological disposition to inebriety, and would succumb to it after taking a single drop of alcohol). Significantly, supporters of the 1898 Habitual Inebriates Act vehemently denied any connection, even by origin, with the Temperance Movement and stressed instead their concern to legislate only for a specific section of the criminal population.[4] They recognized that the condition of the confirmed drunkard was as much a product as a cause of social evils; that habitual inebriety was itself often the result of misery, poverty, and an unhealthy environment. Moreover they recognized that chronic alcoholism, whatever its causes, became a disease which moral treatment alone could not cure.

The British Medical Association (BMA) was pre-eminent in promoting a general reassessment of the habitual inebriate less as sinner than as sick, whether as cause or result of long-term alcoholism. Innumerable articles in the *British Medical Journal* and the short-lived *British Journal of Inebriety*[5] analysed the physical and psychological characteristics of inebriety, examined

[3] R. M. MacLeod, 'The Edge of Hope: Social Policy and Chronic Alcoholism 1870–1900', *Journal of the History of Medicine*, (July 1967), 217.

[4] See, for example, GLRO LCC/MIN/8117, Report by T. L. Murray Browne and Arthur J. Maddison to a meeting of the London County Council (LCC) Inebriates Acts Subcommittee (14 Mar. 1899).

[5] The journal of the Society for the Study and Cure of Inebriety, pub. 1903–4.

its etiology, and prescribed cures ranging widely from hypnosis to treatment by drugs. As founder of the medically orientated Society for the Study and Cure of Inebriety set up in 1884, and elected Chairman of the BMA's Habitual Drunkards Legislative Committee in 1886, Dr Norman Kerr was perhaps the most important proponent of the view that inebriety was best understood as a disease.[6] Suggesting that it had physiological origins and clinical features, he argued that if medical treatment were allowed to replace imprisonment it offered a real possibility of cure.[7]

THE COSTS OF MATERNAL ALCOHOLISM

Towards the end of the century profound fears about the health of the nation, and especially of the degeneration of the 'residuum', led drunkenness to be seen not merely as a sickness but as a hereditary defect which dangerously threatened the very health of the stock (see Chapter 2 above). Fears of degeneration focused on woman, as the more likely parent of hereditary disease and addiction. Dr Norman Kerr assumed that this would be so, insisting: 'that the female parent is the more general transmitter of the hereditary alcoholic taint I have little doubt.'[8] Especially if women were drunk during conception or continued to drink during pregnancy, their inebriety was seen as potentially far more damaging. Numerous studies were devoted entirely to investigating the extent of degeneration attributable to maternal inebriety. For example, in July 1904 the Society for the Study and Cure of Inebriety, at a meeting devoted entirely to women, received papers such as 'Inebriety in Women and its Influence on Child-life' and 'The Causes of Inebriety in the Female and the Effects of Alcoholism on Racial Degeneration'.[9] The Eugenics Education Society regarded the problem as so hazardous they suggested that inebriate women

[6] L. Radzinowicz and R. Hood, *A History of English Criminal Law and its Administration from 1750*, v (1986), 289 ff.

[7] This argument had been put forward at least since the 1870s, see, for example, S. S. Alford, 'The Necessity of Legislation for the Control and Cure of Habitual Drunkards', *Transactions NAPSS 1876* (1877), 383–5. Only once taken up by the BMA did it become widely accepted.   [8] N. Kerr, *Inebriety* (1888), 142.

[9] By F. Zanetti and W. C. Sullivan respectively in the *British Journal of Inebriety* (Oct. 1903).

should be detained until they reached menopause in order to prevent them from bearing potentially defective or addicted children.[10] Others advocated the prohibition of marriage between habitually inebriate 'degenerates' as a marginally less punitive means of preventing them from procreating.

Assessing the impact of alcoholism on women's fecundity was complicated by the obvious difficulty of distinguishing it from other environmental factors: standard of living, diet, and cleanliness. Whilst alcohol alone did not necessarily reduce fecundity, heavy drinking was seen to lead to a whole range of immoral behaviour (promiscuous intercourse, prostitution, etc.) which was liable to promote sterility. Whereas many Victorians had believed that prostitutes were, as if by divine intervention, simply incapable of becoming pregnant, R. W. Branthwaite, the Inspector appointed under the Inebriate Acts, recognized that use of 'preventive measures' and damage of reproductive organs through disease were the real reasons for their apparent sterility.[11]

High rates of stillbirths among inebriate women were taken as evidence of the harmful effects of maternal alcoholism. For example, W. C. Sullivan, the Medical Officer at Holloway Prison, cited a study in Liverpool Prison where over half of the 600 children born to 120 women prisoners were stillborn or died in infancy.[12] That these high rates were due to maternal alcoholism rather than other determining factors does not seem at all clear. When women who had been detained for a substantial part of their pregnancy gave birth to healthier babies than those born outside, this was presented as conclusive proof of the harm done by maternal inebriety. However, this argument failed to recognize that not only was the incarcerated woman kept from alcohol but that she enjoyed a far higher standard of living and medical attention than she was likely to have received outside. Branthwaite hinted at these influences when he argued that the prenatal debility of the inebriate mother did not necessarily lead to stillbirth or infant mortality if conditions of life were otherwise good.[13] Moreover the very

[10] Replies to Questions, Departmental Committee on the Law relating to Inebriates, *PP*, 12 (1908), 981.

[11] Report of the Inspector under the Inebriates Acts for 1909, *PP*, 29 (1911), Pt. 1, 35 ff.   [12] Sullivan, *Alcoholism*, 191.

[13] Report of the Inspector under the Inebriates Acts for 1909 (1911), 38.

facts of infant mortality were extremely difficult to assess given the unreliability of the women's testimony; many, he claimed, could not even remember when their children had died.

Even if the child survived pregnancy it was thought likely to be born puny, possibly mentally deficient, and certainly prone to disease: 'thousands of women are not only destroying their own constitutions, but probably giving birth to neurotic, vicious children, tainted with alcoholism and disease, large numbers of whom should they survive neglected infancy, will grow up weak and rickety.'[14] Inebriate mothers were found to be so debilitated or diseased as to be unable or, worse still, were 'unbothered' to breastfeed their babies, giving them instead poor quality, shop-bought milk which was often contaminated.[15] Even those babies actually breastfed received milk 'poisoned at its very source', whilst others were given alcohol as a medicine or 'pacifier'. Concern about the pre- and postnatal impact of maternal inebriety was compounded by the fear that those who did have children were reproducing faster than the rest of the population. (Excluding those who were childless, the average number of births per woman held in certified inebriate reformatories in 1905 was 6.4, at a time when the average for the population as a whole was only just over 4.) The result was an 'ever-increasing multitude of social failures' who, it seemed to terrified observers, would eventually outnumber the 'fit' population.[16] Thus wider fears of national degeneration became crystallized in the identification of the inebriate mother as a major source of physical deterioration more profoundly marked in each successive child.

This literature was highly moralizing in tone, condemning the inebriate mother for her irresponsibility or deliberate defiance of maternal duties. Alcoholism, it was claimed, weakened maternal instincts and destroyed a woman's ability as homemaker. Although overlaying, that is, the smothering of sleeping infants by drunken mothers, was often found by juries to be an accidental cause of death, it was increasingly deplored as

[14] Zanetti, 'Inebriety in Women', *British Journal of Inebriety* (Oct. 1903), 50–1.

[15] M. Scharlieb, 'Alcoholism in Relation to Women and Children', in T. N. Kelynack (ed.), *The Drink Problem* (1907), 165–8.

[16] GLRO LCC/MIN/8126. LCC Inebriates Acts Subcommittee Papers: Eighth Annual Report of the Lancashire Inebriates Reformatory, Langho, Blackburn, for 1911, 5.

culpable negligence.[17] Women drunks were condemned for
failing to clean, feed, clothe, or care for their infants adequately.
Deliberate cruelty became the subject of considerable concern
as a result of the crusading work of the National Society for the
Prevention of Cruelty to Children and an important focus of
activity under the 1898 Inebriates Act.

Since mid-century, debates on juvenile delinquency had
given prominence to maternal inebriety as a significant causal
factor. For example, the Select Committee on Public Houses of
1852–3 heard evidence that 'not . . . two out of twenty convicts
could claim to have had a good mother'.[18] Mothers drinking
at home during the day were said to provide a more damaging
example than fathers whose drinking was more likely to take
place at night and outside the home. The spectacle of the
drunken mother was decried as incalculably demoralizing; it
could not 'fail to have a debasing influence, as an example,
upon those who are growing up in the midst of it—whose
delicacy is thereby destroyed, and whose sensibilities are
shocked by the repetition of such conduct; and the whole of
whose future lives are tinged and stained by the influence of a
degraded childhood'.[19]

The impact of a woman's alcoholism was not confined to her
influence over children but even her influence over grown men.
Just as criminal women in general were seen as a dangerous
source of moral contagion, so the female inebriate was claimed
to be responsible for the drunkenness of her luckless spouse.
Her failure to provide a haven of domestic comfort 'therefore
impels the workman, whose comfort depends on her, not only
to spend his free time in the public house, but also tends to
make him look to alcohol as a necessary condiment with his
tasteless and indigestible diet'.[20] The drunken wife, it was
argued, gave her husband no incentive to stay at home.

If she fed her addiction in public or took her babies out to
pubs or gin-shops late at night, she was liable to be condemned
as 'not only disgusting . . . but demoralising to the rest of the

[17] See, for example, W. M. Wynn Westcott 'Inebriety in Women and the Overlaying of Infants', *British Journal of Inebriety* (Oct. 1903), 68.

[18] Thomas Wright, former prison visitor in Manchester, evidence to the Select
Committee on Public houses, *PP*, 14 (1854), 382.

[19] H. R. Sharman, *A Cloud of Witnesses against Grocers' Licenses: The Fruitful Source of
Female Intemperance* (n.d.), 4. [20] Sullivan, *Alcoholism*, 111.

community' as well.[21] The rise of the number of women drinking in pubs signified a direct violation of a traditional masculine
bastion and the invasion of an arena which had previously
operated outside and, in many ways, in opposition to the ideal
of domesticity. More than one writer deplored the fact that
'women are not ashamed to be seen drinking in public-houses
as they used to be.'[22] The supposed dangers of women's
growing invasion of the public social sphere were set against
assumptions about their mental frailty: 'Women have shown
an unforeseen facility for adopting masculine vices, without the
saving grace of masculine self-respect. When they give way at
all they are lost: and the temptation to which they are thus
exposed by the removal of conventional safeguards is much
greater than that which assails men, by reason of the physical
weakness and emotional sensibility peculiar to their sex.'[23]
Such beliefs about women's lack of moral character came, as
we shall see, to be highly influential in the formulation of
responses to inebriety in women.

THE EXTENT OF FEMALE ALCOHOLISM

Assessing the extent of female alcoholism over the century is at
best uncertain and always difficult. Mortality returns are unreliable indicators, better seen as expressions of medical opinion
than as the results of chronic alcoholism. Advances in medical
knowledge or fluctuations in 'fashionable' medical opinion
may have altered considerably the numbers of deaths attributed
to inebriety. The pages of the *British Medical Journal* alone
indicate the huge growth in popularity of alcoholism as a
subject for medical investigation in the last years of the
century. Also, any rise in the number of deaths through
drunkenness does not necessarily indicate changes in contem-

---

[21] T. Holmes to the Departmental C. on the Law relating to Inebriates (1908), 936.

[22] R. Jones, 'Alcohol and National Deterioration', in Kelynack (ed.), *Drink Problem*, 232.

[23] A. Shadwell, *Drink, Temperance and Legislation* (1902), 76. See also M. Scharlieb, 'Alcoholism in Women and Children', in Kelynack (ed.), *Drink Problem*, 173; the paper given by S. Knaggs, 'The Habitual Drunkards Act', and subsequent discussion, *Transactions NAPSS 1883* (1884).

porary consumption but often the longer-term effects from earlier periods of high consumption.

Criminal statistics, too, are difficult to interpret. Conviction rates can hardly be said to represent the actual extent of drunkenness and, although they may give some indication of changes over time, other factors such as shifts in public toler-ance and alterations in policing procedures or sentencing policy may all have had a dramatic impact on numbers sent to prison. Nevertheless, they were one of the main indicators by which contemporaries assessed the rise or fall of inebriety. Roy MacLeod identifies 1832 as the year with the highest level of arrests for drunkenness, 21 in 1,000 of the population per annum, falling dramatically to only 4.9 in 1,000 by 1867. A short, sharp rise from 1867 to 1876, reaching 7.6 in 1,000 in that year, was followed by a gradual slow decline to the end of the century.[24] Unfortunately MacLeod does not divide arrests by sex. The Judicial Statistics show that women formed a growing proportion of those convicted, rising from 20 per cent of summary convictions for drunkenness around mid-century to just under 30 per cent by 1900. It is not clear, however, whether these figures represent an actual rise in female in-ebriety or merely that male inebriety was in decline.

Certainly contemporaries believed that female inebriety was increasing dramatically over the last quarter of the century. One author suggested that the supposed increase was 'A National Shame', and launched an outraged polemic against one of the main 'features of present-day life . . . the *deplorable* increase of drunkenness among women'.[25] Inebriety was widely claimed to be far worse in Britain than anywhere else in the world. Whilst men were thought by some to have become more sober, women were said to 'have more and more taken to regular and frequent drinking' to such an extent that they even outnumbered men in pubs.[26] Seemingly impressive figures backed up such assertions, creating a convincing impression that in every major town and city women of all classes were

[24] MacLeod, 'Edge of Hope', 215.

[25] Anon., 'The Increase of Female Inebriety and its Remedy', *British Medical Journal* (July–Dec. 1892). See also N. Kerr, *Female Intemperance* (1880), 3–8; M. Higgs, *Glimpses into the Abyss* (1906), 294–5.

[26] J. W. Horsley, *How Criminals are Made and Prevented* (1913), 140 and 7 respectively.

drinking far more openly and excessively than they had been twenty years before.[27] By 1905 Philip Snowden MP noted that the 'increase of drinking among women . . . seems to be universally accepted as a fact'.[28] More perceptive observers recognized that, since public opinion was less tolerant of drunken women, they were more likely to be proceeded against and, once known as alcoholics, to be repeatedly reconvicted. In consequence, the proportion of women with many convictions in a year was higher than men and arrest and conviction statistics would over-represent numbers of women. At the Social Science Association meeting in 1883, one speaker pointed out that 'it must be borne in mind, in looking at the statistics, that when a woman took to habitual drinking, she was more often convicted than a man, and the several convictions of one woman would swell up the number, and lead to misapprehension as to the proportion of men and women who are drunkards.'[29]

Indeed, it was shown in 1910 that while women accounted for only 20 per cent of the total number of those convicted of drunkenness, a mere 14 per cent of these women accounted for 43 per cent of all those with 20 or more convictions (defined as habitual drunkards).[30] The figures are skewed even further by the fact that a relatively few women amongst this body of habitual offenders received a massive number of convictions. Though this meant that drunkenness was far less widespread than many assumed, confined in fact to a very small number of perpetual recidivists, such revelations did little to alleviate concern. Attention focused instead on 'the heroine of a hundred convictions, whose life is a perpetual horror and a public scandal'.[31] The really incorrigible inebriate was commonly conceived of as female. And certainly the most infamous characters were, as Radzinowicz and Hood point out, all

[27] See G. Pearson, *Hooligan* (1983) for an illuminating analysis of the tendency over the past 100 years or so to identify social problems as being of recent origin or intensity and for each generation to insist that '20 years ago' such horrors did not exist.

[28] P. Snowden, *Socialism and the Drink Question* (1908), 41. See also Anon., 'Physical Deterioration and Alcoholism', in *Justice of the Peace* (Dec. 1905), 579–80.

[29] Contribution made by Miss A. Bewicke to 'Discussion on the Habitual Drunkards Act', *Transactions NAPSS 1883* (1884), 480.

[30] Horsley, *How Criminals are Made*, 138–9.

[31] Howard Association, *Annual Report* (1906), 24.

women.[32] For example, Jane Cakebread became a notorious
public figure renowned for her frequent appearances in court
and her bizarre behaviour. By 1895 she had achieved a stunning
total of 278 police court appearances.[33] Tottie Fay, similarly
taken up by the press, became the subject of much fascinated
and bemused attention as a result of outlandish court appearances in ball gown and satin slippers until she, too, finally
ended up in a lunatic asylum.[34] Such colourful figures attracted
an uneasy mixture of blatant curiosity and outrage. The
numbers of other women who also notched up two or even
three hundred court appearances without the saving grace of
such eccentricities were more roundly deplored. Public outrage
reached a peak in 1891. A flood of letters to the *Daily Telegraph*
during August and September of that year decried the outrageous phenomenon of the habitual female inebriate. Massive
press coverage had provoked a veritable 'moral panic'.

The problem was, however, hardly new, as the briefest
survey of penological literature readily reveals. In studies of
female offenders throughout the nineteenth century, the drunk
was commonly portrayed as the most hardened and most helpless of all women prisoners. As early as the 1830s, notorious
female drunks had captured the public imagination. In the first
half of the century, women like Mary Moriarty and Eliza
Ellams symbolized the nadir of alcohol-driven depravity and
'the utter hopelessness of reform (especially of the female
character), when once that accursed craving had assumed a
chronic form'.[35] The female drunk was, then, a recurring
figure, a 'folk devil',[36] whose excesses surpassed even those of
the habitual criminal or the vagrant. She was simultaneously
feared and pitied for her patent incapacity to stop drinking and
her progressively wretched condition. In the 1890s a press
campaign promoted public demands for new efforts to be made

[32] Radzinowicz and Hood, *History of English Criminal Law*, v. 301.

[33] Ibid. See also Anon., 'A Public Scandal', *British Medical Journal* (Jan.–June 1895), 941.

[34] T. Holmes, 'Habitual Inebriates', *Contemporary Review* (May 1899), 741.

[35] G. L. Chesterton, *Revelations of Prison Life* (1856), i. 71–8, on drunken women admitted to the House of Correction, Coldbath Fields, during his chaplaincy there during the 1830s.

[36] For the classic case study of modern-day 'folk devils', see S. Cohen, *Folk Devils and Moral Panics* (1972).

to formulate a more positive solution. *The Times*, for example, formerly fiercely opposed to long-term detention, swung in the 1890s to join the wide range of newspapers demanding legislation to establish reformatories.[37]

These demands combined with a deepening recognition of the futility of repeated imprisonment, especially since most sentences for drunkenness were a mere seven or at most fourteen days. The Departmental Committee on the Treatment of Inebriates of 1893–4 heard evidence from a variety of witnesses on the adverse effects of repeated short terms of imprisonment. Many supported the widely held belief that the prison not merely failed to reform but was positively conducive to long-term inebriety, a sanatorium, providing a period of recuperation after which the inebriate was launched back into the world, fit to continue her otherwise fatal indulgence.[38] Other critics, though less impressed by the prison's health-giving qualities, were nonetheless convinced that sentencing the inebriate women to regular periods of very short imprisonment was probably the surest means of compounding their dependence. A few years later Eliza Orme despairingly demanded: 'Could a more clever plan be devised for confirming a habit of drinking spirits than locking a person up in a dull cell for three days, with unsuitable food, and then letting her loose into a city of gin-palaces?'[39] The climate of opinion had clearly changed to a deepening realization that any endeavour was better than to allow the Cakebreads and the Fays to continue their appalling 'existence alternated between the streets, the public-house, and the prison'.[40]

In any case such women greatly disturbed the smooth running of the prison. Older drunkards, especially, suffered a variety of medical problems, engendered by long years of drinking, which took up the time and energies of the prison medical officer. If admitted whilst still drunk they all too often disrupted the entire order of the prison by singing and shouting. And even if sober, craving more drink, they were liable to

[37] MacLeod, 'Edge of Hope', 233.

[38] Among others, Albert De Rutzen, Police Magistrate, Westminster, and Canon Acheson, former Chaplain Chester Prison. Departmental Committee on the Treatment of Inebriates, Minutes of Evidence, *PP*, 17 (1893–4), 620, 679.

[39] E. Orme, 'Our Female Criminals', *Fortnightly Review*, 69 (1898), 792.

[40] Holmes, 'Habitual Inebriates', 740.

become difficult to control or even violent. Increasingly, policy makers recognized that traditional forms of punishment were simply inappropriate. New medical views of inebriety evoked considerable sympathy for the female drunk shuttled endlessly about the criminal justice system for crimes which owed more to pathology than moral depravity. In 1888 Norman Kerr took up the case of the archetypal habitual inebriate, a woman imprisoned more than two hundred times for drunkenness and petty theft, and demanded: 'Is it not monstrous to punish such a one as a criminal without any hope of reformation, when if she were treated as a diseased person in an inebriate retreat, there would be a fair opportunity of trying the effect of curative treatment?'[41]

DEVELOPING INEBRIATE REFORMATORIES

To understand the development of public inebriate reformatories around the turn of the century we need to survey preceding legislative experiments directed specifically at the problem of alcoholism, if only to place in context the later, larger endeavour. Although the 1872 Select Committee on the Control and Management of Habitual Drunkards called for legislation to detain criminal inebriates long enough to bring about a cure, the 1879 Habitual Drunkards Act had, in the course of a difficult passage through parliament, been shorn of stringent powers of compulsion. The Act had provided only for the temporary licensing and inspection of private retreats. Inebriates seeking admission were required to prove both their willingness to give up their freedom and their ability to pay. Made permanent by the Inebriates Act 1888, this legislative provision remained little used. The cost made admission impossible for all but the wealthy few, and yet the stigma attached to such incarceration limited its appeal to the very dipsomaniac gentlemen at whom it was primarily directed. Both the 1879 and the 1888 Acts singularly failed to address the plight of the poor, criminal, and especially female, inebriates who, by the 1890s, and in the light of the Cakebread scandal, were seen as the more pressing problem. The Report of the Departmental Committee on the

[41] Kerr, *Inebriety*, 23.

Treatment of Inebriates (1893–4) reflected a new consensus that compulsory commitment to long-term sentences in large public institutions offered the only hope of reform.

The resulting Inebriates Act passed in 1898 added to the existing system of licensed voluntary retreats a two-tier system of certified and state reformatories. The Act applied to two main, not necessarily discrete, groups: habitual drunks; and those who committed serious criminal offences whilst drunk.[42] Both of these could be detained for up to three years, in order to keep them away from drink long enough to fortify their will to abstain. Originally it was intended that state reformatories should be restricted to criminal cases,[43] since it was assumed these would be more difficult to manage. Local authorities and philanthropic bodies would set up certified reformatories mainly to take habitual drunkards, and also some of those convicted on indictment. They received half a guinea per week as a Treasury contribution towards maintenance costs.

Local authorities were slow to respond to the Act. Uncertain as to the numbers liable to be committed they found it difficult to assess the extent of provision needed. Also, as the Inspector appointed under the Act, R. W. Branthwaite, conceded, deficiencies in the drafting hindered action by limiting the ability of local authorities to band together to make provision.[44] Most damagingly for the success of the experiment, they were reluctant to make the commitment or meet the expenses entailed. Many in local government simply felt that provision for the criminal inebriate was the responsibility of the State and did not think that they should have to bear heavy costs simply to relieve the prisons of their worst incorrigibles. It is not surprising, then, that the first certified reformatories were set up and run by philanthropic bodies. In 1899 the Royal Victoria Homes at Brentry and Horfield expanded existing homes and

[42] Section 1 of the 1898 Act pertained to those convicted on indictment of an offence punishable by imprisonment or penal servitude, if committed under the influence of drink or if the offender was found or admitted to being an habitual drunkard; Section 2 applied to those summarily convicted of a range of specified drunkenness offences three times within one year, who were again convicted on indictment or who gave their consent to go to a reformatory. Details from GLRO PC/GEN/1/7, Report of the LCC Subcommittee on the Inebriates Act 1898 (13 Feb. 1899).

[43] That is, under Section 1.

[44] First Report of the Inspector under the Inebriates Acts for 1899, *PP*, 10 (1901), 768–9.

purchased a new site to provide the first national institution for the treatment of 75 women committed under the Act.[45] Duxhurst Reformatory in Surrey, run by Lady Henry Somerset, and St. Joseph's Reformatory at Ashford in Middlesex, run by the Sisterhood of the Good Shepherd, also opened as certified reformatories in 1899, both for women only and the latter exclusively for Roman Catholic women.

Whilst the setting up of reformatories was fraught with antagonisms, there was one issue on which all agreed: that the female inebriate was the greatest problem and must, therefore, be the main focus of their work. Magistrates, viewing women as the intended object of the Act, readily committed them to lengthy sentences but remained loath to detain men for similar periods 'merely for being drunk'. Whereas the drunken woman was construed as a burden and a danger to her family, her male counterpart, however incapable, was assumed to be the breadwinner with a family dependent on his support. In fact, as Inspector Branthwaite pointed out, all too often such men drank their wages away—'the wife keeps the family, and the man is more of a nuisance than a help.'[46] However, philanthropic bodies, assessing the demand on spaces and recognizing the likely impact of these assumptions on committal rates, made little provision for men. The result was that even if a magistrate was willing to commit male inebriates, he had considerable difficulty in finding places to send them.

All these new reformatories accepted cases committed anywhere in the country providing that the local authority (of the area in which the woman was convicted) contributed towards maintenance costs.[47] However, a disappointingly small minority of councils took up the opportunity to participate. At the end of 1900, Branthwaite regretfully reported that only 30 per cent of county councils and 20 per cent of borough councils had so far made any provision.[48] The London County Council

---

[45] A block of cottages on the Brentry estate was designated to hold 30 men but none were sent there in the first year of the Act. Ibid. 766.

[46] Report of the Inspector under the Inebriates Acts for 1901, *PP*, 12 (1902), 751.

[47] Whilst Duxhurst and St Joseph's accepted women simply on the payment of 7*s.* per head per week, the Royal Victoria Homes drew up an elaborate scheme whereby any local council contributing £1000 to the original cost of establishment was entitled to have 7 beds reserved for its use for 25 years at a cost of 3*s.* 6*d.* per head per week.

[48] Report of the Inspector under the Inebriates Acts, *PP*, 12 (1902), 636–7.

(LCC) was the only local authority to set up a reformatory of its own, at Farmfield near Horsley in Surrey. The details of its establishment give some indication of the difficulties and antagonisms involved. The LCC set up an Inebriates Act Sub-committee, under the chairmanship of Dr William J. Collins, a radical medical publicist and reformer, to consider what provision was necessary. Urged from the start by the Home Office to respond immediately with substantial aid to the voluntary bodies already setting up reformatories, the LCC began to investigate the possibility of making its own provision.[49] Disappointed by the quality of advice obtained from the Home Office, the LCC visited Holloway Prison 'with a view to see the class of inmates with whom we should have to deal. The prison officials afforded us all information, but appeared to regard our scheme as Utopian, so impressed were they with the degradation of the "habituals" who haunt their cells.'[50]

The Council's, by now rather cautious, deliberations on the very feasibility of their project were repeatedly interrupted by increasingly terse letters from the Home Office charging them with sabotaging the Act by their 'unwillingness' to co-operate.[51] Evidently under pressure, too, from London magistrates, police, and prison governors, they hurried through arrangements with Duxhurst and St Joseph's reformatories to take women from the London police courts. On advice from the Home Office that 100 men and 200 women were likely to be dealt with in London during the first year,[52] the Council decided that they needed to provide for women far more urgently than for men. The difficulties of finding suitable places for inebriate women in other reformatories, exacerbated by the refusal of many of them to take 'immoral' women, finally convinced the LCC that it would be necessary to make provision of their own. The existing buildings on the Farmfield estate, originally intended to house men, were turned over for women instead. In fact the numbers of men committed under

[49] GLRO PC/GEN/1/7. Report of the LCC Inebriates Acts Subcommittee for 1898 (13 Feb. 1899), 2–3.

[50] W. J. Collins, 'An Address on the Institutional Treatment of Inebriety', *British Journal of Inebriety* (Jan. 1904), 108.

[51] GLRO LCC/MIN/8114. LCC Subcommittee Minutes, esp. 14 July 1899, 13.

[52] In fact, as a later report revealed, the actual numbers were only 2 men and 90 women. GLRO PC/GEN/1/7, Reports of the LCC Subcommittee for 1898 and 1900, at 2 and 5.

the Act remained so low that the LCC was able to accommodate them in other institutions and never found it necessary to set up a reformatory for men. When Farmfield opened with room for thirty women in August 1900 the LCC became the first local authority to establish a reformatory of its own. Nonetheless, its public reception was largely hostile. Farmfield was condemned by some as inadequate (given the number of women who might have been committed had there been more places), and yet by others as wildly extravagant. Only Branthwaite defended the LCC, rejecting the criticisms as unjust and praising their cautious approach.

Instructed to make their institutions 'as little like prisons as possible',[53] the managers of certified reformatories found their attempts to do so thwarted by the most incorrigible of prison habituées. In the first year, seven women were so completely disruptive to the routine of the reformatories holding them that they had to be discharged. Struck by the hopelessness of their task, the reformatories argued that they could not operate without separate provision for the classification and removal of the 10 or 15 per cent of cases that they deemed 'utterly impossible to deal with'.[54] The LCC was perhaps the most vociferous of critics. In condemning the Home Office for its failure to provide for these intractable cases, it fired back many of the criticisms originally used against its own work.[55]

Eventually the government accepted the need for state provision and in 1901 decided to set up two reformatories: one at Aylesbury for women, and a smaller one, in part of Warwick prison, for men. Finally opened late in the autumn of 1902, and housed initially in temporary buildings, they supplied the penal component necessary to the whole system. Originally it had been intended that state reformatories should only take committals under Section 1 of the Act (those convicted on indictment of criminal offences committed under the influence

[53] Report of the Inspector under the Inebriates Acts for 1899, *PP*, 10 (1901), 740.

[54] GLRO PC/GEN/1/7. Report of the Lady Superintendent, Farmfield, for 1900, at 11.

[55] Denouncing delays in the setting up of state reformatories, it dismissed claims that it was not yet clear how much accommodation would be needed, pointing out that certified reformatories had been set up with no better information. See GLRO PC/GEN/1/7, Prefatory Account by Collins, 2. Also Report of the LCC Subcommittee for 1900, 6, referring to a circular by the Secretary of State, 10 Jan. 1899.

of alcohol) on the assumption that these would be the more difficult to manage. The experience of certified reformatories, however, soon indicated that the largest group of refractory women fell within those committed under Section 2—habitual drunkards.[56] It was agreed, therefore, that in addition to those sent to them directly under Section 1, Aylesbury and Warwick should take any intractable cases then held in certified reformatories. The Superintendent at Farmfield, who had found her efforts repeatedly sabotaged by the few women who 'caused constant irritation and annoyance and kept the whole institution in a state of ferment and unrest', thankfully applied for their transfer to Aylesbury.[57]

The prophecy that women would be the main subjects of the reformatory experiment proved, in practice, to be self-fulfilling. In 1904 women represented 91 per cent of admissions to certified reformatories.[58] Female admissions to state reformatories up to that year outnumbered men by two to one.[59]

AIMS AND PURPOSES OF THE REFORMATORY REGIME

Under the 1898 Act, a Departmental Committee was set up to translate the intentions of the legislators into practices and regulations by which reformatories might be run.[60] Recognizing that there was no known cure for inebriety, the committee suggested that the exact mode of treatment be left to the discretion of reformatory superintendents and management boards. They were insistent, however, that the whole purpose of the regime ought to be reformatory and not punitive, relying on moral suasion rather than penal sanctions to maintain order.

The formula adopted owed more to the reformatories established after mid-century for juvenile delinquents than to earlier endeavours at reform within the prison. It reflected con-

[56] Report of the Inspector under the Inebriates Acts, 64–6, in Report of the Prison Commissioners for the year ended March 1902, *PP*, 46 (1902).

[57] GLRO PC/GEN/1/7. Report of the Lady Superintendent, Farmfield, for 1901, at 7.

[58] Report of the Inspector under the Inebriates Acts for 1904, *PP*, 11 (1905).

[59] T. Hopkins, 'The State Drunkard', *Law Times* (Aug. 1904), 321.

[60] Report of the Departmental Committee on Regulations under the Inebriates Acts 1898, *PP*, 12 (1899), *passim*.

temporary beliefs about the causes and character of female drunkenness and the ideal social role to which such women should be restored. Although the non-punitive intentions of the reformatory regime constituted a radical departure from past prison treatment of inebriates, in many ways its character and purpose were reactionary. Deeply anxious about the environment in which alcoholism (amongst a whole host of other vices) appeared to flourish, it recognized no other solution than to recreate, in microcosm, a past world in which the destitution and demoralization of the urban slum simply did not exist.

Given the physical condition of many women arriving at these reformatories, the first requirement was a healthy, plentiful diet and ample exercise in the fresh air. Only then could the central purpose of the regime—to restore 'blunted moral sense' —be attempted.[61] Great efforts were considered necessary to reverse years of demoralization.[62] Physical and moral recovery was seen as a slow and difficult process, continually thwarted by the women's own lack of will or destroyed by the corrupting influence of other more hardened inmates. As in the prisons described in earlier chapters, constant vigilance was considered essential to prevent other inmates detracting from the personal influence of the staff (or 'sisters' as the staff at Farmfield were known).

Less optimistic than the ultimate aim of reform, but nonetheless central, was the purpose of containment, which it was hoped would minimize the pernicious influence of inebriates, prevent them from doing physical harm to others (especially their children), and, more generally, clear the streets of 'shocking spectacles'. On a more pessimistic note still, reformatories appeared to be the only way to avert inebriates from their otherwise 'inevitable end' in criminality or lunacy.[63] Certified reformatories set out, therefore, with a series of somewhat divergent purposes: from mere confinement to the higher, but less readily attainable, goal of moral reform. The ultimate goal was, quite simply, to create of the enfeebled and degraded drunk a model of healthy, domesticated femininity.

[61] GLRO PC/GEN/1/7. Report of the Chaplain for 1903, at 10.

[62] For a succinct discussion of these aims see GLRO LCC/MIN/8124, Report on the History and Working of Farmfield by Elizabeth Forsyth and Charles F. Williamson (16 Dec. 1907).

[63] Report of the Inspector under the Inebriates Acts for 1900, *PP*, 12 (1902), 640.

The purpose of the State Reformatory at Aylesbury was less straightforward. As we have seen, it was conceived by the Home Office as a means of syphoning the refractory element from certified reformatories, in order to leave the latter free to concentrate on reforming the supposedly more amenable. It need do no more than sustain a level of discipline sufficiently strict to secure and contain. This view of the 'hopelessness' of attempting reform was reaffirmed repeatedly throughout the Inspector's reports. Yet it was not a view shared by those actually working at Aylesbury, who propounded a more active regime not dissimilar in intent from that pursued in certified reformatories. Surviving institutional records are infused with the language of reform, of 'raising up' and of encouraging 'self-respect'.[64]

These conflicting intentions, and the dichotomy between avowed purposes and the actual regime, reflected deeper confusions about the very problem being tackled. Significantly, all the reformers recognized that they had little influence beyond the reformatory itself. Accordingly, all attention and aspirations were directed on to the character of the female drunk as primary cause and, therefore, site of reform. As the Medical Officer of Farmfield put it, the women under his care 'require to be remodelled'.[65]

RECREATING A RURAL IDYLL

Whereas local prisons were generally located in the heart of the areas from which they recruited their inmates, the inebriate reformers turned their backs on the appalling environs of the urban slum and built their sanctuaries in the safety of the countryside. Not satisfied even then that they could isolate their charges from the encroachments of the outside world, they set up reformatories on large estates which effectively cocooned them in fields and woodland (as did those later setting up mental hospitals). Farmfield was set in a massive 374 acres of land. The Royal Victoria Homes occupied two neighbouring

---

[64] See *Minute Book of the Visiting Committee, Aylesbury* (1903–17), *passim*. This was kindly lent to me by the present Governor of HM Youth Custody Centre, Aylesbury.
[65] GLRO LCC/MIN/8122. Medical Officer's Report for 1905.

estates at Brentry and Horfield large enough for the cottages for male inebriates to be safely distanced from the main mansion for women by woodland. Even the smaller St Joseph's Reformatory was surrounded by a cordon of 45 acres. That such locations signified a rejection of the city as the site of all social ills was explicitly recognized: ' "Back to Nature" seems a natural resort for these social wrecks, mostly denizens of the festering courts of the great city.'[66] If, as we have observed, woman's rightful place was considered to be in the home, the apotheosis of this ideal sited the 'home *in* a rural village community'. And as Leonore Davidoff has suggested, 'the more cut-off, the more "total" this situation, the greater the likelihood that the definition will remain coherent and thus order and stability maintained.'[67]

At Farmfield two existing mansions were adapted to expand the reformatory to house over a hundred women in small self-contained groups. The women slept in 'brightly furnished' single bedrooms, allowing a degree of privacy regarded as essential to their self-respect. Reminiscent of the continental 'farm school' model adopted by juvenile reformatories fifty years before, this so-called 'cottage system' served exactly the same purpose. It emulated a small village community of 'families' living independently but under the benign care of a maternal hierarchy and meeting together for communal activities such as worship and work. The inebriate woman, like the delinquent child, could, it seemed, best be reformed by relocating her in the heart of the family she had lost. Cleaning, laundering, sewing, and cooking took on an inordinate importance not merely as a means of employment or even of maintaining the institution, but as an invaluable element in the slow process of domestication.

Accommodation at Aylesbury was markedly more prison-like: barrack-style buildings, divided into cells on long, im-

---

[66] Collins, 'Institutional Treatment', 114.

[67] Leonore Davidoff has rightly pointed out that the domestic and rural idyll were ideal types not necessarily reflective of woman's actual existence in either home or village community but suggestive of valued qualities such as deference to male authority (the head of household or lord of the manor), limited aspiration, and maternal contentment. In reality rural life had a far uglier underside and was in any case continually eroded by the harsh intrusions of the outside world. L. Davidoff *et al.*, 'Landscape with Figures: Home and Community in English Society', in A. Oakley and J. Mitchell (eds.), *The Rights and Wrongs of Women* (1976), 144.

personal corridors. Yet even in this austere setting similar importance was attached to the inculcation of domesticity. So impressed were the Visiting Committee (the board of prison visitors) with the importance of this purpose that they proposed to introduce 'explicit teaching on the obligations of social, family, and individual life, lectures on temperance, hygiene, care of children, and domestic economy, practical teaching in elementary cooking, laundry, and garden work'. This concise catalogue of domestic and maternal skills was endowed with several ambitious purposes which the regime seemed curiously inadequate to fulfil, not least 'to make the women wage-earners, and to raise them from their present condition of childishness or savagery'.[68] This lack of fit between avowed purposes and regime reflected a more profound dilemma. Although the ideal aim was to restore a woman to a family of her own, it was far more likely that she would have to support herself. Though domestic service was a poor substitute for her own home, it at least provided the relative safety of someone else's. Moreover, the extraordinary moralizing powers bestowed upon domestic duties suggested that their pursuit could social-ize the supposedly infantile and civilize the 'savage' in inebriate women.

If domestic arrangements within the reformatory were con-trived to recreate an ideal home, outdoors, efforts to reconstruct a rural idyll were even more artificial. At Farmfield a 'model dairy', flower gardens, large vegetable garden, and extensive fields were all cultivated according to an idealized notion of rural English life. Women were taught butter-making, bee keeping, poultry rearing, fruit and potato picking, and, in a parody of pastoral perfection, hay-making. As Dr Collins openly boasted: 'everything is done to take the patients back to simple natural life such as is typified in the English country homestead.'[69] Nowhere was it seriously suggested that these agricultural skills enhanced the women's future prospects; they were merely justified as a healthy distraction from cravings for alcohol. It seems that the reformers could only repudiate, tem-porarily, that morass of problems they could not even begin to tackle, but to which most women would have to return. Only in

[68] Minute Book, Aylesbury. Copy of Report to the Home Secretary (1907–8), 101.
[69] Collins, 'Institutional Treatment', 109.

moments of rare honesty did they address the actual prospects
for the women's future employment. Daunted by their vision of
'the strain and temptation of a single woman's life in the
industrial world', they recoiled hastily to the sanctuary of the
countryside. The LCC was vehemently opposed to instructing
Farmfield inmates in any industrial trade on the grounds that it
would equip the women to return to the untold dangers of
unsupervised employment in trade workrooms and factories.[70]

At Aylesbury outdoor work was severely constrained by the
size of site, which was far too small for any agricultural development. Needlework and light domestic chores failed to exhaust
the energies of the often violent women who were sent there.[71]
So, in poor emulation of their rural sisters, Aylesbury women
were encouraged to cultivate a small plot of land into a pitiful
imitation of a landscape garden.[72] So convinced were the Visiting Committee of the central role that might be played by
agriculture that they sought to acquire additional land on
which to take up market gardening. This, they argued, would
be not only 'highly beneficial to health' but actually 'conducive
to the general welfare and advancement of the object of the
Institution'.[73] Painfully aware that they possessed no cure for
inebriety, the reformatory clung desperately to an ideal
stretched even further to the bounds of plausibility in the
austere setting of Aylesbury than elsewhere.

ATTEMPTING MORAL REFORM

The inebriate reformatories possessed no coherent programme
for moral reform, merely a series of disjointed, sometimes
contradictory tactics intended to 'raise' or 'remodel'. Often
these revealed assumptions and preoccupations otherwise

---

[70] GLRO LCC/MIN/8126. 'Proposed Trade Instruction of Inmates at Farmfield',
Report by F. A. Durham to the LCC Subcommittee (11 Feb. 1911).

[71] Reports of the Prison Commissioners for the year ending March 1902,
Appendix 21, 'Extracts from the Governor's Report, Aylesbury', *PP*, 46 (1902), 622.

[72] It remains today a tiny garden complete with diminutive pond, stepping stones,
and a solitary willow tree, set incongruously against the grim façade of the cell blocks
which now house HM Youth Custody Centre, Aylesbury.

[73] Minute Book, Aylesbury. Copy of the Report of the Board of Visitors to H. J.
Gladstone (31 Mar. 1908), 100.

hidden. For example, great emphasis was placed on the improvement of physical appearance, not merely as indicative of recovery from the appalling ravages of alcoholism but as a statement of the woman's moral health. At Aylesbury the Visiting Committee clearly regarded the ugly and ill-fitting prison-like garb as a hindrance to reform and deliberated lengthily on how to improve it. They stressed the importance of 'encouraging . . . self-respect'; of enabling the women to dress with 'decency', to emulate the dress styles of 'poor women' outside; and suggested that the women be encouraged 'to do their hair in any becoming and neat way they please'.[74] Abandoning Victorian attempts to suppress vanity in women prisoners, they sought instead to promote an active interest in 'feminine pursuits'. Similarly, at Farmfield women were encouraged to make their own dresses in the evenings and praised for their attempts to emulate ladylike demeanour and attire.[75] Such superficial alterations to appearance were clearly seen as reflective of a more profound feminization of character.

By limiting the numbers admitted (most reformatories were kept to well below a hundred inmates) and by breaking up inmates into small groups, reformatory managers sought to preserve a sense of individuality. Each woman could then be known personally. In this way the reformers hoped to dispense with the rulebook and with punitive sanctions in favour of more subtle manipulations—tact, personal influence, and moral guidance. As in prisons, staff were to set themselves up as models of virtue and propriety, and women who attempted to emulate them were rewarded with a series of petty privileges. At Aylesbury a liberally adapted version of the 'mark system' (see Chapter 5 above) substituted the formal structure of rewards with promises of trivial 'luxuries, such as biscuits and jam'.[76] At Farmfield women could earn regular gratuities, though marks were deducted for 'imperfect work, misconduct, unpunctuality, insolence or disobedience'.

Good behaviour was also encouraged by the promise of various treats. Dances and singing evenings, lantern shows and

[74] Ibid. Report on Inmates' Clothing (12 Nov. 1907), inserted at 91.

[75] The very rules demanded that inmates keep themselves 'clean' and 'decent', their possessions and bedding 'neat'. GLRO PC/GEN/1/7, General Regulations, Farmfield, 5.

[76] Minute Book, Aylesbury. Report for 1908, 123.

piano recitals were held throughout the year, whilst at Christmas a ventriloquist and a conjuror entertained.[77] The overall impression is more akin to a rather jolly girls' school than a reformatory supposedly for the worst incorrigibles of London's prisons. The description of Farmfield (when it was later restricted to women classified as 'reformable') as the 'finishing school' of the reformatory system only confirms this analogy. The apparent belief that the demoralized drunk would be restored to useful citizenship by such diversions seems somewhat bizarre. For they were far removed from the likely realities of the women's lives on release, where a taste for such drawing-room entertainments would surely seem a remote guard against the temptations of the city.

This policy designated the women as low and degraded, and attempted to raise them up by exposure to the values and activities of a higher class. Such an aim was even more explicitly expressed in relation to a more conventional agent of reform— that angel of the nineteenth-century prison—the Lady Visitor. At Aylesbury Lady Visitors were encouraged to visit and talk with the women; as in prisons, it was hoped this would lead to 'the growth of woman-like self-respect and to the awakening of higher hopes, from being brought into contact with superior minds and personal ideals'.[78] Yet the moralizing purposes of the regime sat uncomfortably alongside persisting conceptions of inebriety as a disease: for example, at Farmfield the women were incongruously referred to as 'patients'. The regime was clearly incoherent, both in its initial premises and in the means adopted to carry them out. Although well-intentioned, the regime could not succeed in reconciling such widely differing aims.

Punishments for offences against rules and regulations were even more difficult to 'fit' alongside the moralizing purpose of the regime. The premise that women were susceptible to moral suasion demanded that reformatory managers try to enforce discipline by 'kindness and reason' before turning to traditional penalties. Staff met infractions with a finely balanced mixture of quiet condemnation, expressions of disappointment, and

[77] GLRO LCC/MIN/8120. Report of the Lady Superintendent (24 Jan. 1903).
[78] Extracts from the Chaplain's Report, Aylesbury, 606. In Reports of the Prison Commissioners for the Year ending March 1904, *PP*, 37 (1905).

determined efforts to elicit remorse and promises of better behaviour. Whereas the prison regime relied on punitive sanctions against those who failed to comply, the inebriate reformatory employed emotional blackmail, cajolery, and pious exhortation. This strategy relied on considerable intimacy between staff and inmates, so that the former could establish a quasi-maternal stance over their wayward daughters. That this approach tended to infantilize is evident, for example, in the official description of one refractory woman as 'a very bad girl' and her behaviour as 'naughty'.[79]

When women were so violent or determinedly resistant as to defy all attempts at persuasive discipline, the authorities turned in desperation to more traditional methods of punishment— dietary restriction or confinement in solitary cells. Yet even here the maternalistic moralizing desire to secure submission can still be seen at work. Women at Farmfield were not usually sentenced to a fixed period in solitary confinement (as they were in prison) but were incarcerated with the promise that they would be let out only when their behaviour improved. Denied even the right of fixed terms of punishment, they were held at the whim of the staff, whom they had to convince of their penitent desire to be 'good'.

The regime, as we have seen, made strenuous demands on the staff, placing them in a pivotal position on which the success of reform depended. It did not, however, provide them with any formal instruction or training—a fact deplored by Lady Somerset 'as one of the greatest secrets of weakness of the present system'.[80] Nor, apart from rather general criteria, such as the desirability of employing older women over younger and inexperienced staff, was there any clear idea of the type of women suited to the job. In fact, nearly all the women employed at Farmfield were formerly hospital or asylum matrons, and significantly not prison staff. Lady Somerset, at Duxhurst Reformatory, deplored the fact that though their task was 'one of the most difficult works that can be imagined', women who had failed in every other occupation were seen as quite fit to become reformatory matrons.[81] As a result the work was often

---

[79] GLRO PC/GEN/1/7. Report of the LCC Subcommittee for 1901, 2.

[80] Evidence to the Departmental C. on the Law relating to Inebriates (1908), 926.

[81] Ibid.

beyond their capabilities. One Lady Superintendent at Farmfield completed her probationary period but declared: 'I have
no desire to be appointed permanently as I find the worry and
anxiety more than I can bear and find that my health is beginning to fail.'[82] That many matrons at Farmfield felt similarly
dissatisfied is revealed by complaints continually made about
the lengthy hours (fourteen hours per day for three days per
week and ten hours per day for the remaining four), the 'great
strain' their work inflicted, and the loneliness of their isolated,
rural life. Given the considerable difficulties experienced in
trying to recruit staff, these grievances had to be taken
seriously. Surprisingly generous efforts were made to improve
their working conditions and opportunities for enjoyment. For
example, staff at Farmfield were allowed to dine together
instead of with inmates, provided with a circulating library,
and even with reduced rates on the local taxi service. Yet some
staff were simply unequal to the demands made of them. One
cook at Farmfield repeatedly came home drunk, armed with
bottles of spirits. Despite this being a particularly serious
offence (usually averted by demanding that all staff were total
abstainers), she was given many warnings before eventually
being forced to resign.[83]

LIFE INSIDE: INMATE RESPONSE AND RESISTANCE

What sorts of women were admitted to inebriate reformatories?
What backgrounds did they come from? And how did they
respond once there?

Thomas Holmes, writing in 1899, had estimated that 80 per
cent of those qualifying as habituals would be 'women of the
streets', 'the great bulk vice dominated'.[84] The remainder
would be made up in equal parts of demented old women from
the workhouse, and otherwise respectable old women. In fact,
as reformatory managers soon recognized, ascertaining the
true background of the women committed was extremely difficult. Most claimed to be laundresses but, as the Superintendent

---

[82] GLRO LCC/MIN/8120. Letter to LCC Subcommittee (10 Mar. 1903).
[83] Ibid., LCC Subcommittee Papers (1903–4), and LCC/MIN/8115, Minutes
1903–6 (3 Apr. 1903), 75.   [84] Holmes, 'Habitual Inebriates', 744.

at Farmfield pointed out, many had been too often drunk to be capable of holding down any job permanently. They made this claim, as they had done dozens of times before, partly to suggest their previous respectability but primarily in order to secure the most popular institutional jobs.[85] Such ruses, and their general unwillingness to admit that they had had no occupation at all, effectively rob their minutely detailed 'history sheets' and statistics on previous occupation of any real validity.

One gains only the haziest impression that most came from poor working-class areas where they survived by periods of casual employment regularly interspersed with periods of imprisonment for drunkenness. In 1909 Inspector Branthwaite provided detailed figures on 1,031 cases of whom 84 per cent were women (roughly representative of the average proportion of cases admitted to inebriate reformatories). He found that the then average age of women in reformatories was over 38 years. Since they had, on average, become addicted to drink by the age of 26, they had logged over 12 years of inebriety by the time they were first admitted.[86] Although two-thirds (68 per cent) were or had at some time been married, most were living apart from their husbands or were widowed. But only just over half were categorized as 'immoral', this label being confined to those who had actually been convicted of prostitution or of soliciting (or for whom there was very reliable evidence of prostitution from other sources). This figure was, therefore, a highly conservative estimate of the likely number of female drunks who had, at some time, earned their living from prostitution. By the time they were admitted to reformatories the scars of such careers were only too apparent: 'The condition of the majority . . . is deplorable, and provides very bad ground for work. They are all the oldest offenders—dirty, insufficiently clothed, and degraded mentally and morally, having sunk to the lowest depths.'[87] Although in subsequent years it became possible to forestall women before they had reached this state,

[85] GLRO PC/GEN/1/7. Report of the Lady Superintendent, Farmfield, for 1900, at 10.

[86] From Report of the Inspector under the Inebriates Acts for 1909, *PP*, 29 (1911), Pt. 1.

[87] GLRO PD/GEN/1/7. Report of Lady Superintendent for 1901, 6.

there was little apparent change in attitude toward them. They continued to be vilified as 'utterly depraved'.

The reformatory experiment, it seems, had envisaged an altogether different client: downcast but not dead to appeal, penitent, and willing to reform. On this assumption, the certified reformatories allowed their inmates considerable freedom. In stark contrast to the prison, inmates suddenly found themselves allowed to move about with relatively little supervision. At Farmfield, though doors were locked at night, there were no bars on windows or any substantial barriers to prevent escape. The more daring were only too quick to take the first opportunity to run away. One woman, subsequently recaptured, reasoned with disarming simplicity 'that she didn't want to be shut up in a Reformatory as she hadn't done anything wrong'.[88] Often, craving alcohol, women sneaked off across the fields to satisfy their thirst in the first pub they came to, or returned to their old haunts for an ecstatic binge with their friends. Some women openly declared their defiance by escaping at every opportunity, even though they were continually recaptured and returned.

Repeatedly criticized by the Home Office for the apparent laxity of discipline that this continual flight seemed to suggest, Farmfield tightened its security, but to little effect. In 1904, a year in which just 33 women were received, 18 ran away, rousing a storm of protest from the Home Office.[89] When it failed to contain its inmates, the reformatory fell back on traditional sanctions and sent escapees to prison, hoping that this would provide the penal element that it, patently, could not. Yet, as a deterrent to other would-be runaways, imprisonment was found to be singularly ineffective. To the women of Farmfield: 'Holloway has no fears . . . indeed some of them who have been frequently there enjoy paying a visit.'[90] Whether by running away or less overt subversion of the intentions of the regime, the women greatly undermined its power over them. Only the more secure state reformatory at Aylesbury seems to have continued to fulfil its official purpose by being

[88] GLRO LCC/MIN/8119. Superintendent's Report, Farmfield (14 June 1902).
[89] GLRO LCC/MIN/8121. Letter from Henry Cunynghame, Under-Secretary of State, Home Office, to LCC (11 July 1904).
[90] GLRO LCC/MIN/8121. Report of Superintendent (1 July 1904).

feared and loathed by all: 'there is only one place they have great dislike to and that is "The State".'[91]

Even those whom Farmfield managed to restrain from running away were not always compliant. Although most eventually settled down, a minority remained refractory and a few were dangerously disturbed. It is these latter groups who, taxing the ingenuity, courage, and endurance of staff, occupy the pages of officers' reports and minutes of committee meetings. Most of those reported for violent behaviour were young women in their twenties, perhaps less inured to the confines of institutional life than older recidivists. Their behaviour varied from insolence and bad language to extremes of violence that seemed, to fearful observers, to border on the maniacal. Operating on the premise that sympathetic guidance would be sufficient to control these supposed moral delinquents, staff valiantly attempted to plead with the women to stop destroying their rooms and tearing their dresses. The gap between the presumed subjects and the actual recipients of this strategy, and the consequent futility of this endeavour, is well documented in the punishment records of Farmfield. For example, one Mary Ann Bennett 'screamed, kicked and used very bad language from 3 pm to 12 mid. refused all food. throwing it at the Sisters when they took it to her',[92] or again, Linda Bush 'threw all her food at the Sisters & bit Sister Flora's hand, besides other things too bad for repetition'.[93] When all efforts at persuasion finally failed, the authorities abandoned all pretence to 'treatment' and reverted to those punitive sanctions used by prisons. Typically this was the 'solitary' cell, where the galling failures of the regime could be placed out of sight and the very threat they represented to the ideal of penitent compliance removed. Yet even here, and in long-standing prison tradition, some women would continue their defiance. Inmate Susan Ashington, placed in the solitary cell for fighting with another woman, 'in less than five minutes . . . had broken the cell door, every pane of glass in the windows, had taken the iron stays that held & supported the seat & had broken the seat into splinters'.[94] Whilst a few undoubtedly acted out of

[91] Ibid.
[92] GLRO LCC/MIN/8117. Superintendent's Report (Oct. 1900), 2.
[93] GLRO LCC/MIN/8118. Superintendent's Report (22 Feb. 1901).
[94] GLRO LCC/MIN/8119. Superintendent's Report (19 Feb. 1902).

conscious determination to set the regime awry, many were probably to some degree disturbed, mentally deficient, or even more seriously afflicted, for example, by schizophrenia.

As we have seen, Aylesbury Reformatory was set up after demands by certified reformatories for an institution to cater for that minority deemed totally unmanageable. The establishment of Aylesbury was agreed to be fully justified by the conduct of the first women admitted. As the Governor described them: 'the majority are violent-tempered and quarrelsome women. Intemperance has greatly weakened their moral self-control, and rendered them suspicious, vindictive, and untruthful.'[95] Although he admitted that their conduct was partly produced by their incapacity for self-discipline, nonetheless he attributed it also to a conscious, vicious determination to make life unbearable for those in authority. Eventually the persistent violence of a hardened few made it necessary to build a detached refractory block. Here, all pretence at reform was given over to the lesser purpose of 'humane control'.[96] Yet even this more modest goal was continually undermined by their violence, for the more refractory women, as at Farmfield, tore up their clothes and destroyed their cells in outbreaks of 'violent passion'.

In smashing up cells and furniture women expressed extreme frustration against the regime. More self-controlled women took advantage of a more subtle, yet entirely official, means of protest. At Aylesbury women were entitled to lay their grievances before the monthly meeting of the Visiting Committee. They were thus provided with a ready means of criticizing those in authority. It was a channel they were only too quick to exploit. Interpreting their complaints is extremely difficult: were they genuine accusations, malicious attempts to slander individual officers, or merely the confused cries of paranoiacs convinced that they were the victims of cruel persecution? Some complaints were evidently taken very seriously. When three women independently made complaints of ill-treatment by specific officers, the Visiting Committee set up an official

[95] Report of Prison Commissioners for the year ended Mar. 1902, Appendix 21, Governor's Report, Aylesbury, 621.
[96] Report of the Inspector under the Inebriates Acts for the year ending March 1904, *PP*, 37 (1905), 69–70; see also Anon., 'State Inebriates', *Law Times* (Dec. 1906), 113.

enquiry.[97] More commonly women were reprimanded for making unsubstantiated and purely subversive accusations. For example, A. Hardgraft complained that a matron had tantalized her, only to be warned 'that her complaint was frivolous and that the Board might have punished her'.[98] Some women, patently suffering from a paranoia verging on insanity, were treated more leniently. Their grievances were heard but quietly ignored, as in the following case: 'Caroline Tehan . . . stated that she had been kicked and beaten by Attendant Miss Keena, that the Medical Officer examined her the day before to see how much knocking 'bout she could stand, and that she is afraid to go to bed, lest the Officers take advantage to attack her. She said she had written to the Pope of the matter.'[99]

Whilst these women apparently saw themselves as victims, the very fact of their ability to complain furnished them with a considerable degree of power. Obliged to take all but the most far-fetched accusations seriously, the Visiting Committee effectively placed the staff concerned on trial to prove themselves innocent of the charges laid against them. Although in the vast majority of cases the officers were subsequently exonerated, the very threat of such accusations must have in some degree counterbalanced the inequalities of power between inmates and staff. The latter were painfully aware that their slightest slip was liable to be brought against them. The dangers of thus empowering inmates were recognized by the Visiting Committee in 1906 when one member, the Duchess of Bedford, proposed that they visit women on the refractory wards only once every three months (instead of monthly). Justified as a means of curtailing the disruption and excitement concomitant on such accusations, the move was clearly intended to limit inmate powers of subversion.

Although there was no procedure by which inmates at Farmfield could make complaints against one another, there is much evidence that the women became embroiled in disputes. All the women were committed by London courts with the result that 'a larger number of the women sent to Farmfield were well known to one another, and removal from habitual environment is robbed of half its value if old associations are revived by

[97] Minute Book, Aylesbury (1 June 1906), 57.
[98] Ibid. (3 Mar. 1905), 32.   [99] Ibid. (13 Jan. 1911), 164.

personal communication in the reformatory.'[100] Old friendships and enmities were revived and grew all the more intense
in the close confines of the reformatory where petty jealousies
or personal dislike quickly exploded into violence. The records
reveal a continual round of verbal abuse, fights, and attacks,
whose suppression absorbed the energies of the staff. The
Superintendent freely admitted: 'The greatest task in this most
difficult work is to keep the peace between the women. Their
temper at times is quite uncontrollable, and sometimes almost
amounts to madness.'[101] Below the official authority of the staff
there lay a more nebulous hierarchy of powers by which
inmates bargained, coerced, and sanctioned one another.
Punishment records provide only the most partial insight into
an apparently complex network of controls by which the inmates themselves regulated day-to-day life. For example, in
May 1901 one 'Eliza Wannell was punished for having . . .
taken the law into her own hands & given Mary Ann Bennett a
good beating'.[102] On occasion even a single woman could wield
a reign of terror over the rest. To take one extreme example:
Mary Miles 'has violent outbursts of temper & has on 3 occasions threatened to kill 3 different women . . . For 1 day & 1/2
she carried about a huge piece of iron in her pocket to kill a
woman . . . all the patients live in dread of her.'[103] In this and,
more commonly, rather more subtle ways, the women subverted and even repudiated order, so that even the most compliant women were distracted from the goals of the reformatory
by more compelling diversions.

At Aylesbury, while most inmate disputes were probably
resolved informally, an official grievance procedure allowed
women to make complaints against one another. Though far
less numerous than complaints against staff, most appear to
have been of threatened or actual physical violence and they
were certainly taken rather more seriously. Again some were
no doubt false accusations, but most cases seem to have had
some foundation. For example: 'J. Parkinson. Complained of
being hit over the head with a pan by another inmate who still

[100] Collins, 'Institutional Treatment', 114.
[101] GLRO PC/GEN/1/7. Report of Lady Superintendent for 1902, at 9.
[102] GLRO LCC/MIN/8118. Report of Lady Superintendent (3 May 1901).
[103] GLRO LCC/MIN/8120. Report of Lady Superintendent (23 Jan. 1903).

threatens her'; 'D. Macmahon. Complained of the inmate Sarah Jones. Says that Jones has threatened to "do for her" and that she is terrified'; 'Kate Driscoll and . . . Eliza Shingate complained of each other. The Governor was asked to keep these two apart as far as possible.'[104] The disruptive effects of these antagonisms directly distracted from the purpose of reform. The more refractory women, by their irritating or even threatening behaviour, upset and on occasion evidently 'terrified' the other women. Preoccupied by the danger to their physical safety, or embroiled in dispute, the women were scarcely susceptible to the endeavours of the reformers.

CONTACT WITH THE WORLD OUTSIDE

So far we have examined the kinds of women coming into inebriate reformatories and their impact on the intentions of the regime. We need now to examine how the world outside, and particularly the women's families, impinged on life inside; and finally to investigate the women's fate on release. Although most women were committed to the full three-year sentence allowed under the 1898 Act, discharge on licence was possible after only nine months. At Farmfield visits from family and friends were permitted once a month; unlimited ingoing and outgoing mail was allowed subject to censorship; and women were even taken out by staff to attend to family business.[105] Although Aylesbury was considerably more secure and isolated, letters and visits were allowed on the same basis as in prisons. Inebriate reformatories were, therefore, by no means 'total institutions' cut off from the outside world. Certified reformatories, especially, allowed continual two-way communication that set against the intended influence of the regime the demands, worries, and diversions of the woman's life outside.

Perhaps the most poignant concern was that of the incarcerated mother for her children. As we have seen, the supposedly malign influence of the drunken mother had been a major impetus to the setting up of reformatories. Once she was incarcerated, however, her children were no longer feared to be

---

[104] Ibid. 7 Dec. 1906, 67; 12 Apr. 1910, 149; 15 Dec. 1911, 185.

[105] GLRO PC/GEN/1/7, General Regulations, Farmfield, and also GLRO LCC/MIN/8116, Minutes of LCC Subcommittee on regulations concerning visitors.

victims and became instead potential agents of her reform. The mother was continually reminded of the plight of her offspring. Photographs of her children were specially obtained to appeal to maternal instincts in the belief that 'in this . . . lies a redemptive power'.[106] Small infants and babies born to the women inside the reformatory were often allowed to remain there until old enough to be sent to relatives or the workhouse. Indeed, Lady Somerset argued strongly that all children should be allowed to stay with their mother throughout her sentence. Denying that any harm would be done to the children by incarceration, she stressed that the mother would then be able to 'do well by her children' rather than becoming a remote stranger.[107]

When children were cared for outside the reformatory, women were allowed to keep in contact and helped to resolve domestic problems, to settle custody, ensure their safe guardianship, etc. In short, women were encouraged to develop a strong sense of maternal responsibility and every effort was made to ensure that their home and family remained intact. Occasionally they were allowed out on day trips to visit their children, particularly in times of crisis or ill health.

Confounding earlier assumptions that inebriate women were uncaring mothers, inmates' pleas and letters indicate they felt their separation very deeply and worried incessantly about the welfare of their children. Women at Farmfield continually petitioned the LCC Inebriates Acts Subcommittee to intervene on their behalf, to allow them out to visit, or to release them back into their families. Inmate Kate Buttle wrote repeatedly to the Chairman, seeking to be reunited with her husband and five children: 'I feel my position very acutely I am sure my husband would willingly take me home, only of course he has felt very indignant at my behaviour . . .' She stressed her willingness to do any suitable kind of work 'which may enable me to see my children as I am sure the anxiety I feel concerning them is undermining my health'.[108] Unhappily, in this case, as

---

[106] B. Waugh, 'The Restoration of the Female Inebriate', *British Journal of Inebriety* (Oct. 1903), 73–5.

[107] Evidence of Lady Somerset to the Departmental C. on the Law relating to Inebriates (1908), 925–6.

[108] Respectively GLRO LCC/MIN/8119, Letter from Buttle (13 Nov. 1902), and LCC/MIN/8120, Letter of 25 Feb. 1903.

on numerous other occasions, the husband refused to take back his estranged wife. The likelihood of this eventuality was greatly increased by the Licensing Act (1902) which made habitual drunkenness by either spouse legal grounds for judicial separation. Significantly the Act also allowed the husband to apply to have his drunken wife committed (with her consent) to any retreat licensed under the Inebriates Acts: a power not available to the wife. This clause was bitterly criticized since it allowed men simply to discard their wives, often in order to take up with another woman.[109] Even husbands who were originally well-intentioned found that the stigma attached to inebriety made the return of a wife to a once respectable household unthinkable. Whatever the reason, the prime aim of the reformatory—to restore women to domesticity—was effectively confounded. All too often, then, a woman's successful return to respectable society was thwarted by forces that were beyond the scope of the reformatory's control. Her training for domesticity and inculcation in wifely duties and maternal devotion remained only as cruel reminders of the home she had lost.

RELEASE

If reformatories were forced to recognize the constraints on their power to reform their inmates, once women were discharged the limits to their influence were even greater. By releasing women on licence before the expiry of their sentence, some measure of control was possible, in so far as any woman who showed signs of relapsing during this period could be recalled to complete the rest of her term. However, after the sentence had expired, continuing care was extremely expensive and its provision *ad hoc* and inadequate. Wide-ranging efforts were made by philanthropic bodies such as the Church Army, the Women's Total Abstinence Union and the Aftercare and Inebriates Association. But, once a woman had been discharged (that is, after the period of licence), reformatories were not empowered to finance further aid. Any further provision they made relied on government contributions. This legal obstacle was particularly ironic, given that it was generally

[109] T. Holmes, *Known to the Police* (1908), 28–31, 53–8.

agreed that after-care was crucial to the reformatory endeavour. It was vital to ensure that women were provided with surroundings and employment conducive to their continued abstinence. As the Superintendent of Farmfield argued: 'Without something of this kind being done our work is unfinished, as we leave these women just when they most need guidance.'[110]

- The LCC set up an After-care Subcommittee to investigate the provision needed. It appointed a 'travelling sister' to accompany women on their release, to find them jobs (usually in domestic service) or lodgings, and to visit them thereafter. Spending three or four days a week travelling around London, she tried to maintain regular contact with them and to ensure that temptations to drink were minimized. Where practical, women were diverted away from their former neighbourhood, from the pernicious influence of old drinking companions, and, in recognition of the potentially corrupting effects of inmate association, from one another. Following the example set by prisons, 'mark money' earned inside was not usually handed out in a lump sum but in small amounts via an intermediary body, usually one of the after-care agencies. This strategy overrode the temptation to spend it all on drink and obliged women to maintain contact. Where their control was most tenuous, then, reformatories wove a web of links and bonds by which to maintain some possibility of exercising their moralizing influence.

After-care was not limited only by the difficulties of funding and organization but by the unwillingness of many women to be 'aided'.[111] Once their period of licence had expired, or if they were not discharged until the end of their sentence, any further contact with the reformatory was entirely voluntary. Whilst some agreed to be placed under the aegis of one of the many after-care associations, many others clearly resented what they saw as an unwarranted extension of their punishment.

[110] GLRO LCC/MIN/8118. 'The After-care of Patients Discharged or on License and as to the Disposal of Mark Money', Reports of Lady Superintendent (4 Oct. 1901).

[111] Finding that women continually refused to disclose their addresses, the LCC decided to make its After-care Subcommittee responsible for paying out mark money, thereby forcing them to do so. GLRO LCC/MIN/8116, Minutes (2 Apr. 1912).

Their refusal to co-operate posed a particularly acute problem at Aylesbury. If at all amenable, Aylesbury women were transferred back to certified reformatories, later to be released on licence. Only the most refractory remained at Aylesbury until the end of their sentence, after which point the State had no more control over them unless they agreed to submit to the supervision of the Inebriates Reform and After-Care Association, which took on only those women who declared themselves desirous of aid. A Visiting Committee report for 1907–8 noted pessimistically: 'The Board has made it a rule to interview the women about to be discharged. Many have declined proffered help and they have left with the gratuity money in their hands ready . . . to succumb to the first temptation, with consequent prompt recommittal to prison.'[112]

Certain types of women were identified as particularly liable to return to drink, especially those with a family history of inebriety, brought up to alcoholism, or deemed mentally defective but not so insane as to secure a place in an asylum. In addition to these innate or acquired characteristics, those very environmental factors which had been so salient in causing female alcoholism now came back into play. The records of Farmfield's 'travelling sister' catalogue the plight of women, stigmatized by the very sentence that had been intended to reform, and released back into the miserable slum dwellings which had first driven them to drink. The older, more degraded women found it especially difficult to get honest work. Even those who were found places in domestic service were often so lonely or bored that they turned again to drink in desperation.

Once a woman lost her lodgings or job, her return to the streets, to prostitution, and to drunkenness seemed, to contemporaries, all but inevitable.[113] Women found drunk whilst still on licence were generally returned to the reformatory to complete their sentence. Those falling beyond their term of licence were simply committed to prison to begin once more the slow

[112] Minute Book, Aylesbury. Report of the Visiting Committee, Aylesbury, to the Home Secretary (1907–8), 101.

[113] See evidence given by witnesses to the Departmental C. on the Law relating to Inebriates, Replies to Questions (1908), and esp. that of J. T. T. Ramsay, Chairman of the Lancashire Inebriates Acts Board, and Frank Austin Gill, Medical Superintendent, Lancashire Certified Inebriate Reformatory, at 994–8 and 1040.

round of accruing sufficient proof of their habitual inebriety to
enter the reformatory process all over again. The likelihood of
relapse was considered so great that even early release on
licence was condemned as frustrating the very purposes of
reform.[114] Only whilst the women remained carefully cloistered
in the confines of the reformatory was there any real hope of
keeping them from drink. Once released the pressures and
temptations which faced them made their permanent reform
extremely unlikely. A follow-up study of Farmfield's first 600
cases revealed that at most only 19 per cent were 'doing
well'.[115] The remaining 80 per cent or so who fell back into
alcoholism constituted a staggering indictment of the reformatory experiment and gravely undermined any hope of its future
success.

Of the minority who permanently reformed a few kept in
contact with the reformatory. In return the Superintendent
wrote letters, sent presents and flowers, and even visited
women in their homes. Their own letters provide an alternative
view from that already gleaned from the punishment books and
grievance records. These few women had been effectively
socialized into respectable femininity, and were infused with
a sense of gratitude and a genuine affection for members of
staff who helped them to sobriety: 'They constantly write
especially if they are in trouble and want advice, express their
thanks for their sojourn there and are always pleased to speak
of the progress they are making.'[116] In trouble or lacking
employment they would return to the reformatory, seeking
refuge there as if by right. For example, Annie Perry, discharged on licence, found that the position secured for her
provided no proper sleeping accommodation. She reappeared
the very same day, explaining 'that in the circumstances she
considered she was entitled to return to Farmfield'.[117] Women
wrote complaining of unhappy situations in domestic service,
giving news of their families, and seeking information about

[114] See e.g. an article headed 'Frustrated Cures: How Inmates of Inebriate Homes
are Rashly Released', *Daily Chronicle* (Dec. 1906).
[115] Anon., 'The Treatment of Inebriates', *British Medical Journal* (1911), 962.
[116] GLRO LCC/MIN/8125. Report on the History and Making of Farmfield
(16 Dec. 1907).
[117] GLRO LCC/MIN/8116. Minutes of LCC Subcommittee (21 June 1913), 420.

former staff and inmates. Their letters indicate that these few at least had accepted and internalized their role as dependent daughters of the institution or, more specifically, of the Lady Superintendent, to whom they could continue to turn for comfort or direction.

Most of the letters are informal, even affectionate, in tone. Among them is this pathetically penitent letter sent to the Superintendent by a violent woman, transferred to Aylesbury as 'unmanageable': 'you see i have lost all through my temper i must consider myself dead to the world for two years and six months . . . and if i am for saken bye you there is no hopes for me at all . . . with fondest love dont forsake me all together please write to me.'[118] A few are unrestrainedly effusive declarations of love which are difficult to decode. A series of undated letters from former inmate Lizzie Wannell to the same Lady Superintendent, were it not for the fact that they were presented to the Inebriates Subcommittee presumably as evidence of the woman's successful reform, would seem to be an open expression of homosexual love. Beginning 'My darling Pretty Girl' they ramble in illiterate scrawl over several pages. Wannell demanded 'the next time you come to see me in my little humble home I shall want you all to myself and no one else shall have a bit of you . . . my darling Matron . . . Please dont forget to answer wish I was close to you.' Later letters go on to report her refusal to marry her suitor 'Jack'—'I wont tell him I am not tired off single life yet and sick of man'—and conclude dramatically, 'I dont know wich I love Best you owe Jack.'[119] However one reads these letters they indicate, at very least, the intense emotional bonds forged between women within the closed community of the reformatory. Released into a hostile, often lonely world outside, they clung to these remnants of affection. Their letters are indirect testimony of the poverty of their lives outside. More strikingly, the fact that so few women actively sought to maintain contact or looked back on the institution with any warmth draws attention to the fact that the rest presumably did not.

[118] GLRO LCC/MIN/8120. Letter from Annie Deacon (22 May 1903).
[119] GLRO LCC/MIN/8120. Undated batch of letters in LCC Subcommittee Papers (1903–4).

FROM REFORM TO CONTROL

The failure of reformatories to cure more than a small minority caused deep disillusionment amongst local authorities and philanthropic bodies.[120] Whilst the Home Office argued that they should take all classes of inebriate, the institutions used their powers under the 1898 Act to restrict access to those they deemed potentially reformable. They argued that certain groups of women, and especially 'the mentally weak', 'the hopelessly diseased', and 'the oldest and most depraved drunkards' were completely unsuited to reformatory treatment.[121] Further, their very presence could only hinder the progress of those who might otherwise benefit. The Home Office suggestion that different institutions take different classes of women, grouped 'according to conduct and prospect of reformation',[122] was limited partly by the paucity of reliable case history information. Also, unsurprisingly, few institutions were willing to volunteer to take only known irreformables. The LCC, in fact, made repeated bids to have Farmfield redesignated 'as a superior reformatory to which only selected cases shall be sent'.[123] They argued that if reformatories were obliged to take all classes of women, even the old and more hardened, then the scope of their exercise would necessarily be limited: 'the treatment must be regarded as almost entirely penal or as a matter of police administration with a view to clearing the streets of an objectionable and disorderly class.'[124] Such a role contradicted the 1898 Act which had designated the state reformatories as places of penal containment for incurables specifically so as to free the certified reformatories to concentrate on attempting reform. The Home Office, however, insisted that the purposes of deterrence and containment were

---

[120] For example, at St Joseph's the nuns found themselves totally incapable of dealing with the more turbulent cases sent to them. Having tried limiting admissions to supposedly amenable cases but with little success, they eventually decided to close down. Report of the Inspector under the Inebriates Acts for 1903, *PP*, 11 (1905), 7–8.

[121] GLRO LCC/MIN/8121. Report by LCC Clerk, G. L. Gomme, 'Classification of Inebriates' (25 July 1905).

[122] GLRO LCC/MIN/8115. Letter from the Secretary of State to the LCC (20 Oct. 1903).

[123] Ibid. LCC Subcommittee Minutes (25 July 1905); and Minutes (1903–6), *passim*.　　　　　　　　　　　　　　[124] Ibid. Minutes (1 May 1903), 83.

equally important, and that certified reformatories must share these less exalted and less rewarding duties with Aylesbury.

This insistence on the importance of 'cautious control' of all, over and above that of the moral reform of the few, subtly but profoundly redefined the purpose of reformatories. In 1906 Branthwaite revealingly admitted: 'It is proving extremely unfortunate that the word "Reformatory", was ever selected to apply to institutions intended for the reception of committed inebriates.'[125] This growing pessimism about the reformability of the female inebriate reflected a wider disillusionment with the reformatory movement. In the first years of the experiment many had argued that, although corrupt, female habitual inebriates were curable.[126] Subsequent experience, however, severely eroded this early faith in moral reform. Whilst early reports had stressed the 'moral degradation' and 'shame' of women committed, a few years later this vocabulary had subtly but markedly altered: the tone of moral condemnation had been gradually replaced by pseudo-psychiatric discourse. Female inebriates were increasingly described as 'ill-balanced', 'border-land cases', 'neurotic', 'semi-lunatic', or, most commonly of all, as 'feeble-minded'. In 1903 Branthwaite had remarked of women at Aylesbury: 'Our greatest difficulty is to discriminate between insubordination which is due to madness and that which results from pure badness.'[127] Yet only two years later he appears far more certain of the distinction: 'There is hardly a woman at Aylesbury who is drunkard from pure vicious indulgence, practically all of them are so because of their low standard of mental development.'[128]

This shift from a primarily moral to a psychiatric conception of inebriety was in large part due to a wider recognition that sections of the population, though not certifiably insane, were

[125] Report of the Inspector under the Inebriate Acts for 1906, *PP*, 35 (1907), 603.

[126] For early assurances that women were indeed reformable see, for example, the editorial 'Inebriety in Women', *British Journal of Inebriety* (July 1903), 2–3; also early Inspector's Reports, especially that for the year 1903, appendix N, 'Concerning the Curability of Women', in which Branthwaite repudiated the common view of the female inebriate as 'an impression which is not only unjust and cruel, but entirely erroneous', at 145.

[127] Report on State Inebriate Reformatories for year ending March 1903, *PP*, 35 (1904), 69.

[128] Report of the Inspector under the Inebriates Acts for year ending March 1905, *PP*, 16 (1906), 64.

to some degree mentally weak or incapable (see Chapter 7 below). As such they could not be deemed responsible in the same way as other delinquents; for them 'the word morality has no meaning'.[129] Whereas, formerly, the 'immorality' of female inebriates had been deplored, many were now seen to be mentally incapable of supporting themselves by any other means than prostitution. Similarly whereas feeble-mindedness had earlier been designated an attribute or result of prolonged inebriety, increasingly the debate centred on whether it was not in fact the very cause.

The argument was applied with particular force to women, who were considered to be constitutionally more prone to mental disease, to hysteria, and to neuroses. Moreover, the finding that women admitted to reformatories were not so physically ravaged by alcoholism as had first been expected seemed to suggest that they got drunk on very small quantities of alcohol. This only confirmed suspicions of prior mental weakness. The growing redefinition of inebriety as a product of feeble-mindedness was found particularly pertinent to those confined at Aylesbury, of whom not more than 2 or 3 per cent were designated as being 'of absolutely sound mental capacity'.[130]

The Departmental Committee on the Law relating to Inebriates (1908) accepted that female inebriety was 'undoubtedly a constitutional peculiarity' and even that many inebriates were 'mentally defective'. However, its members continued to favour treatment by 'cultivating self-control'. And though they baulked over the label 'disease' because it tended to exonerate, they later argued that inebriates should be treated less punitively on the basis that they were indeed less responsible.[131] These apparent inconsistencies not only reflected ambivalence in attitudes towards the female drunk but also irresolvable contradictions within this medicalized view of deviance.

Above all, this redefinition was disquieting in that it left no possible role for 'reformatories' other than that of humane containment. So long as drunks were regarded as diseased they

[129] Ibid. 12.

[130] Statement by W. H. Winder, Superintendent, Aylesbury, 'Regarding State Inebriate Reformatories and the Inmates Detained Therein', Royal Commission on the Feeble-Minded, Minutes of Evidence, *PP*, 38 (1908), 251.

[131] Departmental C. on the Law relating to Inebriates, (1908), 825, 835, 836.

could conceivably be 'cured', or if they were moral degenerates they could be reformed. However, if they were recognized to be feeble-minded there was little hope for any improvement. Certified reformatories were loath to accept this purely supervisory role and the LCC and the Lancashire Inebriates Acts Board, in particular, were insistent that the long-term care of mentally defective inebriates must be a state responsibility. The two major commissions of 1908, the Departmental Committee on Inebriates and the Royal Commission on the Feeble-Minded, however, moved in the opposite direction, recommending that the Inebriates Acts should be extended to oblige all reformatories to deal with all classes of inebriate, including the mentally defective. Confirming this directive, Branthwaite announced in the following year: 'The nearer an Inebriate Reformatory resembles a mental hospital in all its arrangements, the better will be its suitability for the work it has to do.' [132] The recognition that a large proportion of inebriates were irrevocably feeble-minded undermined the very *raison d'être* of the inebriate reformatory and created widespread disillusionment. The LCC reports on Farmfield became markedly more pessimistic about the possibility of reforming more than a tiny proportion of the women held there. At a time of more general public reaction against incarceration, the LCC questioned whether the huge costs of containment could possibly be justified. [133]

Relations between the LCC and the Home Office had been poor since the very beginning of the experiment and they now deteriorated badly. In 1911 the LCC mounted a nationwide campaign demanding amendment of the law. Chief among their demands were that the government make permanent provision for the feeble-minded and that it bear the whole cost of accommodation and maintenance of all inebriates committed by the courts. [134] The repeated failure of the government to pass long-promised legislation on inebriety prompted threats by the LCC not only to close Farmfield but to end contracts with

[132] Report of the Inspector under the Inebriates Acts for 1909, *PP*, 29 (1911), Pt. 1, 62.

[133] See esp. GLRO PC/GEN/1/7, Reports of the Public Control Committee (1910–13).

[134] GLRO LCC/MIN/8126. 'Inebriates Acts—Proposed Amendment of Law', Report by the Clerk to the Inebriates Acts Subcommittee and to the Public Control Committee (17 Feb. 1912).

numerous other reformatories to take cases from London police courts. The failure of parliament to pass a new Inebriates Bill before the end of the 1914 session was, it seems, the breaking point. On 11 July 1914 the LCC stopped administering the 1898 Act. Without their co-operation none of the cases committed in London, a large proportion of the total, could be dealt with under the Act and it simply became unenforceable. By 1921 the inebriate reformatories had all closed down.

The reformatory experiment failed because it had envisaged female subjects who were quite different from those actually committed by the courts. Having defined the female inebriate primarily as a moral offender, reformatories operated on the premise that, by providing a sufficiently propitious environment and benign moral influences, her cure could be achieved. Finding, instead, women resiliently resistant to the intentions of the regime, or so feeble-minded as to be irredeemable, the very momentum of the endeavour collapsed.

7 NOT BAD BUT MAD: THE DISCOVERY OF 'FEEBLE-MINDEDNESS'

This book has traced a gradual, at times uneven, transition in explanations of and responses to crime. Beginning with the mid-nineteenth century, it opened with moralistic understandings of crime and quasi-religious attempts to reclaim the offender. It concludes with the rise of secular, 'scientific' explanations of deviance, which, in their turn, led to medically orientated attempts to replace punishment with treatment and containment. At the close of the nineteenth century, 'feeble-mindedness' came to be seen as a major social problem, a source of social dangerousness, and the causal link in a wide range of deviant behaviour. This chapter examines the campaign for and enactment of legislation which introduced policies intended to replace penal responses with provisions whose primary purpose was eugenic.[1] Strikingly, as in the case of habitual inebriety, the majority of those labelled as feeble-minded were women. We will examine why psychiatric diagnoses were found to be particularly plausible in explaining female deviance and explore why the consequences of mental deficiency in women were seen to be infinitely greater than in men.

As we observed at the beginning of the previous chapter, alongside drunks, the feeble-minded were considered to constitute the core of prison failures. Those designated as feeble-minded were that dangerous subsection of the mentally deficient who were not so impaired as to show external signs of deficiency but who were, nonetheless, incapable of surviving in normal society. These confirmed recidivists were consequently redefined as helpless defectives rather than as deliberate sinners.

The 'discovery' of feeble-mindedness largely explained the apparent failure of the prison and the reformatory to control their most recalcitrant inmates. It rediagnosed them as blighted by a mental incapacity for making correct moral judgements

---

[1] Or, more properly, accorded with 'negative eugenics' in that they sought to limit damage done to the quality of the nation's stock by breeding amongst the weak and defective.

and, therefore, as incurable inadequates. In turn, this diagnosis removed those thus labelled from the sphere of criminality by redefining them as irresponsible. Punishment and reform could then be deemed not to have failed but merely to be inappropriate to their condition.[2] Their reform within the penal system was abandoned in favour of life-long care which, in many cases, entailed permanent segregation in specialist institutions.

If the discovery of 'feeble-mindedness' tended to exonerate the prison, it did so at the cost of amplifying a myriad of other contemporary anxieties. Pervasive concern with mental heredity became intricately linked with ascendant theories of Social Darwinism. And these crystallized long-standing concerns about urban degeneration in the formal development of the Eugenics Movement. The Eugenicists feared that the rising numbers of urban slum dwellers were breeding generations of physically enfeebled and, above all, degenerate mental defectives.[3] Their fears struck a cord with those increasingly concerned about Britain's 'great power' imperial status. By the end of the nineteenth century, Victorian optimism had been shaken by internal economic crisis and the rise of foreign competition—both commercial and, potentially, military. 'Hereditarianism', with its concomitant theories of degeneration, provided ready explanations for the relative decline of Britain's international status. Similarly, domestic social problems such as unemployment and poverty (compounded by crime and alcoholism) could be understood as the product of a general decline in the national character. The late Victorian preoccupation with the perceived degeneration of urban slum dwellers focused on the feeble-minded as the potential cause of immeasurable damage to the mental, moral, and physical health of the race.

Historians, anxious to assert medicine's increasing influence in informing and directing social policy, have often failed to note the influential role of biological science (and particularly hereditarian theory) in both medical and wider social discourse. Biology, medicine, and social theory combined to designate the apparently defective and the supposedly atavistic

[2] A similar argument is advanced in J. F. Saunders, 'Institutionalised Offenders', Ph.D. thesis, University of Warwick (1983), 279.

[3] For further discussion of middle-class perceptions of slum dwellers, see G. Stedman Jones, *Outcast London* (1971).

as central concerns of late Victorian pessimism.[4] Nor did the intrusion of medicine entirely supersede traditional forms of explanation. Degenerates continued to be defined in moral as well as in medical terms. Nowhere was this combination of the moral with the medical more potent than in the identification of certain groups of women as not only degenerate but also 'feeble-minded'. Feeble-minded women were thought to be innately promiscuous and, despite their weak constitutions, highly fertile. If allowed to continue to breed unchecked, the mentally and physically unfit would increasingly outnumber and eventually overwhelm the fit.[5] The physical threat they posed was greatly exacerbated by their apparent 'moral insensibility'.[6] As the social historian Harvey G. Simmons has argued:

the problem of mentally deficient women was . . . complex and has serious ramifications. Not only were they seen as the biological source of mental deficiency, they also posed a deep threat to existing middle-class and respectable working-class notions of sexuality and familial normality. This explains the near hysteria which characterizes discussions about the social problem of mentally defective women.[7]

Indeed, the tone of much Eugenicist literature was highly sensationalist in its anticipation of the terrible damage that would result from inaction. It gave dire warnings of the need to curb these women.[8] Images of vast numbers of degenerates multiplying in city slums not only captured the public imagination but provided a vehicle for the expression of long-standing fears of the crime, violence, and alcoholism seen as consequent on urban migration. The heyday of this 'negative eugenics' coincided with the rise of late nineteenth-century criminology. In their language, ideas, and target populations (that is, habitual criminals, inebriates, and the mentally defective), negative

[4] For England see R. Smith, *Trial By Medicine* (1981); D. Garland, *Punishment and Welfare* (1985); M. Freeden, *The New Liberalism* (1978). For parallel discussion of France during the same period see R. Nye, *Crime, Madness and Politics in Modern France* (1984); I. Dowbiggin, 'Degeneration and Hereditarianism in French Mental Medicine 1840–90', in W. F. Bynum *et al.* (eds.), *The Anatomy of Madness* (1985), vol. i.

[5] Typical is A. F. Tredgold, 'The Feeble-Minded: A Social Danger', *Eugenics Review* (July 1909), 97–104.   [6] See Ch. 1 for the history of this concept.

[7] H. G. Simmons, 'Explaining Social Policy: The English Mental Deficiency Act of 1913', *Journal of Social History*, 11:3 (1978), 394.

[8] See, for example, works by R. R. Rentoul, *Proposed Sterilisation of certain Mental and Physical Degenerates* (1903) and *Race Culture; or Race Suicide?* (1906).

eugenics and criminology overlapped. Moreover, the two programmes mirrored one another in their proposals, they enjoyed wide cross-party political support, and several leading proponents figured in both movements. The result was two-fold: that deficiency was increasingly seen to be a crime and, conversely, that much habitual criminality came to be considered as the product of inherited mental defect. Henry Maudsley, for example, argued in his early works: 'this hereditary crime is a disorder of mind, having close relations of nature and descent to epilepsy, dipsomania, insanity, and other forms of degeneracy.'[9] Over the latter part of the nineteenth century these views gained considerable currency, so that it became almost a truism that, as unfit bred unfit, habitual criminals bred habitual criminals. Just how far this unquestioning faith in hereditarian explanation was a passing vogue becomes evident if one examines criminological literature at the very end of the period. A marked scepticism is discernible, for example, in Dr James Devon's critique of hereditary explanations of crime, in which he argued that the reasoning involved was all too often spurious since in many cases a criminal's parentage was impossible to ascertain.[10]

Late nineteenth-century Eugenicists addressed the problem of degeneracy in its widest possible impact. The Edwardians, disillusioned with the preoccupation with physical degeneracy so popular in Continental Europe, focused instead on mental defect as the root cause of a whole range of pathological conditions.[11] Mental degeneracy, they argued, could be observed in a significant proportion of the casual poor, slum dwellers, and the prison population. Moreover, it provided quasi-scientific endorsement to very widespread concerns about the 'wastrel' and the 'loafer'.[12] If such groups could be seen to be suffering from hereditary disorders, control could be justified to prevent transmission from one generation to the next in the interests of the 'efficient' society.

[9] H. Maudsley, 'Lecture 2—On Certain Forms of Degeneracy of Mind', in his *Body and Mind* (1870), 66.

[10] J. Devon, *The Criminal and the Community* (1912), 18 ff. The writings of Henry Maudsley also become markedly more sceptical than his earlier work discussed above.

[11] G. R. Searle, *Eugenics and Politics in Britain 1900–1914* (1976), 30–2; Simmons, 'Explaining Social Policy', 391.

[12] See J. Harris, *Unemployment and Social Policy* (1972), especially ch. 5.

The growing currency of feeble-mindedness as a social problem relied on two main factors: first, interest was highly politically motivated, and second, it was a focal point for eugenicism, psychiatry, and criminology (all dominant influences on turn-of-the-century social policy). It thus gained an importance that endorsement by any one of these movements alone could not have achieved. Harvey G. Simmons clearly indicates its stature, though, perhaps, overstates the case: 'Gradually, the realization began to dawn that medical science might have discovered in the phenomenon of feeble-mindedness an explanation for some of the major social problems of the time.'[13] A more sceptical interpretation might suggest that the considerable currency attained by the concept relied less on its inherent credibility as the sole solution to contemporary social problems, and much more on its general compatibility with the dominant strains of contemporary social analysis.

BACKGROUND

It has been argued that one of the few times the Eugenics Movement occupied a national political stage was in its promotion of legislation for the feeble-minded.[14] To understand why this campaign, above all others, should have 'exploded into the public arena',[15] we need briefly to examine the development of concern about female insanity.[16]

Significantly, an increasing proportion—and by the second half of the century the majority—of certified lunatics were women. By 1871, women made up well over half the pauper lunatics in England, and by the end of the century they predominated in all types of lunatic asylum, public and private.[17] One of the few historians to examine the preponderance of women in asylums is Elaine Showalter.[18] She examines this increasing imbalance in the sex ratio and, failing to discover

[13] Simmons, 'Explaining Social Policy', 390–1.
[14] Freeden, *New Liberalism*, 190.   [15] Ibid.
[16] For more general discussion of psychiatric interpretations of female deviance, see Ch. 2 above.
[17] Figures given in E. Showalter, 'Victorian Women and Insanity', in A. Scull (ed.), *Madhouses, Mad-doctors and Madmen* (1981), 315.
[18] See her article, ibid., and more importantly *The Female Malady* (1987).

any medical explanation, suggests various reasons why women should have been more readily perceived, certified, and committed as insane. Whereas doctors were generally reluctant to certify a man as insane, especially if he had a family to support, Showalter argues that committing a woman provoked far less concern. She even suggests that a spell in an asylum was considered an acceptable means of quelling a rebellious, aggressive, or sexually active daughter or wife.[19] Similar attitudes applied to sending women to inebriate reformatories, but Showalter's further claim that the asylum was used against women as 'an deserves efficient agency of socio-sexual control' to be more fully explored.

It is worth noting both that poverty was seen as a major moral cause of insanity and that women were the major recipients of poor relief. Given that they were already living 'on the State' and hence were already designated as social problems, they were more likely to be committed to institutions than those who were not. This view is confirmed by Andrew Scull's conclusion that the Victorian asylum was primarily a means to the 'custodial warehousing of . . . the most difficult and problematic elements of the disreputable poor'.[20] Other reasons for the higher numbers of women in asylums were that women patients tended to outlive men, and that they seem to have been less likely to be discharged as 'cured'. Moreover, the common opinion among doctors that madness was, indeed, primarily a 'feminine malady' encouraged them to incarcerate marginal cases among women more readily than among men.[21]

The assumption that mentally defective women were more likely to 'need' incarcerating than mentally defective men gradually became reflected in institutional provision. In 1859 the Commissioners of Lunacy required that small asylums be limited to one sex. Given the preponderance of potential female clients, more asylums were subsequently designated for women than men. This created more space for them and so tapped the large numbers of women hitherto cared for in the home but considered, to varying degrees, to be mentally deficient. Even in larger asylums, with strictly segregated facilities for both sexes, the greater part was generally given over to women,

[19] Showalter, 'Victorian Women and Insanity', 318.
[20] A. Scull, *Museums of Madness* (1979), 219.   [21] Showalter, *Female Malady*, 55.

again facilitating their committal. For social and institutional, as much as for medical, reasons women came to form the majority of those labelled and incarcerated as insane. Whilst mentally deficient women were rather more likely, therefore, to find themselves incarcerated in lunatic asylums, those who remained outside were also regarded as a greater problem than men. In the latter decades of the century, mentally deficient women still at large, or held in inappropriate institutions, occupied increasing public and official attention.

In addition, certain Victorian psychiatric constructs were found to apply particularly to women. For example, the condition known as 'moral insanity' (now more commonly referred to as 'psychopathy') equated mental health with virtue. The sane person could be expected to make virtuous moral choices, and be punished if he or she did not do so. The 'morally insane' showed themselves incapable of making and still less of abiding by these choices. Though they might show no other sign of mental illness or defect, the 'morally insane' were identifiable by the very fact of their persistently anti-social behaviour. In practice 'moral insanity', without ever being clearly defined, was widely used to denote a perceived moral incapacity— whether manifested merely as impropriety or as outright perversion.[22] And it was in its application to the area of sexual non-conformity, especially in women, that the term gained its greatest credibility.

So long as the boundaries between mental illness and immorality were not clearly drawn, and so long as the definition of medical categories such as moral insanity remained based largely on moral assessments of deviance, psychiatric discourse was strongly resisted by the legal profession. According to the psychiatrists, the extent of a woman's derangement determined her degree of responsibility; if she was diagnosed as morally insane legal notions of responsibility must be seen to be confounded. However, the courts recognized that to accept the designation moral insanity would be to confound all distinction between badness and madness.[23] To redesignate habitual

[22] For more detailed discussion of the development of this concept, see N. Walker and S. McCabe, *Crime and Insanity in England*, ii (1973), ch. 9, 'From Moral Insanity to Psychopathy'.

[23] For discussion of their resistance see, for example, H. Maudsley, *Responsibility in Mental Disease* (1874), 171–3.

criminals as true moral imbeciles rendered legal jurisdiction and penal response entirely inappropriate. It would become necessary instead to replace them with medical diagnosis and treatment. Diagnoses such as moral insanity were, therefore, highly contentious. Yet since women were seen as innately less responsible, and naturally predisposed to mental illness, they were rather more readily fitted into such definitions than men.

THE DISCOVERY OF FEEBLE-MINDEDNESS

The exact origins of the term feeble-mindedness are difficult to determine with any certainty. Simmons asserts that the term was first used by Sir Charles Trevelyan (member of the council of the Charity Organisation Society) in 1876 in a motion proposing that feeble-minded children ought not to be associated with adult idiots.[24] Certainly it does not seem to have been in general usage much before the last quarter of the nineteenth century (previously the condition would have been commonly referred to as 'improvable idiocy'). It had no precise medical definition and its exact meaning continued to be debated right up to the passing of the 1913 Mental Deficiency Act. Generally the term seems to have been applied to that anomalous and previously undesignated grey area between normal intellectual capacity and indisputable idiocy. It identified a population who had never been capable of complete independence but who, given appropriate care and treatment, were thought capable of improvement sufficient to be able to care for themselves. The degree to which improvement was thought possible depended largely on the interpretation given to the term. It could be seen to describe a somewhat blighted mental state in people who were otherwise in every way normal.[25] Or it could represent only the most clearly visible characteristic of an entirely distinct group whose very constitutions were pathological.[26] The very vagueness of the term and, up to 1913, the almost deliberate failure to debate its precise definition, allowed a whole array of

[24] Simmons, 'Explaining Social Policy', 388.
[25] For example, W. A. Coote, 'The Feeble-minded and Rescue Work', in the 'Mental Deficiency' section, *Report of the National Conference on Destitution 1911*, 607–9.
[26] As Dr Tredgold argued in his article, 'The Feeble-Minded', *Contemporary Review* (June 1910), 718.

political, social, and medical commentators to employ the single term to convey a variety of differing, even contradictory, meanings.[27] Any precise definition given to feeble-mindedness was undoubtedly less important than its elastic capacity for embracing a whole array of existing social problems. It re-designated them as biological in origin and therefore demanded medical intervention.

Feeble-mindedness was taken as direct evidence of the pro-liferation of the unfit and family histories were traced back to reveal generations of feeble-minded parents producing more and more defective children. It was widely asserted that feeble-mindedness was untreatable and that such parents were in-capable of producing anything other than defective offspring.[28] It was taken up by a variety of social policy bodies (including the Charity Organisation Society, the London School Board, and the National Vigilance Society), whose respective concerns included the problem of pauperism, mental defect in school children, and the vulnerability of defective girls and women outside institutional care. Despite such wide-ranging concern it took twenty years from the Charity Organisation Society's original debate on the problem of feeble-mindedness[29] for the National Association for Promoting the Welfare of the Feeble-Minded to be set up.

Two women, Miss Mary Dendy and Mrs Hume Pinsent, both long concerned with educational provision for defective children, and latterly with the problem of 'wayward', feeble-minded young women, were pre-eminent in publicizing, organizing, and generally campaigning for recognition of the problem.[30] Their mutual preoccupation with the problem of

[27] For discussion of the various definitions proffered see Simmons, 'Explaining Social Policy', 391–3. Also L. Radzinowicz and R. Hood, *A History of English Criminal Law* (1986), v. 326–7. For discussion of the way in which social problems are generated and then amplified into major policy issues, see J. Goldthorpe, 'The Development of Social Policy in England 1800–1914', *Transactions of the Fifth World Congress of Sociology* (1964), iv.

[28] For example, Tredgold, 'The Feeble-Minded' (1910), 718–19; Anon, 'The Problem of the Feeble-Minded', *British Medical Journal* (Apr. 1910), 1070. See also K. Jones, *Mental Health and Social Policy 1845–1959* (1960), 49.

[29] In a special subcommittee of the society on the need for provision for improvable idiots, 1876–7. Details in Jones, *Mental Health*, 45.

[30] Miss Dendy was a member of the Manchester School Board; she had set up the Lancashire and Cheshire Society for the Permanent Care of the Feeble-minded and founded the Colony at Sandlebridge. Mrs Pinsent had served on the Birmingham

feeble-minded women and girls was widely influential. Indeed, Simmons argues that concern about feeble-minded women was the main reason for the elevation of feeble-mindedness into what he terms a 'major social problem'.[31] We need, therefore, to establish why it was that feeble-mindedness in women was considered to have more serious social implications.

A Royal Commission set up in 1904 to investigate the extent, nature, and costs of feeble-mindedness seemed to take for granted that not only were feeble-minded women the greater problem but that they were more likely to suffer exploitation themselves. It concluded that 'the evil plight of feeble-minded women naturally excites the most general commiseration.'[32] And this was so, even though there was little evidence that such women outnumbered men (on the contrary, Dr A. F. Tredgold quotes a sample of 4,291 feeble-minded persons surveyed by the Royal Commission of whom 2,179 were men and 2,112 women).[33] This assumption was widely reported, with the result that most press coverage focused on women. For example, in 1909 *The Englishwoman* ran an influential series under the general heading, 'Women and the Nation: The Outcasts'.[34]

This concentration on feeble-minded women can, in part, be attributed to the campaigning efforts of Dendy and Pinsent. In addition, large numbers of middle-class women increasingly felt that tackling such problems fell within their rightful sphere of influence. *The Englishwoman* impressed on its readers that it was their duty to concern themselves with the plight of their less fortunate or 'outcast' sisters.[35] Growing numbers of women were interested in penal reform and administration, 'the drink question', and the setting up of inebriate reformatories. But it was in relation to the feeble-minded that they were

Special Schools Subcommittee and had instituted the setting up of an After-care Committee (to supervise children after they had left special schools) and a Girl's Night Shelter. On the basis of her experience she went on to campaign for state intervention on behalf of the feeble-minded.

[31] Simmons, 'Explaining Social Policy', 388.
[32] Report of the Royal Commission on the Care and Control of the Feeble-Minded, *PP*, 39 (1908), 308.   [33] A. F. Tredgold, *Mental Deficiency* (1908), 148.
[34] See 'Women and the Nation', especially 'Introductory' by J. Haslam and 'The Feeble-Minded' by M. Meredith, *The Englishwoman* (May 1909), 406–22.
[35] Ibid. 406–9; see also M. H. Steer, 'Rescue Work by Women among Women' in Baroness Burdett-Coutts (ed.), *Woman's Mission: A Series of Congress Papers on the Philanthropic Work of Women* (1893).

most influential. At the 1911 Conference on the Prevention of Destitution, W. H. Dickinson, MP, noted: 'With a few exceptions the pioneers of this movement have been women guardians and women educationalists.'[36] Their observation of feeble-minded women in workhouses, lying-in hospitals, and prisons spurred them to action and so ensured that wider perceptions of the problem focused on women.

THE COSTS OF FEEBLE-MINDEDNESS IN WOMEN

One cannot, however, attribute the predominant concern with feeble-minded women to the fact of philanthropic women's activism alone. In order to explain this, one needs to examine how feeble-mindedness in women was defined and perceived, and its social costs assessed. For the very definition of feeble-mindedness in women was informed as much by moral as by physical considerations. Assessments of a woman's moral conduct, her apparent degree of self-control—or lack of it—were taken as evidence of her mental state. For example, Mary Dendy argued that a pauper woman becoming pregnant with an illegitimate child should be regarded as having shown herself to be feeble-minded by the very irresponsibility of her actions. Significantly, it was not only the woman's perceived impropriety which defined her as feeble-minded but also her poverty that rendered her a legitimate object of public speculation and judgement.

That the definition of feeble-mindedness in women was as much morally as clinically determined became clear in debates leading up to the long-delayed enactment of legislation for the feeble-minded in 1913. During the course of these debates, most of which related to women, an amendment was proposed to extend the legal definition of feeble-mindedness to include 'sexually Feeble-minded persons, that is to say, female persons who do not belong to any class of the above mentioned classes, of mentally defective persons, but who are feeble-minded and on account of their mental condition fail to exercise due self-control or due self-protection with respect to sexual

---

[36] W. H. Dickinson, 'Homes and Colonies for the Feeble-minded', a paper given at the National Conference on the Prevention of Destitution (1911), *Report*, 638.

immorality'.[37] This amendment demanded that deviation from established notions of sexual propriety alone defined a woman as mentally deficient, without further evidence of any mental defect. Feeble-mindedness, in women, became a description that could be applied not only on the basis of psychiatric judgement but on that of morally determined criteria. This use of apparently objective, scientific labels obscured just how far highly subjective assessments of character and behaviour persisted. And yet if the basis of the assessment was primarily judgemental, its effect tended to be to exculpate rather than to condemn. Women who were promiscuous or who became pregnant outside marriage could, according to the amendment, be reassessed as not deliberately sinful but merely as sick victims of their defective constitution. Although this amendment was eventually defeated, the fact that it was seriously debated indicates how far the definition of feeble-mindedness in women was informed by other considerations than the findings of psychiatric diagnosis. In the Act as it was passed, these can be seen, for example, in the clause suggested by Dendy by which any woman pregnant with, or giving birth to, an illegitimate child whilst on poor relief was made liable to compulsory segregation in an asylum as mentally defective. There need be no proof of her supposed feeble-mindedness other than her poverty and her misfortune in becoming pregnant outside marriage.

By far the largest single group of feeble-minded women were assumed to have become prostitutes, not because they were deliberately sinful but because, as Tredgold put it: 'many of these women, sometimes even mere girls, are possessed of such erotic tendencies that nothing short of lock and key will keep them off the streets.'[38] Consequently they were seen to corrupt and degrade both themselves and those who were quick to take advantage of their 'facile disposition'.[39] This tendency to confound mental deficiency with sexual immorality in women greatly exacerbated its potential for damage. Many articles in the *Eugenics Review* and the *British Medical Journal* focused on the

---

[37] Mr Chancellor MP, speaking on 28 July 1913, Hansard, 5th ser. (1913), vol. 56, cols. 147–8.

[38] Tredgold, 'The Feeble-Minded' (1910), 720; see also Rentoul, *Proposed Sterilisation*, 17.   [39] Tredgold, *Mental Deficiency*, 292.

fertility of feeble-minded women. Interpretations ranged between ostensibly scientific analyses of relative rates of fecundity
in normal and in feeble-minded women to value-laden assessments of the greater sexual activity of the latter group.

More commonly, writers confused the scientific with the
moralistic in outraged revelations of the much higher rates of
fertility among the feeble-minded. Invariably writers provided
figures (often from the most dubious sources) to support their
shocking claims that feeble-minded women were both highly
sexually active and, given their inability to employ those crude
methods of birth control available, 'apt to be prolific'.[40] Most
commonly quoted were surveys such as those by Dr Ettie
Sawyer or Dr Tredgold, showing a higher rate of reproduction
among such women. Sawyer conducted research in London
which showed that whilst the average normal family produced
5 offspring, the feeble-minded produced 7.6. More strikingly
still, Tredgold arrived at figures of an average of 4 children for
the normal family, but 8.4 (including the stillborn) for the
feeble-minded family.[41]

Comparisons of relative birth-rates were used to substantiate
a highly dramatic vision of hoards of feeble-minded women
producing multitudes of children, many mentally defective,
often to a far greater degree than themselves. Medical discourse and moralistic judgement intermingled to make confused and completely unsubstantiated claims. For example, Dr
Clouston (Physician Superintendent of the Royal Edinburgh
Asylum), in a discussion on feeble-minded paupers, asserted
that 'when illegitimate children are born by such young
women, the chances are enormously in favour of their turning
out to be either imbeciles, or degenerates, or criminals.'[42]
Significantly, he did not feel it in any way necessary to clarify
why the 'illegitimate' offspring of the feeble-minded should be
more susceptible to their mother's mental condition than those
born within marriage.

[40] Quote from Anon., 'Problem of the Feeble-minded', 1070. See also W. A. Potts,
'Causation of Mental Defect in Children', *British Medical Journal* (Oct. 1905), 946–7
and Tredgold, 'The Feeble-minded' (1910), 717–22.

[41] Here quoted by A. H. P. Kirby, 'The Feeble-minded and the Voluntary Effort',
Eugenics Review (July 1909), 87.

[42] In evidence to the RC on the Feeble-minded, Report, Pt. 4, 'Mental Defect and
Crime', *PP*, 39 (1908), 120.

The poor genetic make-up inherited by these children was seen to be exacerbated by the feeble-minded mother's inability to instruct, moralize or chastise. It was feared that they were liable to grow up completely out of control, lacking innate self-discipline and any knowledge, let alone understanding, of the moral codes to which they were expected to conform. Unrestrained by maternal strictures, these children were thought to drift uncontrolled into idleness, misery, or vice.[43] Often such women simply lacked the basic qualities necessary to bring their children up with any humanity. Many women charged with neglect or cruelty were subsequently found to be 'mentally incapable of taking care of them'.[44] Whilst some women thus charged were, on assessment, judged to be so defective as to be certifiable, far more were simply deemed feeble-minded. They abused their children, not out of malice but because they lacked the wit to realize the cruelty of their actions. The social costs of feeble-mindedness in women, from their supposed fecundity, through their incapacity as mothers, to neglect or even cruelty to children, were considerable. And if diagnosis of feeble-mindedness tended to exculpate, it did not lessen the social misery caused.

There was much concern, even sympathy, for the plight of the women themselves. Their lives were often wretched, existing in poverty and filth on the margins of society. They were easy prey, incapable of escaping the attentions of men who sought to exploit their weak will. Their vulnerability was of particular concern to the growing numbers of social feminists. Ursula Roberts, for example, commented: 'Feeble-minded girls are peculiarly liable to be seduced. In some cases there is a positive tendency to immorality, in more it is a negative quality that ruins them—they are too feeble to make any resistance to the demands of unprincipled men. Once they are seduced, the downward path is easy.'[45] Much of the evidence given to the Royal Commission on the Feeble-minded (1908) emphasized the need to extend control over feeble-minded women to save them from sexual violation. The dangers of attack, and the

---

[43] See National Association for Promoting the Welfare of the Feeble-minded, *First Annual Report* (1897), 6.   [44] Devon, *Criminal and the Community*, 56–8.

[45] U. Roberts, *The Cause of Purity and Women's Suffrage*, Church League of Women's Suffrage Pamphlet 5 (no date, *c*.1911–12), 5–6. See also similar arguments in Meredith, 'The Feeble-minded', 421.

need therefore to protect such women, were invoked to justify incarceration. Similarly, in parliamentary debates, their evident vulnerability combined with concern over the consequent dangers of prostitution, the spread of 'loathsome diseases', and their own repeated pregnancies fuelled demands for intervention. For example, Dr Chapple, MP, argued that: 'To obtain control early of such a woman is humanitarian, and it is the bounden duty of the State to mitigate the sufferings of such women by getting them early under proper control. They are not in prison when under control.'[46] Left unprotected, such women all too often sank into extreme poverty from which they were quite incapable of raising themselves. Note, for example, Thomas Holmes's horrified description of female vagrants outside a London workhouse: 'What do we see? Squalor, vice, misery, dementia, feeble minds and feeble bodies. Old women on the verge of the grave eating scraps of food gathered from the City dustbins. Dirty and repulsive food, dirty and repulsive women!'[47]

DISILLUSIONMENT WITH PENAL AND REFORMATORY REGIMES

Clearly the very incapacity of feeble-minded women to live up to social norms of propriety and decency shocked those who ventured into their sphere. But of more pressing concern, because they came to the immediate and unavoidable notice of those in positions of authority, were those feeble-minded women who repeatedly reappeared in reformatories, rescue homes, workhouses, and prisons.

By far the highest proportion of feeble-minded women in any inmate population was to be found in inebriate reformatories. Whilst the majority of habitual inebriates were found, whether as cause or product of their addiction, to be below average in mental capacity, the most chronic and incurable drunks were almost all feeble-minded.[48] The records of inebriate reformatories examined in the previous chapter reveal how their

---

[46] Dr Chapple, 17 May 1912, Hansard, 5th ser. (1912), vol. 38, col. 1516.
[47] T. Holmes, *London's Underworld* (1912), 57.
[48] Kirby, 'The Feeble-minded and the Voluntary Effort', 90.

managers were persistently challenged by the disciplinary problems provoked by these inmates. Because they were, quite simply, incapable of understanding the strategies designed to cure them of alcoholism, they remained immune to reform. Finding the very purpose of their institutions perverted by the more pressing need simply to maintain order, reformatory managers became preoccupied with seeking means to segregate or remove such 'incorrigibles'.

Given their vulnerability and incapacity for self-protection, it was not found surprising that a very high proportion of feeble-minded women also turned up in Magdalen and Rescue Homes, often seeking shelter to give birth to their illegitimate babies. And this may, in part, explain why observers mistakenly tended to confound illegitimate pregnancies with feeble-mindedness. For example, in Birmingham, Dr Potts found that in the city's Magdalen and Rescue Homes 37 per cent of the inmates were feeble-minded. Nationwide research into all the larger Magdalen homes in England elicited 100 responses, revealing that of 14,725 women accommodated over a three-year period, 2,521 were classed as feeble-minded, and that these women alone were known to have added 1,000 illegitimate children to the population.[49]

The workhouse also attracted large numbers of feeble-minded women who tended to drift towards it in search of sustenance.[50] Workhouse maternity wards, in particular, were a magnet for the many women who found themselves pregnant with no means of supporting themselves or their babies. As a result, workhouses were doubly burdened with feeble-minded women and their children. The workhouse was evidently ill suited to the containment or management of such women, and its disciplinary regime ill adapted to allow for the uncontrolled vagaries of their behaviour.

The prison was less appropriate still. Estimates of the numbers of feeble-minded women in prison varied considerably and most figures given were agreed to be under-representative. Perhaps surprisingly, given the greater concern with feeble-mindedness in women, the proportion of male prisoners assessed

[49] Both pieces of research quoted in Kirby, ibid.
[50] For a breakdown of the various factors which drew the feeble-minded to the workhouse, see Tredgold, *Mental Deficiency*, 282–3.

as weak-minded was slightly higher than that of women. The Prison Commission Medical Inspector, Dr Herbert Smalley, noted in his annual report for 1905 that approximately 2 per cent of male but only 1.1 per cent of female prisoners were feeble-minded. Significantly, he was sceptical of the validity of these figures. He argued: 'I am not convinced that more male criminals are feeble-minded than female, but I attribute the preponderance rather to the greater leniency that is shown to females at the various courts of justice, and to the fact that discipline in the prison is rather more elastic amongst females.'[51] Offences committed by feeble-minded women were mostly of such a petty order (minor theft, begging, or drunkenness) that the courts were inclined to let such women go unpunished more readily than mentally defective men, who tended to face more serious charges, often for sexual assault.[52] Those feeble-minded women who were committed tended to receive lighter sentences in recognition of their condition. For example, in Holloway Prison the sentences for similar offences for sane women averaged six weeks to a year, but those for feeble-minded women were generally only a month or less—yet again reducing their proportion of the daily prison population. In any case, the very brevity of most sentences meant that many women were released before being detected as feeble-minded: especially since the more flexible regime in female prisons allowed greater play for eccentricities of behaviour.[53] This enabled them to go unnoticed or, at least, to avoid formal categorization as feeble-minded by prison medical officers. According to Dr Smalley's evidence, those women who were officially recognized as mentally defective were far more likely than men to be certified as insane and removed to a lunatic asylum, or, if found to have a history of alcoholism, to an inebriate reformatory.[54]

All these factors combined to reduce the numbers of women in prison who were recognized as feeble-minded. Dr G. B. Griffiths, Deputy Medical Officer and Deputy Governor of

[51] Report of Prison Commissioners for 1905, Report by the Medical Inspector, *PP*, 50 (1906), 39.

[52] H. Smalley, RC on the Feeble-Minded, Minutes of Evidence, *PP*, 35 (1908), 289.   [53] See Chs. 4 and 5 above.

[54] Report of Prison Commissioners for 1905, Report by the Medical Inspector, Herbert Smalley, 39.

Holloway, admitted the difficulties of assessing the extent or degree of feeble-mindedness among women held in his own prison, given the lack of information available about prisoners on commitment. Apart from the few details provided by magistrates, or revealed by the prisoners themselves, the only information readily available was that obtained by immediate observation. Many officials, including Griffiths, Smalley, and Mary Gordon, on her appointment as Lady Inspector of Prisons, argued that far larger numbers of women prisoners were feeble-minded than official figures indicated.[55] And evidence to the Royal Commission on the Feeble-minded suggested that as many as 10 per cent of the daily average prison population were feeble-minded (suggesting that they were an even higher proportion of receptions, given that the feeble-minded were mostly committed for minor offences and, therefore, to shorter sentences).

Most importantly, the feeble-minded seemed to be responsible for an amount of crime and to pose problems of policing and prison discipline out of all proportion to their number. Many writers were adamant that the feeble-minded woman, in particular, lacked sufficient control and capacity for self-management to survive unaided outside the prison.[56] Held for only a few weeks or months she was released unimproved and unaided into a world whose demands she could not meet. All too often she sank to the social depths, 'the lowest haunts', resorting to prostitution or petty theft to survive, prey to the temptations of alcohol. Returned to the prison on a new charge after the shortest of intervals, she was liable to enter into a round of repeated short-term sentences from which she was unlikely ever to escape. Mary Gordon argued that the surest proof of the real extent of mental defect among women prisoners was to be found in prison case histories. These revealed lives of destitution and crime, often begun early in their teens. She lamented: 'Girls of this description often wander away or live

[55] H. Smalley and G. B. Griffiths to RC on the Feeble-Minded, Minutes of Evidence, 289 and 301. Also Report of Prison Commissioners, Report of the Lady Inspector of Prisons for 1912, *PP* (1912), 36.

[56] For example, C. H. Melland, 'The Feeble-minded in Prisons', in the 'Mental Deficiency' section, *Report of the National Conference on the Prevention of Destitution 1911*; also Devon, *Criminal and the Community*.

in common lodging houses, deserting or perforce being abandoned by their friends. They tramp about, and "sleep out",
but they are often unable to get a living even by prostitution or
theft. They are sometimes almost starving, and suffer every
kind of hardship, and form of exploitation.'[57] Of these the very
worst were 'those unfortunate women who spend a large part
of their lives in local prisons and who, at 50, may be found
living much the same life as at 20 years of age'.[58] Stamped as
recidivist so early in life, the feeble-minded woman found it
almost impossible to escape the dual roles of petty offender and
prison habituée. The problem was not only appalling in the
obvious, immediate costs of such recidivism but was amplified
by the fact that this continued round of imprisonment and
temporary release created 'conditions of emotional stress' that,
over time, aggravated her condition. Her poor early mental
health degenerated into that 'extreme perversity and moral
insensibility of the criminal class of habitués'.[59] J. Bruce
Thomson acclaimed as 'perhaps the most astounding fact
known in all psychological history' his finding that one in every
thirty-six women in the convict class (at the General Prison at
Perth) was insane.[60]

This relationship between mental defect and the astound-
ingly high rates of recidivism notched up by a core of women
prisoners amplified concern about both feeble-mindedness and
habitual crime. However, explanations of their causal relationship remained confused. On the one hand, feeble-mindedness
was certainly seen as a major cause of habitual recidivism.
On the other, it seemed likely that, especially in cases where
the intellect was already weak, mental decline was accelerated
by the impact of repeated imprisonment. Whether feeble-
mindedness was the cause or product of habitual recidivism remains irresolvable. Although concern about feeble-mindedness
was undoubtedly conflated by its association with habitual
criminality, it is difficult to fit beliefs about the dangerousness
of the habitual criminal alongside descriptions of the pathetic,
feeble-minded old women who were the worst recidivists. At

[57] Report of Prison Commissioners, Report by the Lady Inspector of Prisons for
1912, 37.   [58] Ibid.
[59] J. B. Thomson, writing of weak-minded women with 10, 50, and even 100
previous convictions in Scottish prisons, 'The Psychology of Criminals', *Journal of
Mental Science* (Oct. 1870), 338.   [60] Ibid. 344–5.

Aylesbury Convict Prison, for example, one visitor found only a 'general air of apathy' and 'a large number of rather miserable old women, such as may be seen in any workhouse', with 'dull, stupid, almost feeble-minded faces'.[61]

If the feeble-minded seemed to pose social problems each time they were released from prison, they were no less a source of anxiety when incarcerated. Penal administrators had always maintained that women prisoners, as a class, were of weaker intellect and less able to control their passions than men. Consequently they were not only seen as less reformable but less able to comply with the strictures of the penal regime. And it was such assumptions about their weaker mental health that had necessitated and justified the development of a less rigid, more adaptable regime in women's prisons. Even within this more flexible regime, feeble-minded women proved incapable of complying with those demands that were made upon them, inevitably undermining any attempt to impose uniform discipline. Often they were simply indifferent to systems of rewards and punishments and, unintentionally, obliged officials to mitigate the severity of the regime.

These problems were not new. Evidence given to the Royal Commission into the Working of the Penal Servitude Acts (1878–9) by officials at Millbank and Woking Female Prisons revealed that the many 'half-witted women' held caused great difficulties for the implementation of discipline.[62] The Lady Superintendent of the latter prison deemed punishment to be both ineffective and unfair, given that such women had no understanding of right and wrong. Whilst isolation was the policy commonly adopted to deal with men suspected of being mentally defective (in part as a test against impostors), Dr Gover, the Medical Officer at Millbank, warned against such treatment of women on the grounds that 'women are so impulsive, and a woman placed in a separate cell may do herself mischief.'[63] Whilst men actually found to be 'weak-minded' were generally segregated in a separate wing, the smaller numbers of women made such costly, specialist provision less

---

[61] 'X.Y.Z., 'England's Convict Prison for Women', *The Englishwoman* (Apr.–June 1912), 268–9.

[62] R. M. Gover, Medical Officer at Millbank, and Mrs Sarah Gibson, Lady Superintendent at Woking Female Prison, to Royal Commission into the Penal Servitude Acts, Minutes of Evidence, *PP*, 37 (1878–9), 204 and 554.   [63] Ibid. 204.

easily justifiable in female prisons. The Royal Commission (1878–9) consequently recommended complete segregation of feeble-minded men yet demanded only that women be 'kept as separate as possible'.[64]

In subsequent years Dr Gover remained vehemently opposed to the segregation of feeble-minded women prisoners. He claimed that those involved in the management of mentally defective prisoners recognized the need to formulate differentiated treatment appropriate to each sex. In a letter to the Chairman of the Prison Commissioners, Sir Edmund Du Cane, he argued:

As regards women, it is advisable, as far as possible to avoid associating together those who are labouring under the same form of mental defect or disease, and this particularly applies to women who are partially or wholly imbecile. When a number of such women are placed together, the result is, in many cases, a development of the hysterical tendency or element to a point which is injurious, and which renders the subject of it difficult of management.[65]

He even suggested that madness became 'contagious' in women confined closely together. Rejecting segregation on these grounds, he recommended instead that weak-minded women merely be monitored. That such provision proved to be inadequate is evident from the number of circulars and standing orders subsequently issued dealing specifically with the problem of those who, though not certifiably insane, were deemed to be 'unfit' for ordinary prison discipline.[66]

By the turn of the century, specific provisions were drawn up for the classification and separation of the feeble-minded away from other prisoners, under the supervision of specially selected officers. Special provisions were also made for diet; for 'untasked' work; for punishment to be inflicted only with the assent of the Medical Officer; and, perhaps most importantly, for discharge into safe custody or under supervision. The aim of these provisions, applied to men as well as women, was to mitigate the severity of the regime as far as possible in respect

[64] Ibid. 43.

[65] Letter from Dr Gover to E. Du Cane, 9 Jan. 1880, PRO HO45 9588/89824.

[66] Report of the Prison Commissioners for the year ending March 1902, appendix 17a, Copies of Circulars and Standing Orders, *PP*, 46 (1902).

of the feeble-minded. At the same time it sought to minimize the extent to which their very presence in prison tended to undermine the punishment of other prisoners. Nonetheless, management of the feeble-minded in prison clearly remained problematic and their presence there unsatisfactory. The Duchess of Bedford (Vice-President of the Association of Lady Visitors and Vice-President of the Rescue Work section of the Pimlico Ladies' Association), in evidence to the Royal Commission on the Feeble-minded, argued that such women needed specialized provision and, therefore, must be regarded as a class apart from ordinary prisoners. Unlike many less optimistic observers, she argued that feeble-minded women were to some degree improvable and that if they were placed under the influence of a well born and highly educated lady (presumably such as herself) they could be raised by their association. Here again one sees a role being claimed for women as superintendents or on the management committees of institutions in which feeble-minded women and girls were held.[67]

The Duchess of Bedford was far from alone in pressing for the segregation of the feeble-minded. The prevailing view both inside and outside prison administration was that the feeble-minded should not merely be segregated within the prison but that they should be removed altogether from what could only be seen as a totally inappropriate environment. The general consensus in favour of complete segregation in specialist institutions was clearly reflected in evidence given to the Royal Commission on the Feeble-minded. As one observer noted: 'with regard to the segregation of mentally defective women the evidence was specially conclusive.' She went on to argue: 'They are a source of great trouble to those who have to deal with them, as they are not amenable to ordinary prison discipline and need to be treated differently from other prisoners.'[68]

Even before the Commission finally reported (in 1908, four years after it had first begun to take evidence) the shift in favour of separating feeble-minded women from other prisoners was beginning to lead to administrative reorganization. In

[67] Her Grace Adeline, Duchess of Bedford, to RC on the Feeble-minded, Minutes of Evidence, *PP*, 36 (1908), 328–9. Such arguments were clearly very similar to those used more generally in support of female superintendence and the role of Lady Visitors in women's prisons. See my discussions in Ch. 3 above.
[68] Meredith, 'The Feeble-minded', 412 and 413.

1907 Aylesbury Convict Prison for Women became the first female prison to segregate its feeble-minded in an entirely separate wing ('D' Hall), where they were supervised by specially selected officers and 'treated in all respects with exceptional leniency'.[69] Their separation was reported to be a great improvement: 'These women have always exercised a very disturbing influence amongst the other prisoners, and their elimination has rendered it possible to maintain a much higher state of discipline than had hitherto been the case.'[70] The removal of that group who were 'all more or less noisy, irritable, and difficult to manage' obviously allowed for the much smoother running of the rest of the prison and was subsequently hailed as an 'unqualified success'.[71]

By contrast, the continuing lack of specialist provision for the feeble-minded in local prisons was found increasingly intolerable. The report of the Prison Commissioners for 1909 declared that such prisoners, 'especially in the case of women, constitute one of the saddest and most unprofitable features of prison administration',[72] and called for a purging of prisons of those who properly required medical rather than penal discipline. It was hardly surprising that the final report of the Royal Commission should declare itself firmly in favour of separate, long-term detention. To a significant extent these demands for the removal of the feeble-minded from the prison system, by rejecting the penal in favour of the medical sphere, constituted a tacit decriminalization of those who could be judged mentally defective.

SPECIALIST PROVISION FOR THE FEEBLE-MINDED

If the difficulties of managing the feeble-minded in prisons, workhouses, and reformatories provided negative reasons for their removal, more positive inducements were suggested by the example of existing voluntary provision. Since its establish-

---

[69] Report of Prison Commissioners for 1907, Appendix 21, Extracts from the Reports of Officers of Convict Prisons: Aylesbury Prison, *PP*, 52 (1908), 415.

[70] Ibid.

[71] Report of Prison Commissioners for 1909, Appendix 21, Extracts from the Reports of Officers of Convict Prisons: Aylesbury Prison, *PP*, 45 (1909), 485.

[72] Report of Prison Commissioners for 1909, 158.

ment in 1897, the National Association for Promoting the Welfare of the Feeble-minded had encouraged the proliferation of small homes. Significantly, all the leading organizers and the majority of the Association's subscribers were women, with the result that nearly all these homes were set up for the care of girls and women.[73] The Association provided a vital central organizing service: collecting and disseminating information, encouraging the establishment of homes, setting up appeals, collecting funds, and making grants available for their maintenance. Their primary concern was to protect feeble-minded women from exploitation or, failing that, to care 'for the morally as well as mentally deficient . . . who by reason of their feeble intellects, have fallen into sin, but are not depraved'.[74]

The years around the turn of the century saw the proliferation of homes, each set up for a few dozen women inmates. Sandlebridge, organized by the Lancashire and Cheshire Society for the Permanent Care of the Feeble-minded, of which Miss Mary Dendy was the head, was by far the largest of these and certainly the most influential. By 1911 the Chairman of the National Association, W. H. Dickinson MP, reported that since the first one was established in 1887 (in North London by a Miss Alexander) the number of homes for women had risen to thirty-three, though these provided for 'only a small fraction of the women in dire need of help'.[75] Nevertheless these homes, run on lines of charitable refuges rather than as penal institutions, constituted a pre-existing skeleton provision to which penal administrators turned in making their case for the removal of the feeble-minded from prison.

Although they took evidence from these voluntary homes, it is significant that the Royal Commission turned to America in seeking a model for state-controlled provision. They visited hospitals for the insane, training schools, and colonies in the hope that these might provide models for the development of similar provision in Britain. They found considerable diversity of provision between states; in some there were institutions capable of holding up to 2,000 inmates. In general they were

[73] See the National Association for Promoting the Welfare of the Feeble-minded: *First Annual Report* (1897); *Fifth Annual Report* (1901); and *How to Help the Feeble-minded* (1899).
[74] Ibid., *How to Help*, 21.   [75] Dickinson, 'Homes and Colonies', 639–40.

impressed by the special institutions already set up and held them up as 'examples that in many respects, may be imitated in this country with great advantage'. Significantly they held up the Newark State Custodial Home for Feeble-Minded Women as representing the most advanced and successful example of existing provision. This institution for 600 women of childbearing age had as its primary aim their segregation and detention 'in order to prevent the propagation of persons of feeble-mind with its attendant evils to the community'.[76] What impressed the members of the Commission most was that at least a quarter of the women had been brought before the courts on some criminal charge and sent by a magistrate to the custodial home without first being convicted.[77] Newark State seemed, therefore, to operate as a ready alternative to imprisonment. It effectively replaced the unremitting cycle of short sentences, so deplored in England, with secure, long-term, and ostensibly non-punitive detention. Moreover the American experience confirmed that curing or even simply improving the feeble-minded was at best a slim possibility. It thus provided a strong case for indefinite, even permanent, detention, the force of which clearly impressed the visiting members of the Commission.

CAMPAIGNING FOR LEGISLATION

After four years of investigation, resulting in a massive seven volumes of evidence, the Royal Commission's overriding conclusion was that the feeble-minded must be prevented from breeding. During the later years of the nineteenth century, sterilization of the 'unfit' and the prevention of propagation by those who were chronically diseased or mentally defective had become the standard means advocated to maintain the quality of the population. Up to this point such propositions had remained largely on a theoretical level. However, the concern to limit the numbers of those deemed feeble-minded in the interests of 'national efficiency' forced the eugenics debate into the public arena for the first time.[78]

[76] Quoted in Radzinowicz and Hood, *History of English Criminal Law*, vol. 5, 327.
[77] J. Wormald and S. Wormald, *A Guide to the Mental Deficiency Act 1913* (1914), 44.
[78] Freeden, *New Liberalism*, 190–2.

The Commission's report insisted that its primary aim was not to purify the race, but to protect the feeble-minded from exploitation and destitution. However, its most important recommendations all related to segregation and detention as means of preventing procreation.[79] It proposed the setting up of a single authority—the Board of Control—under whose aegis local authorities would be empowered to register and detain mental defectives, and to maintain, care for, and educate those held. The existing system of voluntary provision would not be superseded but would be expanded with the help of state subsidies and the encouragement of profit-making endeavours intended to minimize costs.

These proposals were clearly intended to pacify the workhouse and, in particular, the prison authorities—at whose instigation the whole enquiry had first been initiated. But they also reflected a high level of Eugenic influence. Although the Commission was dubious of proposals to restrict marriage between the feeble-minded, and refused even to consider the highly controversial option of sterilization, it was, nonetheless, preoccupied with the procreative potential of feeble-minded women. 'Life-long care', with its connotations of benign protection, was a far less controversial means of securing what was, at base, a eugenic aim.[80] Significantly, the effect of the concern to prevent procreation ensured that political debate, legislative proposals, and eventual provision all focused on the feeble-minded woman as the greatest social problem. Eugenic influence was reflected in the close co-operation between the National Association for Promoting the Welfare of the Feeble-minded and the newly established Eugenics Education Society (set up 1907). In 1910 they joined together to campaign for legislation and were widely supported, most influentially by *The Times* newspaper. The campaign generated public meetings nationwide and hundreds of resolutions deploring the inadequacy of existing provision and pressing the government to legislate. By the end of 1912 the Home Office alone had received 800 resolutions from public bodies pressing for

[79] Report of the RC on the Care and Control of the Feeble-Minded.
[80] For discussion of the relationship between eugenics and the feeble-mindedness debate, see Jones, *Mental Health*; Searle, *Eugenics and Politics*, esp. ch. 9; M. Freeden, 'Eugenics and Progressive Thought', *Historical Journal*, 22:3 (1979); Radzinowicz and Hood, *History of English Criminal Law*, 322–38.

legislation.[81] As a consequence the issue became a recurring subject of parliamentary debate from 1910 onwards.

The costs of feeble-mindedness in women formed the basis on which the extension of state control to segregate and detain could be justified. Areas which had previously been considered private concerns of the domestic domain and, therefore, beyond legitimate state intervention—such as birth control, motherhood, and the welfare of children—were increasingly redesignated as social issues, of legitimate public concern. Where the woman was deemed to be unfit for motherhood, or inadequate or neglectful in her maternal duties, the State claimed a right to intervene.[82] Its new role would expand to protect those incapable of self-restraint or unable to resist attacks upon their virtue, and to manage or to supervise those who were unable to look after themselves. Further, it sought to detain those whose intellectual inadequacies and consequent lack of moral sense made them prone to crime.

The inebriate reformatory experiment had provided an important precedent. It had established the right of state control to outweigh the rights of the individual in cases where he or she was deemed incapable of living up to the demands that freedom entailed. The 'liberty of the subject' was seen to be overridden by the demands of 'decency, morality, and good order' in the case of 'those who cannot or will not conform to our social requirements'.[83] The sphere of penality was thus transmuted from the classical equation of punishment proportionate to a criminal offence, to indeterminate sentences in welfare-oriented institutions for those deemed morally irresponsible. Like habitual inebriates, the feeble-minded could be seen to be so deficient as to fall below minimum standards of citizenship and so be liable to be denied their status and rights. Feeble-minded women were particularly susceptible to this infantilization. Yet many observers argued that their vulner-

[81] Jones, *Mental Health*, 62.

[82] See, for example, Mr Astor's speech on the need to provide specialist institutions for the containment of feeble-minded women, especially those pregnant through 'immorality', 5 June 1912, Hansard, 5th ser. (1912), vol. 39, cols. 164–6. See also Freeden, *New Liberalism*, 177–9.

[83] GLRO LCC/MIN/8126. LCC Inebriate Acts Subcommittee Papers 1910–12. Frank A. Gill, 'The Lancashire Inebriates Reformatory, Langho, near Blackburn', Seventh Annual Report to the Lancashire Inebriates Acts Board, 1910, 8–9.

ability and their all too frequent exploitation made a cruel mockery of abstract rights of the 'liberty of the subject'. The archetypical feeble-minded woman, presenting herself time after time at the lying-in hospital or workhouse maternity ward, by her very actions testified to her incapacity to survive unaided and so justified state intervention.

Proponents of state responsibility for the incapable and defective argued that such legislation was exculpatory—signifying a recognition that the sinner was, in fact, sick.[84] Their arguments did not go unopposed. Traditional liberals saw the freedom of the individual being heedlessly superseded by proposals to control and, worse still, to contain the feeble-minded. L. T. Hobhouse, for example, argued that the proposed definition of feeble-mindedness as 'those incapable of competing on equal terms with their fellows' would have the effect of making non-conformity, infirmity, old age, and dependence into crimes. This would render the 'unfit' liable to what were effectively penal sanctions.[85] Others, though they shared the eugenic premise that certain members of society were indeed 'unfit', questioned the desirability of taxing the fit for the support of those who were not. In the case of feeble-minded women, the costs of maintenance were greatly increased by the need to isolate them for the whole of their childbearing life. Many also feared that instead of ensuring a decline in the numbers of feeble-minded, such provisions could well serve only to maintain them. This was a long-standing basis for opposition to any provision. For example, the work of the Association for Promoting the Welfare of the Feeble-Minded had been criticized because it 'seems to tend towards the preservation and the multiplication of the feeble-minded rather than to their decrease'.[86] There was also considerable opposition on the grounds that the proposed legislation challenged the domain of the courts. Whilst the Home Secretary, Winston Churchill, suggested that decisions on permanent detention be based on medical rather than judicial grounds, he recognized the potential dangers of indeterminate sentencing. He argued that indeterminate sentences must be imposed solely on

---

[84] For example, Dr Chapple, 17 May 1912, Hansard, 5th ser. (1912), vol. 38, col. 1515. [85] Discussed in Freeden, *New Liberalism*, 191–2.
[86] Anon., 'The Care of the Feeble-Minded', *The Humanitarian* (July 1898), 64.

medical criteria and not because the individual was, for example, politically troublesome or 'inconvenient'. Josiah Wedgwood, a radical Liberal, on the other hand, violently opposed legislation on the grounds that the role of the courts in deciding what was criminal would be superseded. He denounced the prospect of 'specialists' extending their jurisdiction over those who had committed no crime but who could be judged to be in some way 'abnormal'. The likely result, according to Wedgwood, was that anyone deemed to deviate in any way from an arbitrarily defined norm would be labelled feeble-minded and incarcerated.[87]

In particular, Wedgwood attacked the way the debate concentrated on women and especially on poor women. He argued that legislation would penalize them unjustly, and insisted: 'This Bill is eminently a Bill which we as men have no right to pass. It is a Bill which affects principally women, because under it its practical effect will be to put feeble-minded women into gaol much more than men.'[88] Incarcerating women for what was hitherto considered anti-social, but certainly not punishable, behaviour was, according to Wedgwood, an unacceptable infringement of personal freedom inflicted by one sex upon those of the other who had least power to resist. Wedgwood waged a one-man campaign against the Mental Deficiency Bill. However, over the period up to 1913, the concept of feeble-mindedness was progressively narrowed and the Bill's scope so limited that opposition was largely eroded. The final Bill, whilst it certainly satisfied the eugenicists, so underplayed its eugenic purpose as to be palatable to most former critics.

THE MENTAL DEFICIENCY ACT 1913

The Act finally passed in 1913 divided mental deficiency into four categories: idiots, imbeciles, feeble-minded, and moral imbeciles, all of whom had to be proved to have suffered their condition from birth, or at least from an early age.[89] To come

[87] J. Wedgwood, 17 May 1912, Hansard, 5th ser. (1912), vol. 38, cols. 1470–7.
[88] Ibid. col. 1476.
[89] For contemporary description and commentary on the details of the Act, see H. Davey, *The Law relating to the Mentally Defective* (1913); Wormald and Wormald,

under the jurisdiction of the Act they had also to fit one or more of a number of categories of proven social inadequacy, dependency, vulnerability, or deviance. The classifications effectively encompassed a whole array of social outcasts: the neglected or abandoned child, the juvenile delinquent, the vagrant, the drunkard, and the criminal. Notably, that most controversial of categories relating to women who were pregnant with, or already the mothers of, illegitimate children, was narrowed. It finally covered only those who were in receipt of poor relief when pregnant or at the time of giving birth. None the less, the fact that the mother of a legitimate child remained entitled to relief, whilst the mother of an illegitimate child became liable to be labelled feeble-minded and incarcerated, signified a clear moral condemnation.

The Act established a central authority, the Board of Control, to direct provision by local authorities and appointed inspectors to supervise the institutions they set up. Special local authority committees, established under the Act, were obliged to provide both guardianship in the community and institutional care for mental defectives in their area. Provision for the criminally inclined and those with dangerous or violent propensities was to be made in one, or possibly two, state institutions also under the direction of the Board of Control. Institutional care orders were to be initially for one year, subsequently renewable for five-year periods. Significantly, women were to be appointed at every level: from the Board down through the inspectorate to the local authority committees. The novelty of writing this executive role for women into a piece of social legislation is indicative of the extent to which its provisions were considered to pertain primarily to their sex.

A major purpose of the Act was to provide care or safe custody for the feeble-minded then in the community. Above all it was designed to remove them completely from workhouses, prisons, and reformatories where their presence had long been disruptive. It also sought to protect feeble-minded girls and women by providing that attempted or actual sexual assault on them, and their exploitation as prostitutes, became

Guide; E. Fox, 'The Mental Deficiency Act and its Administration', *Eugenics Review* (Apr. 1918), 1.

legal misdemeanours, liable to imprisonment for up to two years (with or without hard labour).[90] Although similar provisions had applied under the Criminal Law Amendment Act 1885, it had been the responsibility of the Crown to prove that the man did know that the woman concerned was feeble-minded. Now, the 1913 Act demanded that the defendant prove that he had no reason to suspect the woman was mentally defective, so shifting the burden of proof on to the man. These provisions represented a considerable extension of protection over feeble-minded women, especially those who had previously been all too easily inveigled into prostitution.

In its early years the implementation of the Mental Deficiency Act was severely hampered on several fronts. Its provisions were complex and difficult to interpret. Local authorities were unwilling to build the necessary institutions or to meet the cost of running them. And, above all, the outbreak of the First World War diverted resources away from its implementation. Shortage of institutional provision was a major hindrance, not easily overcome given the financial constraints of wartime. The Board of Control was forced to limit its activities to urgent cases. These were designated as children, young girls, or women, most of whom were then held in workhouses, special schools, or prisons and urgently required more appropriate, specialist care.[91] Most seriously of all, the Act failed to establish preventative provision—the mentally defective came under its aegis only after they had become inebriates, paupers, convicted criminals, or had fallen into prostitution. The Board of Control quickly recognized the damage done by these constraints and bemoaned their inability to act sooner. The Act, they argued, afforded 'no protection for the feeble-minded girl whose sexual tendencies, though well-known, have not so far landed her in the trouble indicated by the section' (Section 2(1)(b) of the Mental Deficiency Act 1913).[92] Nor did it provide for the consequences of this precarious freedom, for it conferred 'no power to detain feeble-minded women with any

[90] For a detailed list of the offences named under section 56 of the Act, see Wormald and Wormald, *Guide*, 81–3.

[91] Third Annual Report of the Board of Control for 1916, *PP*, 143 (17 Oct. 1917), 32–5.

[92] First Annual Report of the Board of Control for 1914: Part One (16 Feb. 1916), 50.

number of illegitimate children unless they were born while the mother was in receipt of poor relief'.[93] The efforts of libertarians to limit the degree to which the Act infringed 'the liberty of the subject' had clearly been successful. Yet their attempt to remove feeble-minded women from the penal sphere and to substitute social welfare provision had deprived the legislation of power, which many on the Board of Control considered vital to its success.

In the longer term, the very purpose of the Act was more profoundly eroded by new findings. It was found that feeble-minded women did not, in fact, necessarily give birth to mentally deficient children. It gradually became clear that the fertility rates of such women had been greatly over-estimated. They did not invariably fall into prostitution. Nor were they inevitably the victims of sexual exploitation. And their mental deficiency was not necessarily synonymous with amorality, nor did it always entail appalling social costs. These findings greatly undermined arguments for permanently segregating feeble-minded women. Moreover, they inevitably challenged the primary causal position previously accorded to the feeble-minded woman in the aetiology of poverty, inebriety, crime, and prostitution. As Simmons concludes in his analysis of the period after 1913: 'The de-dramatization of the role of morality as a cause of social problems also meant that feeble-minded women no longer occupied a central place in the drama.'[94]

Dangerous female defectives, on the other hand, continued to be a subject of major concern. They were the one group who could not be released from care and yet who caused the greatest disruption to institutions attempting to contain them. Ironically, that model institution for female inebriates, the LCC reformatory at Farmfield, which had closed largely because it had proved impossible to manage its feeble-minded inmates (see Chapter 6 above), was reopened as the first state institution for dangerous female defectives. The regime seems hardly to

---

[93] Ibid.

[94] Simmons, 'Explaining Social Policy', 400. As concerns provision for the feeble-minded in general, in the longer term it continued to be limited by the reluctance of local authorities to impose the heavy demands of staff and buildings on the rates. Throughout the 1920s, Board of Control reports criticized local authorities for failing to ascertain the numbers of defectives in their areas in order to evade the requirement to provide supervision. See Jones, *Mental Health*, Pt. 2.

have been changed,[95] but all pretensions to reform were abandoned and replaced only by long-term containment avowedly 'not to punish but to protect'. Removed from the penal sphere, diagnosed as sick rather than sinful, even these most dangerous of feeble-minded women were effectively de-criminalized.

The impact of this changing diagnosis of serious female offenders as mad rather than bad had a significant impact on the rates of those entering the criminal justice system. The proportion of women making up the daily population of convict prisons more than halved over the second half of the nineteenth century. Putting aside the possibility that women were actually becoming less criminally inclined and more prone to mental illness, we must turn yet again to the social construction of images of deviant women for explanation of these trends.[96] Changing opportunities, differential expectations of behaviour, and, not least, major shifts in political, social, and scientific thought had a profound impact on the ways in which female criminality was perceived. Feeble-mindedness had gained considerable popularity as an explanation for female offending, and the failure of the prison to control or to reform those so diagnosed. Whilst the currency of feeble-mindedness itself diminished in subsequent years, explanations of female crime as a product of mental illness, and paradoxically also the presentation of mental illness as an alternative to criminality in women, endured to become prevailing features of criminological thought in the twentieth century.

[95] For description of the regime see Third Annual Report of the Board of Control for 1916 (17 Oct. 1917), 40.

[96] And in doing so, return full circle to Chapter 1. For the continuation of these trends through the twentieth century, see A. Morris, *Women, Crime and Criminal Justice* (1987), ch. 3, 'Theories of Women's Crime'; F. Heidensohn, *Women and Crime* (1985), chs. 5 and 6.

POSTSCRIPT

Historically, changing views about criminal women related directly to prevailing notions of femininity. This finding raises important questions about the extent to which male behaviour was also assessed in relation to parallel ideas about masculinity. Whereas male criminals were generally seen to act out 'manly' traits and aspirations, criminal women not only broke the law but also contravened various moral prescriptions about women which were deemed essential to the moral order of Victorian society. The extent to which female crime was seen to repudiate feminine moral duties partly explains the high level of anxiety it seems to have provoked. In addition this anxiety about female crime was exacerbated by its relationship to prevailing social problems: urban filth and overcrowding; disease and degeneration; the threat posed by urbanization to community and family. By studying attempts to control crime historically, this book has shown how concern about crime is a mirror for the wider problems which preoccupy society at any given time. As such it has obvious implications for present-day criminology.

In its examination of institutional responses, this study has suggested that the attempt to bring about a uniform and deterrent prison system in the first half of the nineteenth century was far less successful than many historians have claimed. Moreover, the problems of imposing such penal regimes in male prisons became even greater in women's prisons where policies had to be adapted to fit assumptions about femininity. Close investigation of the prisons themselves has revealed that the realities of prison life were often starkly divergent from the intentions of the reformers. This book has provided evidence of how official aims and regulations were often greatly distorted in the process of adaptation to the demands of actually running the prison. Moreover, as our attempt to reconstruct experiences of custody has shown, life inside is better understood as the sum of relations between inmates (both warders and prisoners) than as the perfect realization of penal theory. This points to the danger of relying on official reports and external observations alone, but also demonstrates the value of writing, if not a 'history from below', at least a 'history from within'. Institutions did not always live up to the ideals they were established to achieve. Indeed their failings, by creating dissatisfaction with existing provision, were often the very source of new policies.

Around the turn of the century, links drawn between criminality, alcoholism, and insanity crystallized long-standing eugenic fears about deviance in women. Female crime came to be seen less as the deliberate contravention of social norms or laws and instead as the manifestation of innate pathology. Although the intervention of medicine and, in particular, psychiatry tended to de-criminalize female deviancy, paradoxically, in stressing its hereditary aspects they also amplified its perceived social costs. For, through motherhood, criminal women might transmit their pathological qualities to the next generation. Accordingly the new reformatories for alcoholics and asylums for the feeble-minded catered overwhelmingly for women: seeking to cure them or, where this proved impossible, to restrain them for so long as they were able to bear children. Whether women were really so much more drunken or mentally deficient than men is questionable. Whatever the case, it does seem likely that the smaller scale of female imprisonment allowed for experimentation not possible in the much larger male system. For those women who could be seen as sick or mad, rather than bad, punitive responses came to be seen as less appropriate than specialized treatment outside the penal sphere.

There remained, however, large numbers of women who were not certifiable under the 1913 Mental Deficiency Act.[1] These women occupied the 'borderland' of sanity; they were incapable of surviving unprotected in wider society yet imprisonment was entirely inappropriate for them. Repeatedly sentenced to prison for petty offending or public disorder, their continued presence there was a constant cause of disruption and a source of anxiety and shame to Prison Commissioners throughout the inter-war years.[2]

---

[1] Or under the subsequent Mental Deficiency Act 1927.
[2] See A. Smith, *Women in Prison* (1962), 230–6.

Concern that the prison was a wholly unsuitable environment for many women inmates led eventually to the Mental Health Act 1959 which, in repealing the 1913 Act, sought to expedite further removal of mentally deficient women from the penal system. Significantly, the preoccupations motivating this reform seem to have changed little from those which had prompted the original legislation half a century earlier. Note, for example, the observations of one commentator writing just after the 1959 Act was passed:

Social inadequacy and social inefficiency which are typical of mental defect often influence women to commit the types of crime to which their sex is particularly prone . . . The majority of such women are quite unable to take advantage of any training offered in prison. They disrupt the routine, and hamper the progressive schemes for work and education. Even borderline cases who can profit by some form of training require care and protection, not imprisonment.[3]

In the second half of the twentieth century, the view that much female crime is the product of mental deficiency has become commonplace, infiltrating the mainstream prison system itself. Significantly, the most important of prisons for women today, the new Holloway prison, was designed on the assumption that 'most women and girls in custody require some form of medical, psychiatric or remedial treatment . . . [Holloway] will be basically a secure hospital to act as the hub of the female system. Its medical and psychiatric facilities will be its central feature and normal custodial facilities will comprise a relatively small part of the establishment.'[4] Whilst Holloway has remained more punitive than was originally intended, its special facilities for mentally disturbed prisoners are testament to the continuing view of criminal women as mentally ill or inadequate. What began as a view that a proportion of criminal women were inadequate now extends to envelop views of the female prison population as a whole. As the sociologist Carol Smart observed in 1977: 'the assumption underlying this policy is that to deviate in a criminal way is "proof" of some kind of mental imbalance in women.'[5] Psychotropic drugs are

---

[3] Ibid. 231.

[4] James Callaghan, then Home Secretary, speaking in 1968, quoted in A. Morris, *Women, Crime and Criminal Justice* (1987), 109.

[5] C. Smart, 'Criminological Theory: Its Ideology and Implications Concerning Women', *British Journal of Sociology*, 28 (1977), 96.

dispensed in far larger quantities in female prisons than those for men, ostensibly in response to women prisoners' greater mental problems.[6] The tendency to self-mutilate has at times become almost endemic in female prisons but there seems to be no self-destructive equivalent, short of suicide, among male prisoners. The fact that women in prison have a punishment rate roughly double that of men is also cited as evidence of their greater psychiatric disturbance. As a result of all these factors, the expectation that women in prison are likely to be in some way inadequate is now common amongst penal policy makers.

It would undoubtedly be overly simplistic to draw a straight line from the end of the Victorian era to the present day. Nonetheless, it is worth emphasizing how little understanding of female crime has altered in the intervening years when compared to interpretations of male criminality. Since the heyday of biological determinism and psychiatric intervention, those seeking to explain male crime have developed a rapidly changing array of sociological theories. The emergence of feminist criminology offers the prospect of new understandings of female criminality but as yet has had only limited impact on 'mainstream' criminology.[7] For the moment, at least, penal policy continues to be dominated by beliefs about women which were pre-eminent around the turn of the century. To explain the resolute persistence of these ideas we must recognize the extent to which psychological responses to female crime are founded on earlier, deeply moralistic theories about what constitutes normality and what deviance in women. It is only through historical research that we can recognize just how far these Victorian beliefs about women continue to inform penal policy today.

[6] F. Heidensohn, *Women and Crime* (1985), 73–5.

[7] See L. Gelsthorpe and A. Morris, 'Feminism and Criminology in Britain', in P. Rock, *A History of British Criminology* (1988); L. Gelsthorpe and A. Morris (eds.), *Feminist Perspectives in Criminology* (1990).

APPENDIX OF TABLES

A NOTE ON THE JUDICIAL STATISTICS

It is held by criminologists, almost as a truism, that official crime statistics scarcely begin to indicate the 'real' amount of crime, so great is the unreported and unrecorded 'dark figure'. Changes in popular attitudes towards women as criminals, in levels of policing, and in court administrative and sentencing policy combined to affect the number of those officially designated criminal and often to alter the gap between the 'actual' level of crime and that recorded. For those attempting to examine the nature and extent of criminal activity itself, official statistics are, therefore, of limited use.

Given that this book is less a history of female criminality than of the social policy it provoked, the fact that the statistics give a better indication of prevailing perceptions of crime and criminals than of the activity itself is rather less of a problem. Indeed, one could read the figures less as the facts of crime and more as indicators of the perception and definition of crime, of policing, of judicial and penal responses. Many of the objections traditionally raised to using criminal statistics (not least, by J. J. Tobias[1]) then become less pertinent. Criminal statistics may well be social constructs which tell us little about the 'dark figure' of crime but they tell a great deal about the motivations of those charged with data collection and the preoccupations of bureaucrats. Moreover crime figures, irrespective of their accuracy, were important both in informing public opinion and as a basis for formulating policy.

Obviously official statistics deal only with crimes that come to official attention and those criminals known to the police, going to court, or actually reaching prison. However, numbers of apprehensions, committals for trial and convictions do tell a lot about the work of these institutions. These figures can be taken to indicate the general level and type of crime coming to

---

[1] J. J. Tobias, *Crime and Industrial Society in the Nineteenth Century* (1967) and *Nineteenth Century Crime* (1972).

the attention of the criminal justice system and may be used to establish trends over time.[2]

Greater problems are caused by major reorganizations in the presentation of criminal statistics carried out at various times in the nineteenth century. In my period, the most pertinent are those put into effect in 1857 and in 1893. The reorganization carried out in 1857 created major discontinuities, particularly in the enumeration of summary offences. If we confine our analysis to statistics after 1857, they are then recorded with reasonable consistency up to 1893, when they were completely reorganized, so making comparisons across this date extremely difficult. Even within the period, the comparability of statistics over time is often placed in doubt by the changes in the policing, court, and penal practices suggested above.

Distortions were also caused by alterations in enumeration and, more seriously, by changes in the law, especially relating to summary offences. Probably the most significant of legal reforms was the Summary Jurisdiction Act of 1879 which extended statutes of 1847, 1850, and 1855 by bringing under summary jurisdiction: (1) all children under twelve (for all but a few specified, serious offences); (2) all adults pleading guilty to larceny, embezzlement, etc. or consenting to be tried summarily for these types of offences concerning property not exceeding 40 shillings in value.

The effects of this legislation were marked, as Gatrell and Hadden point out:[3] the total number of larcenies tried summarily rose from 42,011 in 1879 to 51,025 in 1880. Other new legislation (for example, the 1872 Elementary Education Act, the 1872 Habitual Drunkards Act, and the 1885 Criminal Law Amendment Act) not only shifted existing offences to other categories but actually added new types of offence to the total. Whilst the movement of offenders under the 1879 Summary Jurisdiction Act is easy to plot, the effects of other legal reforms on the total number of proceedings, or on the categories under which offenders were most likely to be tried, are often not at all clear.

---

[2] See arguments in support of this view by V. A. C. Gatrell and T. B. Hadden, 'Criminal Statistics and their Interpretation', in E. A. Wrigley (ed.), *Nineteenth Century Society* (1972).

[3] Ibid.

When using official crime statistics, or when reading arguments based on them, the various limits to their accuracy and consistency over time must be held constantly in view. The tables used in this book, where not indicated otherwise, are derived from the Judicial Statistics produced annually as parliamentary returns. A number of errors were noted in the original returns and wherever possible these have been corrected. Any remaining gaps in the tables reflect missing figures or inconsistent reporting in the original returns. Particular problems or weaknesses in the data are discussed, where specific tables are referred to, in the body of the book (see Chapters 1, 4, and 5).

TABLE 1a. *The criminal classes (excl. vagrants and tramps) (both sexes expressed as % of total)*

	1860	1865	1870	1875	1880	1885	1890
known thieves/depredators							
male %	77.7	76.3	76.9	75.9	77.1	78.0	78.4
female %	22.3	23.7	23.1	24.1	22.9	22.0	21.6
total	38,094	22,773	22,014	17,632	17,907	15,457	14,189
receivers of stolen goods							
male %	80.4	79.1	75.8	75.8	75.7	73.6	73.4
female %	19.6	20.9	24.2	24.2	24.3	26.4	26.6
total	4,440	3,024	2,602	1,602	1,297	1,171	1,158
suspected persons							
male %	81.6	81.1	80.0	78.3	80.1	79.4	80.3
female %	18.4	18.9	20.0	21.7	19.9	20.6	19.7
total	35,206	29,591	28,371	22,217	20,261	18,599	15,878
TOTALS							
male %	79.6	79.0	78.4	77.2	78.6	78.6	79.3
female %	20.4	21.0	21.6	22.8	21.4	21.4	20.7
total	77,740	55,388	52,987	41,451	39,465	35,227	31,225

TABLE 1b. *Numbers of suspected persons at large (both sexes expressed as %)*

	1893	1895	1900	1905	1910
habitually engaged in crime					
male %	76.8	77.9	77.2	80.3	83.1
female %	23.2	22.1	22.8	19.7	16.9
TOTAL	7,509	5,640	5,256	4,035	3,972

TABLE 2. *Character of persons proceeded against (summarily and by indictment) (expressed as % of men/women proceeded against)*

	1857	1860	1865	1870	1875	1880	1885	1890
known thieves								
% of men	5.9	4.7	4.0	3.5	2.2	2.8	2.7	2.4
% of women	5.2	5.2	4.2	3.8	2.9	3.0	3.1	2.9
prostitutes								
% of women	28.0	24.0	19.1	20.1	16.1	18.6	18.6	12.0
vagrants/tramps								
% of men	4.5	3.7	3.9	5.1	2.7	4.6	4.0	3.4
% of women	5.8	5.0	5.1	5.1	2.6	3.5	3.2	3.1
suspicious characters								
% of men	12.8	11.9	10.2	9.2	6.2	6.8	6.9	5.6
% of women	7.7	8.9	7.0	6.2	4.5	4.4	4.5	4.1
no known occupations								
% of men	1.7	6.0	6.2	6.6	—	—	—	—
% of women	2.0	6.0	7.0	9.0	—	—	—	—

previous good character							
% of men	35.6	37.5	41.9	42.9	49.0	49.8	49.7	53.0
% of women	16.8	18.9	22.3	23.3	28.7	28.9	30.2	34.7
character unknown							
% of men	39.5	36.1	33.8	32.6	33.5	30.9	31.6	30.9
% of women	34.6	32.1	35.3	32.5	34.7	33.9	32.4	35.5
habitual drunks							
% of men	—	—	—	—	6.4	5.0	5.0	4.7
% of women	—	—	—	—	10.2	7.6	8.3	7.8
TOTAL							
male	314,432	323,551	392,504	447,821	541,353	553,964	576,093	623,608
female	86,832	86,229	95,469	105,661	130,582	131,671	127,195	132,131
total	401,264	409,780	478,973	553,482	671,935	685,635	703,288	755,739

Appendix of Tables

TABLE 3a. *Offences determined summarily*

	1857	1860	1865	1870	1875	1880	1885	1890
proceeded against[a]								
male %	78.8	79.4	80.7	81.1	80.7	80.9	81.9	82.6
female %	21.2	20.6	19.3	18.9	19.3	19.1	18.1	17.4
total	369,233	384,918	458,914	526,869	649,827	663,404	684,081	738,061
convicted[b]								
% of men	66.1	69.4	70.8	75.5	80.4	79.5	81.3	83.2
% of women	53.1	55.1	57.3	67.4	72.2	71.4	74.9	76.8
total	233,759	255,803	312,882	389,712	512,425	517,373	548,436	605,921
discharged[b]								
% of men	33.9	30.6	29.2	24.5	19.6	20.5	18.7	16.8
% of women	46.9	44.9	42.7	32.6	27.8	28.6	25.1	23.2
total	135,474	129,115	146,032	137,157	137,402	146,031	135,645	132,140

[a] Both sexes expressed as % of total.
[b] Expressed as % of men/women proceeded against.

TABLE 3b. *Offences determined summarily—those for which women were most commonly convicted only*
(expressed as % of male/female total convictions)

	1857	1860	1865	1870	1875	1880	1885	1890
assaults on peace officers								
% of male	5.1	4.4	4.3	3.3	2.6	2.3	2.3	2.2
% of female	3.1	2.3	2.2	1.6	1.2	1.2	1.2	1.3
assaults, common								
% of male	12.7	12.6	13.6	10.2	8.7	7.2	6.7	5.8
% of female	17.1	17.8	19.5	14.0	12.6	10.5	10.0	9.5
breaches of peace								
% of male		2.6	2.6	2.9	2.7	2.1	2.1	1.8
% of female		2.7	2.8	4.3	4.7	3.6	3.5	2.8
drunkenness/drunk and disorderly								
% of male	18.7	21.7	22.6	26.3	34.4	27.2	28.8	26.5
% of female	21.7	25.5	25.9	36.5	43.9	38.0	36.4	36.9
local acts/by-laws, offences against								
% of male	8.0	8.4	6.1	7.4	6.8	7.0	6.5	7.8
% of female	2.8	3.6	5.1	5.9	6.7	8.5	8.6	9.1

TABLE 3b. (cont.)

	1857	1860	1865	1870	1875	1880	1885	1890
other malicious damage/ trespass								
% of male	3.1	2.8	3.6	3.1	2.7	2.3	2.4	1.7
% of female	3.6	3.3	3.5	2.8	2.1	1.9	1.8	1.8
offences punishable as misdemeanours								
% of male	2.5	1.6	1.6	1.7	1.9	2.4	1.6	1.6
% of female	4.0	2.4	1.5	1.1	1.2	1.2	1.1	1.2
larceny under 5s.								
% of male	4.0	2.7	2.2	2.2	1.3			
% of female	7.6	6.6	4.7	4.1	2.4			
simple larceny (Summary Jurisdiction Act 1879)								
% of male						4.9	4.4	4.1
% of female						6.6	5.9	5.4
other categories larceny, grouped								
% of male	2.9	3.0	2.3	1.7	1.6	0.6	0.6	0.6
% of female	3	6.2	3.7	2.0	1.9	1.2	1.3	1.3

prostitutes							
% of female	13.3	10.7	9.4	9.3	7.7	9.4	9.8	6.7
begging								
% of male	2.0	1.4	1.6	2.9	1.3	3.1	2.7	1.9
% of female	2.7	2.3	2.1	1.0	1.5	1.4	1.1	1.1
other offences								
% of male	27.7	28.1	39.5	38.3	36.1	40.5	41.9	46.0
% of female	21.1	16.6	19.6	17.4	14.1	16.5	19.3	22.9
TOTAL								
male	192,235	212,018	262,214	322,792	421,984	426,837	455,899	507,126
female	41,524	43,785	50,668	66,920	90,441	90,536	92,537	98,795
total	233,759	255,803	312,882	389,712	512,425	517,373	548,436	605,921

TABLE 4a. *Offences determined on indictment—those for which women were most commonly committed for trial only (each sex expressed as % of male/female total commitments)*

	1857	1860	1865	1870	1875	1880	1885
larceny, simple							
% of male	38.5	38.1	38.2	39.0	34.5	32.6	28.5
% of female	37.7	40.8	41.5	42.8	44.2	40.3	37.4
larceny from person							
% of male	7.5	9.0	8.8	7.5	7.2	7.0	6.1
% of female	23.2	22.9	20.7	18.4	16.4	17.0	16.3
larceny by servants							
% of male	4.6	4.9	4.6	4.1	5.0	3.5	2.5
% of female	6.7	6.5	6.7	5.5	6.0	5.0	4.9
receiving stolen goods							
% of male	2.6	2.6	2.4	2.9	2.1	2.0	1.7
% of female	4.5	4.2	4.3	4.6	3.3	3.6	2.6
fraudulently obtaining goods							
% of male	3.8	4.9	3.9	5.6	5.6	7.6	8.0
% of female	3.9	4.4	3.7	6.3	4.7	7.4	7.9
larceny to £5 in dwelling house							
% of male	2.6	2.5	2.0	2.0	1.8	2.2	1.5
% of female	3.0	3.5	2.4	3.3	2.4	2.4	2.5
murder							
% of male	0.4	0.5	0.5	0.3	0.7	0.4	0.7
% of female	1.0	1.0	1.1	0.7	1.3	0.5	0.8

concealing of birth							
% of male	—	—	—	—	—	—	—
% of female	1.6	2.0	2.7	1.9	2.5	2.3	2.0
burglary							
% of male	7.3	6.3	6.8	7.0	5.4	8.0	8.3
% of female	2.2	2.4	2.9	2.8	1.8	3.6	3.3
uttering counterfeit coin							
% of male	3.5	3.0	2.0	1.6	1.5	2.0	2.3
% of female	5.9	4.4	3.1	2.7	1.4	2.2	1.2
keeping disorderly house							
% of male	0.1	0.2	0.2	0.2	0.2	0.2	0.3
% of female	0.8	1.0	1.9	1.0	0.7	1.0	0.7
other offences							
% of male	29.1	28.0	28.4	29.8	36.0	34.5	40.1
% of female	9.5	6.9	9.0	10.0	15.3	14.7	20.4
TOTAL							
male	12,940	11,150	14,432	13,353	10,738	11,345	10,489
female	3,563	3,648	3,900	3,389	3,023	2,763	2,056
total	16,503	14,798	18,332	16,742	13,761	14,108	12,545

TABLE 4b. *Summary of indictable offences—those for which women were most commonly committed for trial only (1857 and 1860 not available; each sex expressed as % of that sex's total commitments)*

	1857	1860	1865	1870	1875	1880	1885	1890
offences against the person								
% of male			12.7	11.6	18.3	15.9	17.2	20.2
% of female			8.0	7.2	11.7	10.5	12.2	13.0
against property with violence								
% of male			12.5	12.6	11.7	15.2	16.2	17.6
% of female			4.3	4.6	3.3	5.1	5.2	6.1
against property without violence								
% of male			66.7	67.9	62.9	61.3	54.6	52.2
% of female			80.2	81.4	77.6	76.2	71.8	69.5
malicious offences against property								
% of male			1.9	1.6	1.3	2.0	2.0	2.0
% of female			0.7	0.1	1.0	0.7	1.5	2.3

forgery and offences against currency							
% of male			3.0	2.7	2.9	3.9	4.3	2.4
% of female			3.7	3.4	2.1	3.1	3.9	2.8
not included above							
% of male			3.2	3.5	2.9	1.8	5.8	5.6
% of female			3.1	2.6	4.4	4.3	5.4	6.3
TOTAL							
male	12,940	11,150	14,432	13,353	10,738	11,345	10,489	9,039
female	3,563	3,648	3,900	3,389	3,023	2,763	2,056	1,722

TABLE 4c. *Disposal of persons apprehended for indictable offences*

	1857	1860	1865	1870	1875	1880	1885	1890
persons apprehended[a]								
male %	73.1	72.6	75.9	76.2	75.9	78.2	81.0	80.7
female %	26.9	27.4	24.1	23.8	24.1	21.8	19.0	19.3
total	32,031	24,862	29,049	26,613	22,108	22,231	19,207	17,678
persons discharged[b]								
% of men	38.1	32.1	27.4	22.3	22.5	21.6	19.5	17.9
% of women	54.9	42.1	39.6	33.0	28.4	27.4	26.3	25.5
total	13,631	8,659	8,814	8,245	6,617	6,481	5,005	4,998
persons committed for trial[c]								
% of men	61.9	67.9	72.6	77.3	77.5	78.4	80.5	82.1
% of women	45.1	57.9	60.4	77.0	71.6	72.6	73.7	74.5
total	18,390	16,203	20,235	18,368	15,491	15,750	14,202	12,680

[a] Both sexes expressed as % of total.
[b] Expressed as % of men/women apprehended.
[c] Including those bailed for trial and those committed for want of sureties; expressed as % of men/women apprehended.

TABLE 5. *Population of local prisons*

	1857	1860	1865	1870	1875	1880
greatest number of prisoners at one time						
male	18677	14989	17957	19537	17458	19190
female	4962	4567	4803	4858	5095	4840
daily average number of prisoners						
male	15002	—	14334	15881	14433	15929
as % of total	79.0	—	78.9	80.1	78.1	80.3
female	3987	—	3827	3949	4054	3906
as % of total	21.0	—	21.1	19.9	21.9	19.7
TOTAL	18989	—	18161	19830	18487	19835

TABLE 6. *Nature of commitments to local prisons (expressed as % of male/female total commitments)*

	1857	1860	1865	1870	1875	1880
awaiting trial at assizes and sessions						
% of male	14.9	14.2	7.4	10.2	9.5	7.6
% of female	12.3	12.9	11.8	7.9	6.5	5.0
total	20,212	16,190	17,953	16,235	14,691	12,350
on summary conviction						
% of male	57.9	56.2	35.7	72.3	74.8	74.8
% of female	71.2	71.9	75.4	82.3	85.3	86.7
total	86,795	70,151	92,665	125,730	132,594	139,546
for want of sureties						
% of male	2.2	2.9	1.2	1.9	19.8	1.3
% of female	2.4	2.8	2.4	2.0	1.8	1.4
total	3,163	3,309	3,073	3,242	3,303	2,367
on remand and discharged						
% of male	9.9	9.1	4.1	6.5	6.5	6.3
% of female	11.6	10.3	8.0	6.2	6.0	6.0
total	14,653	10,964	10,367	10,751	10,900	11,090

for debt and on civil process						
% of male	12.6	13.0	4.7	6.7	3.9	6.0
% of female	2.4	2.1	1.3	1.0	0.3	0.5
total	14,339	11,707	9,443	8,804	4,845	8,187
under the Mutiny Act						
% of male	2.6	4.6	1.0	1.7	3.3	3.5
% of female	—	—	—	—	—	—
total	2,808	3,961	1,940	2,107	3,967	4,619
male commitments	107,384	85,513	102,949	125,796	121,344	131,621
female commitments	34,586	30,769	32,492	41,073	48,956	46,538
TOTAL	141,970	116,282	135,441	166,769	170,300	178,159

TABLE 7. *Previous commitments to local prisons (expressed as % male/female total commitments)*

	1857	1860	1865	1870	1875	1880
those with previous commitments						
% of male	26.4	26.1	29.1	32.3	33.6	33.5
% of female	39.8	42.4	41.3	45.5	51.3	53.0
females as % of total with previous commitments	32.6	36.9	30.9	31.5	38.1	35.8
TOTAL	42,169	35,381	43,964	59,698	65,871	69,062
with more than 10 previous commitments						
% of male	0.7	1.0	1.2	1.7	2.6	2.8
% of female	5.0	8.4	7.3	8.0	11.4	14.5
females as % of total with more than 10 previous commitments	70.1	75.8	65.8	60.1	63.7	65.0
TOTAL	2,464	3,409	3,636	5,469	8,734	10,421

TABLE 8. *Previous occupation of women convicted prisoners (expressed as % of total women)*

	1857	1860	1865	1870	1875	1880
domestic servants	10.4	10.2	9.9	9.2	7.5	7.8
labourers, char/needlewomen	23.9	28.5	24.8	26.1	27.5	24.1
factory workers	6.0	6.2	7.0	8.2	11.5	12.9
mechanics/skilled workers	1.9	2.3	2.3	2.4	2.6	2.3
shopwomen and clerks	0.2	0.2	0.3	0.2	0.2	0.2
shopkeepers and dealers	4.2	3.0	3.8	4.9	4.1	7.7
professional employments	0.1	0.1	—	—	0.1	0.1
prostitutes	—	—	19.2	21.0	18.3	16.3
no occupation	51.6	48.2	32.0	27.9	28.0	28.3
not ascertained	1.7	1.3	0.7	0.1	0.2	0.2
TOTAL	33,746	100,614	32,423	40,983	48,801	46,439

TABLE 9. *Offences for which sentenced to penal servitude (types of offence as % of total offences; both sexes expressed as % of total within type)*

	1870	1875	1880	1885	1890
offences against the person	11.2	16.1	12.7	16.6	21.0
male %	94.0	95.1	94.6	93.6	93.5
female %	6.0	4.9	3.6	6.4	6.5
against property with violence	17.4	14.8	18.0	19.6	22.5
male %	95.8	94.6	96.4	97.0	98.2
female %	4.2	5.4	3.6	3.0	1.8
against property without violence	62.3	61.5	60.2	52.3	47.3
male %	82.1	76.0	78.4	84.5	90.7
female %	17.9	24.0	21.6	15.5	9.3
malicious offences against property	2.7	1.3	2.6	2.0	2.8
male %	95.9	90.9	100.0	95.2	95.0
female %	4.1	9.1	—	4.8	5.0
against the currency	5.3	5.0	6.0	8.1	5.6
male %	85.1	85.2	92.4	94.0	85.4
female %	14.9	14.8	7.6	6.0	14.6
not included above	1.1	1.3	0.5	1.4	0.8
male %	100.0	77.3	100.0	92.9	83.3
female %	—	22.7	—	7.1	16.7

TABLE 10. *Population of convict prisons (both sexes expressed as % of total)*

	1857	1860	1865	1870	1875	1880	1885	1890
received during year								
male (no.)	6,236	5,398	5,309	5,203	5,661	1,642	4,535	1,784
as %	89.2	83.7	86.5	86.5	89.2	85.8	91.9	94.9
female (no.)	758	1,050	830	810	685	272	401	95
as %	10.8	16.3	13.5	13.5	10.8	14.2	8.1	5.1
TOTAL	6,994	6,448	6,139	6,013	6,346	1,914	4,936	1,879
daily average population								
male (no.)	6,540	6,450	5,971	8,028	8,621	9,184	7,640	4,568
as %	87.9	83.8	83.1	87.1	87.5	89.2	91.6	93.8
female (no.)	899	1,249	1,213	1,192	1,236	1,113	699	302
as %	12.1	16.2	16.9	12.9	12.5	10.8	8.4	6.2
TOTAL	7,439	7,699	7,184	9,220	9,857	10,297	8,339	4,870

TABLE 11. *Previous convictions and sentences (taken from Penal Servitude Acts Commission 1878–9—snapshot of convict population on 6 May 1878 only))*

	Total		As % of sex	
	male	female	male	female
no previous convictions of any kind	2,064	124	23.0	10.1
no previous sentence to penal servitude	4,672	635	52.0	51.8
previous sentences to penal servitude				
one	1,620	331	18.0	27.0
two	520	110	5.8	9.0
three	97	26	1.1	2.1
four	9	—	0.1	—
five	1	—	—	—
TOTAL	8,983	1,226	100.0	100.0

TABLE 12. *Prison character (taken from Penal Servitude Acts Commission 1878–9—snapshot of convict population on 6 May 1878 only)*

	Total		As % of sex	
	male	female	male	female
exemplary	1,674	390	18.6	31.8
very good	1,779	146	19.8	11.9
good	2,151	288	23.9	23.5
fair	1,477	140	16.4	11.4
indifferent	1,131	132	12.9	10.8
bad	514	84	5.7	6.9
very bad	257	46	2.9	3.8
TOTAL	8,983	1,226	100.0	100.0

TABLE 13a. *Removals from convict prisons after ending of transportation (1868) (each sex expressed as % of that sex's total)*

	1870	1875	1880	1885	1890
discharged on ticket-of-leave					
male %	80.8	78.3	76.7	75.2	78.2
female %	84.3	78.7	79.9	76.1	75.2
discharged at end of sentence					
male %	6.9	11.8	12.3	12.7	16.1
female %	6.5	13.3	11.3	56.8	20.4
died					
male %	7.0	7.6	7.3	3.4	2.9
female %	7.8	7.3	4.7	2.6	1.8
otherwise removed					
male %	5.3	1.2	3.7	8.7	2.8
female %	1.4	0.7	4.1	4.5	2.7
TOTAL					
male	1,426	1,724	1,707	1,903	1,327
female	293	286	319	268	113

TABLE 13b. *Removals from convict prisons—total figures for period 1868 to 1876 (taken from Penal Servitude Acts Commission 1878–9)*

	male	female	total
no. licensed	10,565	2,095	12,660
as % of total removed	88.1	86.2	
no. of licences revoked	838	304	
as % of licences	7.9	14.5	
no. discharged by expiration of sentence	1,423	336	1,759
as % of total removed	11.9	13.8	
TOTAL	11,988	2,431	14,419

BIBLIOGRAPHY

I. Primary Sources
   (A) Archival
   (B) Annual Reports
   (C) Reports of Royal Commissions and Committees
   (D) Miscellaneous Parliamentary Papers
   (E) Home Office Printed Memoranda
   (F) Printed Sources
II. Secondary Sources
   (A) Doctoral Theses
   (B) Unpublished Papers
   (C) Published Sources

I. PRIMARY SOURCES

(A) ARCHIVAL

Greater London Record Office

GLRO MJ/OC 21. Middlesex Sessions Records—Visits of Mrs Fry and Other Friends to the House of Correction, Middlesex (1825–6).

GLRO MA/RS/1/550–725 (175 vols.). House of Correction, Westminster—Visiting Justices Reports, Annual Returns and Accounts etc. (1846–78).

GLRO WA/G/5–14. Minute Book of Westminster House of Correction—Minutes of Meetings of the Visiting Justices (1849–78).

GLRO MA/RS/1/742. Reports of the Governors and Chaplains of the Houses of Correction, at Cold Bath Fields and Westminster, on the Causes of Crime (1850).

GLRO ACC 1159/18. Minute Book of Middlesex Society to Promote the Reformation and Employment of Discharged Criminals (1856–79).

GLRO MA/G/Gen/1244. Rules and Regulations for the House of Correction at Westminster (1861).

GLRO MA/RS/2/71–90. Reports of Middlesex Prisons (1862–77).

GLRO MA/G/Gen/1252. Middlesex House of Correction at Westminster amended proposed Regulations and Rules for the Government of the Prison (1865).

GLRO MF 487. Particulars of Officers at House of Correction, Westminster (1877).

GLRO A/LWC/1. London Diocesan Council for Penitentiary, Rescue and Preventative Work—Minute Book (1889–1912).

GLRO LCC/MIN/8114–8115. London County Council (LCC) Inebriates Acts Subcommittee Minutes (1899–1904).

GLRO LCC/MIN/8116–8128. LCC Inebriates Acts Subcommittee Minutes (1904–14).

GLRO PC/GEN/1/7. LCC Inebriates Acts Subcommittee Misc. Printed Reports (1900–13).

HM Youth Custody Centre, Aylesbury

Aylesbury State Inebriate Reformatory Minute Book (1903–1917) (by permission of the Governor).

London School of Economics and Political Science

Jebb Papers (Boxes 4–12).

Public Record Office, Kew, London

Papers held at HO12; HO14; HO20; HO21; HO22; HO24; HO45; HO144; Pri Com 2; Pri Com 7.

(B) ANNUAL REPORTS

Reports of the Surveyor-General, *PP* (1847 to 1857–8). Irish University Press reprints 1970.

Reports of the Inspectors appointed to visit the different Prisons of Great Britain, *PP* (1847–77).

Reports of the Directors of Convict Prisons (RDCP), *PP* (1852–95).

'Judicial Statistics', *Parliamentary Papers* (hereafter *PP*) (1857–1913).

Reports of the Commissioners of Prisons and Directors of Convict Prisons, *PP* (1896–1914).

Reports of the Inspector under the Inebriates Acts, *PP* (1899–1902).

Reports of the Inspector under the Inebriates Acts, 1879 to 1899, *PP* (1900–2).

Reports of the Inspector under the Inebriates Acts, 1879 to 1900, *PP* (1902–11).

Reports of the Board of Control, *PP* (1914–16).

(C) REPORTS OF ROYAL COMMISSIONS AND
COMMITTEES

Report from the Select Committee on Inquiry into Drunkenness, *PP*, 8 (1834).

First Report from the Select Committee of the House of Lords appointed to inquire into the present State of the Several Gaols and Houses of Correction, *PP*, 11 (1835).

Report from the Select Committee on Prison Discipline, *PP*, 17 (1850).

Report from the Select Committee on Public Houses, *PP*, 3 (1852–3).

Report from the Select Committee on Public Houses, *PP*, 14 (1854).

Reports from the Select Committee on Transportation, *PP*, 17 (1856).

Report from the Select Committee on Transportation, *PP*, 13 (1861).

Report from the Select Committee of the House of Lords on the present State of Discipline in Gaols and Houses of Correction, *PP*, 9 (1863).

Report of the Commissioners appointed to inquire into the Operation of the Acts relating to Transportation and Penal Servitude, *PP*, 21 (1863).

Report from the Select Committee on the Control and Management of Habitual Drunkards, *PP*, 9 (1872).

Report from the Select Committee of the House of Lords on Intemperance, *PP*, 11 (1877).

Report of the Committee into the Dietaries of the Prisons (1878), *Home Office Printed Memoranda*, iv. 427.

Report of the Commissioners appointed to inquire into the Working of the Penal Servitude Acts, *PP*, 37–8 (1878–9).

Report from the Select Committee of the House of Lords on Intemperance, *PP*, 10 (1878–9).

Report of the Commission appointed to inquire into the Subject of Criminal Lunacy, *PP*, 32 (1882).

Report of the Departmental Committee on the Treatment of Inebriates, *PP*, 17 (1893–4).

Report of the Departmental Committee on Prisons, *PP*, 56 (1895).

Report of the Departmental Committee appointed to revise the Criminal Portion of the Judicial Statistics, *PP*, 108 (1895).

Reports of the Departmental Committee to advise as to the Regulations to be made under the Inebriates Act 1898, *PP*, 12 (1899).

Report of the Departmental Committee appointed to inquire into the Operation of the Law Relating to Inebriates and to their Detention in Reformatories and Retreats, *PP*, 12 (1908).

Royal Commission on the Care and Control of the Feeble-Minded, *PP*, 35–9 (1908).

(D) MISCELLANEOUS PARLIAMENTARY PAPERS

A Bill for Providing Places of Confinement in England or Wales for Female Offenders under Sentence or Order of Transportation, *PP*, 3 (4 Aug. 1853).
Copy of Correspondence between the Secretary of State for the Home Department and the Directors of Convict Prisons, on the subject of the Recommendations of the Royal Commission on Penal Servitude, *PP*, 49 (1864).
Women and Children in Public-Houses: Information obtained from certain Police Forces as to the Frequenting of Public-Houses by Women and Children, *PP*, 89 (1908).

(E) HOME OFFICE PRINTED MEMORANDA

Memorandum respecting Decrease in Crime, and especially in the Crimes affected by the Habitual Criminals Act, 1869 (22 Oct. 1872), xxiv.
Habitual Drunkards Act 42 and 43 Vict. Cap. 19. Model Rules for Retreats (1881) and *Additional Model Rules for the Management of Retreats licensed under the above Act* (Aug. 1883), xxv.
Suppression of Vice (July 1894), xvi.
Accommodation at Metropolitan Police Courts—Report by the Chief Magistrate (23 Feb. 1895), xviii.
Confidential. Capital Cases. Infanticide. (Mar. 1900), xvii.
Strictly Confidential. Memorandum on Home Office Practice in dealing with Criminal Lunatics (1 Nov. 1913), xxiv.

(F) PRINTED SOURCES

ACTON, WILLIAM, *Prostitution: Considered in its Moral, Social, and Sanitary Aspects in London and other large Cities and Garrison Towns* (2nd edn., London, Frank Cass, 1972; 1st edn. 1857).
ADAM, HARGRAVE, *Woman and Crime* (London, T. Werner Laurie, 1914).
ALFORD, STEPHEN S., 'The Necessity of Legislation for the Control and Cure of Habitual Drunkards', *Transactions of the National Association for the Promotion of Social Science (NAPSS) 1876* (1877).
A.M.M., 'The Decrease of Crime', *The Englishwoman's Review*, no. 301 (15 Feb. 1890).

AMOS, SARAH M., 'The Prison Treatment of Women', *Contemporary Review*, 73:390 (June 1898).

ANON., 'The Care of the Feeble-minded', *The Humanitarian*, 13:1 (July 1898).

ANON., 'Certified Inebriate Reformatories', *Justice of the Peace*, 63:2 (14 Jan. 1899).

ANON., 'Drawing-Room Alcoholism', *The Englishwoman's Review* (NS), 6 (Apr. 1871).

ANON., 'Drunken Women', *British Workman*, 8 (1862).

ANON., 'Feeble-Minded Women and Degeneracy', *British Medical Journal*, 1 (6 May 1911).

ANON., 'Female Convicts', *Victoria Magazine*, 2 (Nov.–Apr. 1863–4).

ANON., 'Female Habitual Drunkards', *The Englishwoman's Review*, 9:64 (15 Aug. 1878).

ANON., 'Female Inebriety', *British Medical Journal*, 11 (July–Dec. 1892).

ANON., 'Female Life in Prison', *The English Woman's Journal*, 10:55 (1 Sept. 1862).

ANON., 'The Foundation of the Society for the Study of Inebriety', *British Journal of Inebriety*, 1:1 (July 1903).

ANON., 'Gone to Jail', *All The Year Round*, 7:171 (2 Aug. 1862).

ANON., 'Habitual Drunkards', *Law Times*, 115 (16 May 1903).

ANON., 'Habitual Inebriates in England', *Justice of the Peace*, 69:29 (22 July 1905).

ANON., *How to Help the Feeble-Minded*, issued by the National Association for Promoting the Welfare of the Feeble-Minded (1899).

ANON., 'The Increase of Female Inebriety and its Remedy', *British Medical Journal*, 11 (July–Dec. 1892).

ANON., 'Indiana Female Prison', *The Englishwoman's Review* (NS) (15 Dec. 1877).

ANON., 'Inebriety in Women', *British Journal of Inebriety*, 1:1 (July 1903).

ANON., 'Intemperance and the Social Evil', *Temperance Spectator*, 2 (Jan.–Dec. 1860).

ANON., 'Maternal Inebriety', *The Humanitarian*, 16:3 (Mar. 1900).

ANON., 'The Mental and Moral Dignity of Woman', *The Female's Friend* (Jan. 1846).

ANON., 'The Operation of the Inebriates Act 1898', *Law Times*, 107 (27 May 1899).

ANON., 'The Petting and Fretting of Female Convicts', *Meliora*, 6 (1864).

ANON., 'Physical Deterioration and Alcoholism', *Justice of the Peace*, 69:49 (9 Dec. 1905).

Bibliography 333

ANON., 'Prison Mission, Wandsworth', *The Englishwoman's Review*, 12:98 (15 June 1881).
ANON., 'Prison Photographs', *All The Year Round*, 15:373 (16 June 1866).
ANON., 'The Problem of the Feeble-Minded', *British Medical Journal*, 1 (30 Apr. 1910).
ANON., 'A Public Scandal', *British Medical Journal*, 1 (Jan.–June 1895).
ANON., 'Reclamation of Female Drunkards', *The Englishwoman's Review*, 9:66 (15 Oct. 1878).
ANON., 'Reform of Female Drunkards', *The Englishwoman's Review* (NS), 8:55 (15 Nov. 1877).
ANON., 'The Report of the Committee appointed to consider the Eugenic Aspect of Poor Law Reform—Section II The Eugenic Principle and the Treatment of the Feeble-Minded', *Eugenics Review*, 2:3 (Nov. 1910).
ANON., 'State Inebriates', *Law Times*, 122 (1 Dec. 1906).
ANON., 'Statistics of Drunkenness', *The Englishwoman's Review* (NS), 59 (15 Mar. 1878).
ANON., 'The Treatment of Inebriates', *British Medical Journal*, 1 (1911).
ANON., 'Women and Crime', *The Englishwoman's Review* (NS), 26–7:227 (15 Oct. 1895).
ANON., 'Women Convicts', *The Englishwoman's Review* (NS), 18:173 (15 Oct. 1887).
ARMSTRONG, JOHN, 'Female Penitentiaries', *Quarterly Review*, 83 (Sept. 1848).
ARNOT, Revd WILLIAM, 'The Criminality of Drunkenness with the Consequent Rights and Duties of Society in regard to the Criminals', *Transactions NAPSS 1859* (1860).
ASTOR, Lady, *Why Students should be Interested in the Drink Question*, privately printed pamphlet (London, 1922).
ATKINSON, STANLEY, 'The Medico-Legal Relations of Alcoholism', in T. N. Kelynack (ed.), *The Drink Problem* (1907).
—— 'The Care of Children Neglected by Drunken Parents', *National Conference on Infantile Mortality, 23–25 Mar. 1908, Report of Proceedings* (Westminster, P. S. King, 1908).
BARTLEET, R. S., 'Social Results of the Employment of Girls and Women'. *Transactions NAPSS 1868* (1869).
BEBEL, AUGUST, *Woman in the Past, Present and Future* (trans. London, Modern Press, 1885).
BEDFORD, ADELINE M., 'Treatment of Women in Prisons', *International Congress of Women* (1899).

BEDFORD, ADELINE M. 'Fifteen Years' Work in a Female Convict Prison', *The Nineteenth Century and After*, 68:404 (Oct. 1910).

BEGGS, THOMAS, 'What are the Principal Causes of Crime, considered from a Social Point of View?', *Transactions NAPSS 1868* (1869).

BEWICKE, A., Contribution in 'Discussion on the Habitual Drunkards Act', *Transactions NAPSS 1883* (1884).

BOGELOT, Madame, 'Prison Reform Work of St. Lazare, Paris', *Report of the International Council of Women* (Mar.–Apr. 1888).

BOOTH, Mrs BRAMWELL, *Mothers and the Empire and other Addresses* (London, Salvation Army, 1914).

BOSANQUET, CHARLES B. P., *London: Some Account of its Growth, Charitable Agencies and Wants* (London, Hatchard, 1868).

BRANTHWAITE, R. W., 'Inebriety and Mental Defect', *Report of the National Conference on the Prevention of Destitution 1911*.

BREMNER, JOHN A., 'What Improvements are Required in the System of Discipline in County and Borough Prisons?', *Transactions NAPSS 1873* (1874).

British Ladies' Society for Promoting the Reformation of Female Prisoners, *Twenty-first Report of the Committee* (printed by Richard Barrett, 1844).

British Society for Promoting the Reformation of Female Prisoners, *Second Annual Report of the Committee* (printed by William Belch, c.1824).

BUCKNILL, JOHN CHARLES, 'Habitual Drunkenness: A Vice, Crime, or Disease?', *Contemporary Review*, 29 (Feb. 1877).

BURDETT-COUTTS, Baroness (ed.), *Woman's Mission: A Series of Congress Papers on the Philanthropic Work of Women* (London, Sampson Low, Marston, 1893).

BURT, Revd J. T., 'On the Adaption of Punishment to the Causes of Crime', *Transactions NAPSS 1857* (1858).

CARPENTER, MARY, *Juvenile Delinquents: Social Evils, their Causes and their Cure* (London, Cash, 1853).

—— 'Reformatories for Convicted Girls', *Transactions NAPSS 1857* (1858).

—— 'On the Treatment of Female Convicts', *Transactions NAPSS 1863* (1864).

—— *Our Convicts* (2 vols., London, Longman, 1864).

—— *Reformatory Prison Discipline* (London, Longman, 1872).

CARTER, Revd THOMAS, 'On the Crime of Liverpool', *Transactions NAPSS 1858* (1859).

C.E.B., 'A New Danger for Women', *The Englishwoman's Review* (NS), 43 (15 Nov. 1876).

Charity Organisation Series, *The Feeble-Minded Child and Adult* (London, Swan Sonnenschein, 1893).

CHESTERTON, GEORGE LAVAL, *Revelations of Prison Life* (2 vols., London, Hurst and Blackett, 1856).

CHICHESTER, Earl of, *Report and Observations on the Discipline and Management of Convict Prisons by the late Major-General Sir J. Jebb* (London, Hatchard, 1863).

CLAY, Revd W. L., *The Prison Chaplain: A Memoir of the Late Revd John Clay* (London, Macmillan, 1861).

—— 'On Recent Improvements in our System for the Punishment and Reformation of Adult Criminals', *Transactions NAPSS 1865* (1866).

CLAYE SHAW, T., 'Psychology of the Inebriate Mother', *British Journal of Inebriety*, 1:2 (Oct. 1903).

COLLINS, Sir W. J., 'An Address on the Institutional Treatment of Inebriety', *British Journal of Inebriety*, 1:3 (Jan. 1904).

COOTE, WILLIAM ALEX., 'The Feeble-Minded and Rescue Work', *Report of the National Conference on the Prevention of Destitution 1911*.

CRAIG, ISA, 'Emigration as a Preventative Agency', *The English Woman's Journal*, 2:11 (1 Jan. 1859).

CROFTON, WALTER, 'Repression of Crime: on Female Convicts', *Law Times*, 41 (20 Oct. 1866).

—— 'Female Convicts, and Our Efforts to Amend Them', *Transactions NAPSS 1866* (1867).

—— 'Female Criminals: Their Children's Fate', *Good Words for 1873* (1873).

—— 'The Training of Prison Officers', *Transactions NAPSS 1881* (1882).

DALRYMPLE, DONALD, 'What Measures may be Adopted with a View to the Repression of Habitual Drunkenness?', *Transactions NAPSS 1870* (1871).

—— 'Asylum for Drunkards', *MacMillan's Magazine*, 26 (May–Oct. 1872).

DAVENPORT-HILL, FLORENCE, 'Women Prison Visitors', *The Englishwoman's Review* (NS), 16:152 (15 Dec. 1885).

DAVENPORT-HILL, MATTHEW, 'On Irish Convict Prisons', *Transactions NAPSS 1857* (1858).

—— 'Brief Remarks on the Treatment of Criminals under Imprisonment for Life', *Transactions NAPSS 1866* (1867).

DAVEY, HERBERT, *The Law Relating to the Mentally Defective, The Mental Deficiency Act, 1913 (3 & 4 Geo. V, c. 28)* (London, Stevens, 1913).

DENDY, MARY, 'Can the Feeble-Minded Be Made Happy in Confinement? The Experiment at Sandlebridge', in E. Fry *et al.*, *The Problem of the Feeble-Minded* (1909).

—— 'Feeble-Mindedness, Destitution, and Crime', *Report of the National Conference on the Prevention of Destitution 1911*.

336 *Bibliography*

DEVON, JAMES, *The Criminal and the Community* (New York, John Lane, 1912).

DICKENS, CHARLES, *Oliver Twist* (Middlesex, Penguin Books, 1978; 1st edn. 1837–9).

DICKINSON, W. H., *The Treatment of the Feeble-minded* (London, P. S. King, 1903).

—— 'Homes and Colonies for the Feeble-minded', *Report of the National Conference on the Prevention of Destitution 1911*.

DIXON, HEPWORTH, *The London Prisons* (London, Jackson and Walford, 1850).

DU CANE, E. F., 'Address on the Repression of Crime', *Transactions NAPSS 1875* (1876).

—— *The Punishment and Prevention of Crime* (London, Macmillan, 1885).

—— 'An Account of the Manner in which Sentences of Penal Servitude are Carried out in England', in E. Pears (ed.), *Prisons at Home and Abroad: Transactions of the International Penitentiary Congress London, 1872*, printed for private circulation (HM Prison Maidstone, 1912).

EDMUNDSON, MARY, SIBTHORPE, FANNY, and EUSTACE, ELIZABETH J., 'Dublin Prison Gate Mission for Women', *The Englishwoman's Review*, 13 (15 May 1882).

ELLIS, HAVELOCK, *The Criminal* (London, Walter Scott, 1890).

—— *Man and Woman: A Study of Human Secondary Sexual Characters* (4th edn. London, Walter Scott, 1904; 1st edn. 1894).

FARRAR, F. W., 'Drink and Crime', *Fortnightly Review*, 59 (June 1893).

FAULKS, E., 'The Sterilisation of the Insane', *Journal of Mental Science*, 57 (1911).

The Female's Friend. Under the Sanction of The Association Institution, For Improving and Enforcing the Laws for the Protection of Women, i–iv (London, Houlston and Stoneman, Jan.–Apr. 1846).

FIELD, Revd J., *Prison Discipline* (London, Longman, 1846).

FLETCHER, SUSAN WILLIS, *Twelve Months in an English Prison* (Boston, Lee and Shepard, 1884).

FOX, EVELYN, 'The Mental Deficiency Act and its Administration', *Eugenics Review*, 10:1 (Apr. 1918).

FRAZER, CATHERINE, 'The Origin and Progress of the British Ladies' Society for Promoting the Reformation of Female Prisoners, Established by Mrs Fry in 1821', *Transactions NAPSS 1862* (1863).

FRY, Sir EDWARD, *et al.*, *The Problem of the Feeble-Minded: An Abstract of the Report of the Royal Commission on the Care and Control of the Feeble-Minded* (London, P. S. King, 1909).

ЛДHe yкa

FRY, ELIZABETH, *Observations on the Visiting, Superintending, and Government of Female Prisons* (2nd edn. London, John and Arthur Arch, 1827).

—— *Sketch of the Origin and Results of Ladies' Prison Associations, with Hints for the Formation of Local Associations* (London, John and Arthur Arch, 1827).

FULLER, L. O., 'Alcoholism, Crime and Insanity', *Journal of Mental Science*, 55 (Oct. 1909).

FYFFE, C. A., 'The Punishment of Infanticide', *The Nineteenth Century*, 1:4 (June 1877).

GIBB, DAVID, 'The Relative Increase of Wages, of Drunkenness, and of Crime', *Transactions NAPSS 1874* (1875).

GORDON, MARY, *Penal Discipline* (London, George Routledge, 1922).

GREENWOOD, JAMES, *The Seven Curses of London: Scenes from the Victorian Underworld* (Oxford, Basil Blackwell, 1981; 1st edn. 1869).

GREG, W. R., *Why are Women Redundant?* (London, N. Trubner, 1869).

GRIFFITHS, Major ARTHUR, *Memorials of Millbank and Chapters in Prison History* (London, Chapman & Hall, 1884).

—— *Secrets of the Prison-house or Gaol Studies and Sketches*, 2 vols. (London, Chapman & Hall, 1894).

—— *Fifty Years of Public Service* (London, Cassell, 1904).

GURNEY, J. J., *Notes on a Visit made to some of the Prisons of Scotland in Company with Elizabeth Fry* (Constable, 1819).

GUY, W. A., 'On some Results of a Recent Census of the Population of the Convict Prisons in England; And especially on the Rate of Mortality at present Prevailing among Convicts', *Transactions NAPSS 1862* (1863).

An Habitual Drunkard, 'Habitual Drunkenness', *Westminster Review*, 129:5 (May 1888).

HARRIS, VERNON, 'The Female Prisoner', *The Nineteenth Century and After*, no. 363 (May 1907).

HASLAM, JAMES, 'Women and the Nation: The Outcasts. I. Introductory', *The Englishwoman*, 2:4 (May 1909).

HIGGS, MARY, *Glimpses into the Abyss* (London, P. S. King, 1906).

HILL, OCTAVIA, *Homes of the London Poor* (London, Frank Cass, 1970; 1st edn. 1875).

HILL, ROSAMOND, 'A Plea for Female Convicts', *The English Woman's Journal*, 13:74 (Apr. 1864).

HOLLANDER, BERNARD, *Crime and Responsibility* (London, Ethological Society, 1908).

HOLMES, THOMAS, 'Habitual Inebriates', *Contemporary Review*, 75:401 (May 1899).

HOLMES, THOMAS, *Pictures and Problems from the London Police Courts* (London, Thomas Nelson, 1900).

—— 'The Criminal Inebriate Female', *British Journal of Inebriety*, 1:2 (Oct. 1903).

—— *Known to the Police* (London, Edward Arnold, 1908).

—— 'Crime and Mental Defect', *Report of the National Conference on the Prevention of Destitution 1911*.

—— *London's Underworld* (London, J. M. Dent, 1912).

Home Office, *Suffragist Women Prisoners*, pamphlet (Home Office, 1909).

HOPKINS, TIGHE, 'The State Drunkard', *Law Times*, 117 (6 Aug. 1904).

—— *Wards of the State: An Unofficial View of Prison and the Prisoner* (London, Herbert and Daniel, 1913).

HORSLEY, Canon J. W., *Jottings from Jail: Notes and Papers on Prison Matters* (London, T. Fisher Unwin, 1887).

—— *Prisons and Prisoners* (London, C. Arthur Pearson, 1898).

—— *How Criminals are Made and Prevented: A Retrospect of Forty Years* (London, T. Fisher Unwin, 1913).

HORSLEY, Sir VICTOR, and STORGE, MARY D., *Alcohol and the Human Body* (6th edn., enlarged. London, Macmillan, 1920, 1st edn. 1907).

Howard Association, *Annual Reports 1872–1910* (London, Howard Association).

—— 'County and Borough Prisons', a letter from William Tallack to *The Times*, 6 Dec. 1879 (London, Howard Association, 1880).

—— *The Paris Prison Congress 1895: Summary Report* (London, issued by Howard Association, 1895).

HOYLE, WILLIAM, *Crime in England and Wales: An Historical and Critical Retrospect* (London, Effingham, Wilson, 1876).

HUBBARD, LOUISA M., *The Englishwoman's Year Book for 1881* (London, Hatchards, 1881).

HUNT, THORNTON, 'The English Convict System', *Cornhill Magazine*, 3 (1861).

HUTCHISON, EVALINE, 'Women in the Police Courts', *The Englishwoman*, 20 (Oct.–Dec. 1913).

JEBB, Major-General Sir JOSHUA, 'Prison Discipline', *Transactions NAPSS 1862* (1863).

JELLICOE, ANNE, 'A Visit to the Female Convict Prison at Mountjoy, Dublin', *Transactions NAPSS 1862* (1863).

JOHNSON, HARRIET M., *Our Future Citizens*, pamphlet, proof copy (1899).

JOHNSTONE, J., 'Women and the Nation: The Outcasts. I. The Inebriates; II. Inebriety and Women; III. Inebriety and Parenthood,' *The Englishwoman*, 4 (1909–10).

JOHNSTONE, M. F., 'The Life of a Woman Convict', *Fortnightly Review*, 75 (1901).

JONES, ROBERT, 'Alcohol and National Deterioration' in T. N. Kelynack (ed.), *The Drink Problem* (1907).

KELLOR, FRANCES, *Experimental Sociology: Descriptive and Analytical* (New York and London, Macmillan, 1901).

KELYNACK, T. N. (ed.), *The Drink Problem: In its Medico-Sociological Aspects* (London, Methuen, 1907).

KERR, NORMAN, *Female Intemperance* (London, National Temperance, Jan. 1880).

—— *Inebriety: Its Etiology, Pathology, Treatment and Jurisprudence* (London, H. K. Lewis, 1888).

KINGSMILL, JOSEPH, *Chapters on Prisons and Prisoners and the Prevention of Crime* (London, Longman, 1854).

KIRBY, A. H. P., 'The Feeble-minded and the Voluntary Effort', *Eugenics Review*, 1:2 (July 1909).

KNAGGS, SAMUEL, 'The Habitual Drunkards Act', *Transactions NAPSS 1883* (1884).

LANKESTER, EDWIN, 'The Repression of Infanticide', *Transactions NAPSS 1866* (1867).

—— SAFFORD, A. HERBERT, and MAINE, Mrs, various papers under heading 'Can Infanticide be Diminished by Legislative Enactment?', *Transactions NAPSS 1869* (1870).

LANGDON-DOWN, R., 'The Mental Deficiency Bill', *Eugenics Review*, 5:2 (July 1913).

LETTSOM, ELLIOT, 'What are the Principal Causes of Crime, considered from a Social Point of View?', *Transactions NAPSS 1868* (1869).

LLOYD, M. A., *Susanna Meredith: A Record of a Vigorous Life* (London, Hodder & Stoughton, 1903).

LOMBROSO, CAESAR, and FERRERO, WILLIAM, *The Female Offender* (London, T. Fisher Unwin, 1895).

LONDON, JACK, *The People of the Abyss* (London, ARCO, 1962; written 1902).

MADDISON, ARTHUR J. S., *Hints on Aid to Discharged Prisoners* (London, Reformatory and Refuge Union, 1888).

MARTINEAU, HARRIET, 'Life in the Criminal Class', *Edinburgh Review*, 122:201 (Oct. 1865).

MAUDSLEY, HENRY, *Body and Mind: An Inquiry into their Connection and Mental Influence* (London, Macmillan, 1870).

—— *Responsibility in Mental Disease* (Henry J. King, 1874).

MAYBRICK, FLORENCE ELIZABETH, *My Fifteen Lost Years* (New York, Funk & Wagnalls, 1905).

MAYHEW, HENRY, *London Labour and London Poor*, iv, *Those That Will Not Work* (London, Dover, 1968; 1st edn. 1861–2).

MAYHEW, HENRY, and BINNEY, JOHN, *The Criminal Prisons of London and Scenes of Prison Life* (London, Griffin, Bonn, 1862).

MEADEN, W. D., 'Duties of Prison Officers', *Transactions NAPSS 1862* (1863).

MEARNS, ANDREW, *The Bitter Cry of Outcast London: An Inquiry into the Condition of the Abject Poor* (London, Frank Cass, 1970); 1st edn. 1883).

MELLAND, CHARLES H., 'The Feeble-minded in Prisons', *Report of the National Conference on the Prevention of Destitution 1911*.

MERCIER, CHARLES, *Criminal Responsibility* (Oxford, Clarendon Press, 1905).

—— *Crime and Insanity* (London, Williams and Norgate, 1911).

—— *Crime and Criminals: Being the Jurisprudence of Crime—Medical, Biological, and Psychological* (London, University of London Press, 1918).

MEREDITH, M., 'Women and the Nation: The Outcasts. II. The Feeble-Minded', *The Englishwoman*, 2:4 (May 1909).

MEREDITH, Mrs, 'The Formation of a Disqualified-for-Liberty Class of Criminal Female Offenders', *Transactions NAPSS 1870* (1871).

MEREDITH, SUSANNA, *A Book About Criminals* (London, James Nisbet, 1881).

MERRICK, G. P., 'Emigration for Female Prisoners', *The Englishwoman's Review*, 79 (15 Nov. 1879).

—— *Work among the Fallen as Seen in the Prison Cell* (London, Ward, Lock, 1891).

M.M.B., 'Correspondence. Men and Women as Habitual and Occasional Criminals. To the Editor of the "EWR"', *The Englishwoman's Review*, 19:182 (14 July 1888).

MORRISON, W. D., *Crime and its Causes* (London, Swan Sonnenschein, 1891).

MOUAT, FREDERIC, 'Address on the Repression of Crime', *Transactions NAPSS 1881* (1882).

National Association for Promoting the Welfare of the Feeble-Minded, *First Annual Report* (London, 1897).

—— *How to Help the Feeble-Minded* (London, 1899).

—— *Fifth Annual Report* (London, 1901).

—— *Report of Conference with Poor Law Guardians and Others* (London, 28 Feb. 1901).

—— *Annual Conference on After-care Report for 1911* (London, 1911).

National Conference on the Prevention of Destitution 1911, *Report of Proceedings* (London, P. S. King, 1911).

National Conference on the Prevention of Destitution 1912, *Report of Proceedings* (London, P. S. King, 1912).

National Society for the Prevention of Cruelty to Children, *World of Forgotten Children* (London, NSPCC, 1893).

—— *Justice to Children: A Ten Years' Review* (London, NSPCC, May 1894).

—— *Not a Prosecuting Society*, Letter Leaflet Series 30 (Oct. 1894).

NICOLSON, DAVID, 'The Morbid Psychology of Criminals', *Journal of Mental Science*, 19:86 (6 parts, July–Oct. 1873).

NUGENT, Revd JAMES, 'Incorrigible Women: What are We to Do with Them?', *Transactions NAPSS 1876* (1877).

ORME, ELIZA, 'Our Female Criminals', *Fortnightly Review*, 69 (1898).

OWEN, Mrs M. E., 'Criminal Women', *Cornhill Magazine*, 14 (1866).

PEARS, EDWIN (ed.), *Prisons at Home and Abroad: Transactions of the International Penitentiary Congress London, 1872*, printed for private circulation (HM Prison Maidstone, 1912).

PIKE, LUKE OWEN, *A History of Crime in England*, ii, *From the Accession of Henry VII to the Present Times* (London, Smith, Elder, 1876).

PLINT, THOMAS, *Crime in England* (London, Charles Gilpin, 1851).

POTTS, W. A., 'Causation of Mental Defect in Children', *British Medical Journal*, 2 (14 Oct. 1905).

A Prison Matron (attrib. Frederick William Robinson), *Female Life in Prison* (2 vols., London, Hurst & Blackett, 1862).

—— *Memoirs of Jane Cameron: Female Convict* (2 vols., London, Hurst & Blackett, 1864).

—— *Prison Characters drawn from Life with Suggestions for Prison Government* (2 vols., London, Hurst & Blackett, 1866).

RANKEN, W. BAYNE, 'The Origin and Progress of the Discharged Prisoners' Aid Society', *Transactions NAPSS 1858* (1859).

—— 'On Adults after their Release from Convict Prisons', *Transactions NAPSS 1860* (1861).

RENTOUL, ROBERT REID, *Proposed Sterilization of Certain Mental and Physical Degenerates* (London, Walter Scott, 1903).

—— *Race Culture; or, Race Suicide? (A Plea for the Unborn)* (London, Walter Scott, 1906).

RICHARDSON, H. M., 'The Outcasts', *The Englishwoman* (Sept. 1909). Repr. as a pamphlet by National Union of Women's Suffrage Societies.

ROBERTS, URSULA, *The Cause of Purity and Women's Suffrage*, Church League for Women's Suffrage Pamphlet, 5 (no date, *c*.1911–12).

ROWNTREE, JOSEPH, and SHERWELL, ARTHUR, *The Temperance Problem and Social Reform* (London, Hodder & Stoughton, 1899).

RUSSELL, CHARLES E. B., 'Some Aspects of Female Criminality and its Treatment', *The Englishwoman*, 37 (Jan. 1912).

RYAN, MICHAEL, *Prostitution in London* (London, H. Balliere, 1839).

RYLE, REGINALD JOHN, 'The Origin of Feeble-Mindedness', in

National Association for Promoting the Welfare of the Feeble-Minded, *Report for 1911*.

SAFFORD, A. HERBERT, 'What are the Best Means of Preventing Infanticide?', *Transactions NAPSS 1866* (1867).

SCHARLIEB, MARY, 'Alcoholism in Relation to Women and Children', in T. N. Kelynack (ed.), *The Drink Problem* (1907).

SCOUGAL, FRANCIS, (pseud. for Felicia Mary Francis Skene), *Scenes from a Silent World or Prisons and their Inmates* (Edinburgh and London, William Blackwood, 1889).

SHADWELL, ARTHUR, *Drink, Temperance and Legislation* (London, Longmans, Green, 1902).

SHARMAN, HENRY RISBOROUGH, *A Cloud of Witnesses against Grocers' Licenses: The Fruitful Source of Female Intemperance* (London, Lile and Fawcett, no date).

SMITH BAKER, R., 'The Social Results of the Employment of Girls and Women in Factories and Workshops', *Transactions NAPSS 1868* (1869).

SNOWDEN, PHILIP, *Socialism and the Drink Question* (London, Independent Labour Party/Socialist Library, 1908).

STEER, MARY H., 'Rescue Work by Women among Women', in Baroness Burdett-Coutts (ed.), *Woman's Mission* (1893).

SULLIVAN, W. C., 'The Causes of Inebriety in the Female and the Effects of Alcoholism on Racial Degeneration', *British Journal of Inebriety*, 1:2 (Oct. 1903).

—— *Alcoholism: A Chapter in Social Pathology* (London, James Nisbet, 1906).

—— 'The Criminology of Alcoholism' in T. N. Kelynack (ed.), *The Drink Problem* (1907).

SYMONS, JELINGER C., *Tactics for the Times: As regards the Condition and Treatment of the Dangerous Classes* (London, John Olivier, 1849).

TALLACK, WILLIAM, *Penological and Preventative Principles* (2nd edn. London, Wertheimer, Lea, 1896; 1st edn. 1889).

TAYLOR, H. B., 'Influence of Women on Temperance', *The Englishwoman's Review*, 46 (15 Feb. 1877).

THOMSON, BASIL, *The Criminal* (London, Hodder and Stoughton, 1925).

THOMSON, J. B., 'The Hereditary Nature of Crime', *Journal of Mental Science*, 15:72 (Jan. 1870).

—— 'The Psychology of Criminals', *Journal of Mental Science*, 17:75 (Oct. 1870).

TIMPSON, Revd THOMAS, *Memoirs of Mrs Elizabeth Fry* (London, Aylott & James, 1847).

TODD, Miss, 'Prison Mission and Inebriates' Home', *The Englishwoman's Review*, 12:98 (15 July 1881).

TREDGOLD, A. F., *Mental Deficiency* (London, Baillère, Tindall and Cox, 1908).

—— 'The Feeble-Minded: A Social Danger', *Eugenics Review*, 1:2 (July 1909).

—— 'The Feeble-Minded', *Contemporary Review*, 97:534 (June 1910).

TWINING, LOUISA, 'Women as Public Servants', *The Nineteenth Century* (Dec. 1890).

VINCENT, ARTHUR, *Lives of Twelve Bad Women* (London, T. Fisher Unwin, 1897).

VINCENT, C. E. HOWARD, 'Address on Repression of Crime', *Transactions NAPSS 1883* (1884).

WAUGH, Revd BENJAMIN, 'The Restoration of the Female Inebriate', *British Journal of Inebriety*, 1:2 (Oct. 1903).

WEBB, SIDNEY, and WEBB, BEATRICE (eds.), *The Break-Up of the Poor Law: Being Part One of the Minority Report of the Poor Law Commission* (London, Longmans, Green, 1909).

—— *English Prisons under Local Government* (London, Longmans, 1922).

WELLS, Revd ASHTON, 'Crime in Women: Its Sources and Treatment', *Transactions NAPSS 1876* (1877).

WESTCOTT, W. M. WYNN, 'Inebriety in Women and the Overlaying of Infants', *British Journal of Inebriety*, 1:2 (Oct. 1903).

WILSON, JOHN DOVE, 'Can any Better Measures be Devised for the Prevention and Punishment of Infanticide?', *Transactions NAPSS 1877* (1878).

WORMALD, JOHN, and WORMALD, SAMUEL, *A Guide to the Mental Deficiency Act, 1913* (London, P. S. King & Son, 1914).

WRENCH, MATILDA, *Visits to Female Prisoners at Home and Abroad* (London, Wertheim and Macintosh, 1852).

X.Y.Z., 'England's Convict Prison For Women: An Impressionist Report of a Visit to H.M. Prison, Aylesbury, in Nov. 1911', *The Englishwoman*, 14 (Apr.–June 1912).

ZANETTI, FRANCES, 'Inebriety in Women and its Influence on Child-Life', *British Journal of Inebriety*, 1:2 (Oct. 1903).

II. SECONDARY SOURCES

(A) DOCTORAL THESES

CREW, HILARY, 'Money, Morality and Mental Deficiency: An Examination of the Attitudes and Influences that caused Destitute Unmarried Mothers to be certified as "Moral Defectives" under the Mental Deficiency Act of 1913', M.Soc.Sc. (Birmingham, no date).

HAYWARD, A. R., 'Murder and Madness: A Social History of the Insanity defence in Mid-Victorian England', M.Litt. (Oxford, 1983).

JOHNSTON, VALERIE J., 'Diet in Workhouses and Prisons 1835 to 1895', D.Phil. (Oxford, 1981).

MAPPEN, ELLEN F., 'Women Workers and Unemployment Policy in late Victorian and Edwardian London', D.Phil. (Rutgers University, New Brunswick, NJ, May, 1977).

SAUNDERS, JANET FLORENCE, 'Institutionalised Offenders: A Study of the Victorian Institution and its Inmates, with special reference to late Nineteenth-Century Warwickshire', Ph.D. (Warwick, May 1983).

STEVENSON, SIMON JOHN, 'The "Criminal Class" in the Mid-Victorian City: A Study of Policy conducted with Special Reference to those made subject to the Provisions of 34 & 35 Vict., c. 112 (1871) in Birmingham and East London in the Early Years of Registration and Supervision', D.Phil. (Oxford, 1983).

(B) UNPUBLISHED PAPERS

GELLING, ANNE, 'The Didactic Moralists and the Ideology of Femininity' (Nuffield College, Oxford, 1984).

INNES, JOANNA, 'English Houses of Correction and "Labour Discipline" c.1600–1780: A Critical Examination' (Somerville College, Oxford, July 1983).

LESSELIER, CLAUDIE, 'Les Femmes et la prison 1820–1939: Prisons des femmes et reproduction de la société patriarcale' (formerly held in Greater London Council Feminist Library, no date).

SMITH, ROGER, 'Medicine and Murderous Women in the Mid-Nineteenth Century', paper read at the Social History Conference on Crime, Violence and Social Protest (Birmingham, Jan. 1977).

(C) PUBLISHED SOURCES

ABRAMS, PHILIP, *The Origins of British Sociology: 1834–1914* (Chicago, University of Chicago Press, 1968).

ADLER, FREDA, *Sisters in Crime* (New York, McGraw-Hill, 1975).

ALEXANDER, SALLY, *Women's Work in Nineteenth Century London: A Study of the Years 1820–50* (London, Journeyman, 1976).

BAILEY, VICTOR, 'Crime, Criminal Justice and Authority in England', *Society for the Study of Labour History Bulletin* (Warwick) 40 (Spring 1980).

—— (ed.), *Policing and Punishment in Nineteenth Century Britain* (London, Croom Helm, 1981).

BANKS, OLIVE, *Faces of Feminism: A Study of Feminism as a Social Movement* (Oxford, Martin Robertson, 1981).

BANKS, J. A., and BANKS, OLIVE, *Feminism and Family Planning in Victorian England* (Liverpool, Liverpool University Press, 1965).

BASCH, FRANÇOISE, *Relative Creatures: Victorian Women in Society and the Novel 1837–67* (London, Allen Lane, 1974).

BEATTIE, J. M., 'The Criminality of Women in Eighteenth Century England', *Journal of Social History*, 8 (1975).

BEDDOE, DEIRDRE, *Carmarthenshire Women and Criminal Transportation to Australia 1787–1852*, repr. from the *Carmarthenshire Antiquary* (1977).

—— *Carmarthenshire's Convict Women in Nineteenth Century Van Diemen's Land*, repr. from the *Carmarthenshire Antiquary* (1979).

—— *Welsh Convict Women: A Study of Women Transported from Wales to Australia 1787–1852* (Barry, Stewart Williams, 1979).

BENNETT, JAMES, *Oral History and Delinquency: The Rhetoric of Criminology* (Chicago and London, University of Chicago Press, 1981).

BISHOP, CECIL, *Women and Crime* (London, Chatto and Windus, 1931).

BLAND, LUCY, 'In the Name of Protection: The Policing of Women in the First World War', in J. Brophy and C. Smart (eds.), *Women-in-Law* (1985).

BLOM-COOPER, LOUIS (ed.), *Progress in Penal Reform* (Oxford, Clarendon Press, 1974).

BOWKER, LEE H. (ed.), *Women, Crime and the Criminal Justice System* (Toronto, Lexington Books, 1978).

BRANCA, PATRICIA, *Women in Europe since 1750* (London, Croom Helm, 1978).

BRENZEL, BARBARA, 'Lancaster Industrial School for Girls: A Social Portrait of a Nineteenth Century Reform School for Girls', *Feminist Studies*, 3 (Fall 1975).

BRIDENTHAL, RENATE, and KOONZ, CLAUDIA (eds.), *Becoming Visible: Women in European History* (Boston, Houghton Mifflin, 1977).

BRISTOW, EDWARD J., *Vice and Vigilance: Purity Movements in Britain since 1700* (Dublin, Gill & Macmillan, 1977).

BRODSKY, ANNETTE M. (ed.), *The Female Offender* (Beverley Hills and London, Sage, 1975).

BROPHY, JULIA, and SMART, CAROL (eds.), *Women-in-Law: Explorations in Law, Family and Sexuality* (London, Routledge & Kegan Paul, 1985).

BURMAN, SANDRA, and HARRELL-BOND, BARBARA E. (eds.), *Fit Work for Women* (London, Croom Helm, 1979).

BURNS, J. H. (ed.), *The Collected Works of Jeremy Bentham* (London, Athlone, 1970).

BYNUM, W. F., PORTER, ROY, and SHEPHERD, MICHAEL (eds.), *The Anatomy of Madness: Essays in the History of Psychiatry*, i, *People and Ideas*, and ii, *Institutions and Society* (London, Tavistock, 1985).

CARLEN, PAT, *Women's Imprisonment: A Study in Social Control* (London, Routledge and Kegan Paul, 1983).

CHRIST, CAROL, 'Victorian Masculinity and the Angel in the House', in M. Vicinus (ed.), *A Widening Sphere* (1977).

CLARK, ANNA, *Women's Silence Men's Violence: Sexual Assault in England 1770–1845* (London, Pandora, 1987).

CLOWARD, R. A., 'Social Control in the Prison', in R. A. Cloward, *Theoretical Studies in the Social Organisation of the Prison* (1960).

—— CRESSEY, RONALD R., OHLIN, L., and SYKES, G. (eds.), *Theoretical Studies in the Social Organisation of the Prison*, Social Science Research Council, Pamphlet 15 (New York, SSRC, 1960).

COHEN, STANLEY, *Folk Devils and Moral Panics: The Creation of Mods and Rockers* (London, MacGibbon and Kee, 1972).

—— *Visions of Social Control: Crime, Punishment and Classification* (Cambridge, Polity, 1985).

—— and SCULL, ANDREW (eds.), *Social Control and the State: Historical and Comparative Essays* (Oxford, Martin Robertson, 1983).

COLLINS, PHILIP, *Dickens and Crime* (London, Macmillan, 1964).

COMINOS, PETER, 'Late Victorian Sexual Respectability and the Social System', *International Review of Social History*, 8 (Amsterdam, 1963).

—— 'Innocent *Femina Sensualis* in Unconscious Conflict', in M. Vicinus (ed.), *Suffer and Be Still* (1980).

CONWAY, JILL, 'Stereotypes of Femininity in a Theory of Sexual Evolution', *Victorian Studies*, 14:1 (Sept. 1970).

CRESSEY, DONALD, 'Limitations on Organization of Treatment in the Modern Prison', in R. A. Cloward *et al.* (eds.), *Theoretical Studies in the Social Organisation of the Prison* (1960).

CRITES, LAURA (ed.), *The Female Offender* (Lexington, Mass., Lexington Books, 1978).

DAVIDOFF, LEONORE, 'Class and Gender in Victorian England', in J. L. Newton *et al.* (eds.), *Sex and Class in Women's History* (1983).

—— L'ESPERANCE, JEAN, and NEWBY, HOWARD, 'Landscape with Figures: Home and Community in English Society', in A. Oakley and J. Mitchell (eds.), *The Rights and Wrongs of Women* (1976).

DAVIS, JENNIFER, 'The London Garrotting Panic of 1862: A Moral Panic and the Creation of a Criminal Class in Mid-Victorian England', in V. A. C. Gatrell *et al.* (eds.), *Crime and the Law* (1980).

DELACY, MARGARET E., 'Grinding Men Good? Lancashire's Prisons at Mid-Century', in V. Bailey (ed.), *Policing and Punishment* (1981).

—— *Prison Reform in Lancashire 1700–1850: A Study in Local Administration* (Manchester, Manchester University Press, 1986).

DELAMONT, SARA, and DUFFIN, LORNA (eds.), *The Nineteenth Century Woman: Her Cultural and Physical World* (London, Croom Helm, 1978).

DOBASH, RUSSELL P., DOBASH, R. EMERSON, and GUTTERIDGE, SUE, *The Imprisonment of Women* (Oxford, Basil Blackwell, 1986).

DONAJGRODZKI, A. P. (ed.), *Social Control in Nineteenth Century Britain* (London, Croom Helm, 1977).

DONZELOT, JACQUES, *The Policing of Families: Welfare versus the State* (London, Hutchinson, 1979).

DOWBIGGIN, IAN, 'Degeneration and Hereditarianism in French Mental Medicine 1840–90: Psychiatric Theory as Ideological Adaption', in W. F. Bynum *et al.* (eds.), *The Anatomy of Madness*, i (1985).

DUFFIN, LORNA, 'Prisoners of Progress: Women and Evolution', in S. Delamont and L. Duffin (eds.), *The Nineteenth Century Woman* (1978).

DYHOUSE, CAROLE, 'Working-class Mothers and Infant Mortality in England 1895–1914', *Journal of Social History*, 12:2 (Winter 1978).

—— 'Mothers and Daughters in the Middle-Class Home *c.*1870–1914', in J. Lewis (ed.), *Labour and Love* (1986).

DYOS, H. J., 'The Slums of Victorian London', *Victorian Studies*, 11:1 (Sept. 1967).

—— *Exploring the Urban Past: Essays in Urban History* (Cambridge, Cambridge University Press, 1982).

EVANS, ROBIN, *The Fabrication of Virtue: English Prison Architecture 1750–1840* (Cambridge, Cambridge University Press, 1982).

FEE, ELIZABETH, 'The Sexual Politics of Victorian Anthropology', in M. S. Hartman and L. Banner (eds.), *Clio's Consciousness Raised* (1974).

FINNEGAN, FRANCES, *Poverty and Prostitution: A Study of Victorian Prostitutes in York* (Cambridge, Cambridge University Press, 1979).

FORSYTHE, WILLIAM JAMES, *The Reform of Prisoners 1830–1900* (London, Croom Helm, 1987).

FOUCAULT, MICHEL, *Madness and Civilisation: A History of Insanity in the Age of Reason* (trans. London, Tavistock, 1967).

FOUCAULT, MICHEL, *Discipline and Punish: the Birth of the Prison* (trans. Harmondsworth, Peregrine, 1982; first publ. 1975).

FOX, Sir LIONEL WRAY, *The English Prison and Borstal Systems* (London, Routledge & Kegan Paul, 1952).

FREEDEN, MICHAEL, 'Eugenics and Progressive Thought: A Study in Ideological Affinity', *Historical Journal*, 22:3 (Sept. 1979).

—— *The New Liberalism: An Ideology of Social Reform* (Oxford, Oxford University Press, 1978).

FREEDMAN, ESTELLE B., *Their Sisters' Keepers: Women's Prison Reform in America 1830–1930* (Ann Arbor, Mich., University of Michigan Press, 1981).

GARLAND, DAVID, 'The Criminal and his Science: A Critical Account of the Formation of Criminology at the end of the Nineteenth Century', *British Journal of Criminology*, 25:2 (Apr. 1985).

—— *Punishment and Welfare: A History of Penal Strategies* (Aldershot, Gower, 1985).

—— 'British Criminology before 1935', *British Journal of Criminology*, 28:2 (Spring 1988).

GATRELL, V. A. C., 'The Decline of Theft and Violence in Victorian and Edwardian England', in V. A. C. Gatrell *et al.* (eds.), *Crime and the Law* (1980).

—— LENMAN, BRUCE, and PARKER, GEOFFREY (eds.), *Crime and the Law: The Social History of Crime in Western Europe since 1500* (London, Europa, 1980).

—— and HADDEN, T. B., 'Criminal Statistics and their Interpretation', in E. A. Wrigley (ed.), *Nineteenth Century Society* (1972).

GELSTHORPE, LORAINE, and MORRIS, ALLISON, 'Feminism and Criminology in Britain', in P. Rock (ed.), *A History of British Criminology*, special edn. of *British Journal of Criminology*, 28:2, Spring 1988 (Oxford, Clarendon Press, 1988).

—— —— (eds.), *Feminist Perspectives in Criminology* (Buckingham, Open University Press, 1990).

GIALLOMBARDO, ROSE, *Society of Women: A Study of a Women's Prison* (New York, John Wiley, 1966).

GIBSON, HELEN E., 'Women's Prisons: Laboratories for Penal Reform', in L. Crites (ed.), *The Female Offender* (1978).

GOFFMAN, ERVING, *Asylums: Essays on the Social Situation of Mental Patients and Other Inmates* (Harmondsworth, Pelican, 1970).

GOLDFARB, R. L., and SINGER, L. R., *After Conviction* (New York, Simon and Schuster, 1977).

GOLDTHORPE, JOHN, 'The Development of Social Policy in England 1800–1914', *Transactions of the Fifth World Congress of Sociology*, Washington DC, 2–8 Sept. 1962, vol. 4 (International Sociological Association, 1964).

GORHAM, DEBORAH, *The Victorian Girl and the Feminine Ideal* (London, Croom Helm, 1982).

GOVE, WALTER R. (ed.), *The Labelling of Deviance: Evaluating a Perspective* (New York, Sage, 1975).

GRAFF, HARVEY J., 'Crime and Punishment in the Nineteenth Century: A New Look at the Criminal', *Journal of Interdisciplinary History*, 7:3 (Winter 1977).

GROSSER, GEORGE H., 'External Setting and Internal Relations of the Prisons', in R. A. Cloward, *Theoretical Studies in the Social Organisation of the Prison* (1960).

HAHN, NICHOLAS FISCHER, 'Female State Prisoners in Tennessee 1831–1979', *Tennessee Historical Quarterly*, 39:4 (Winter 1980).

—— 'Matrons and Molls: The Study of Women's Prison History', in J. A. Inciardi and C. E. Faupel (eds.), *History and Crime* (1980).

HALL, CATHERINE, 'The Early Formation of Victorian Domestic Ideology', in S. Burman and Barbara E. Harrell-Bond (eds.), *Fit Work for Women* (1979).

HARDING, CHRISTOPHER, HINES, BILL, IRELAND, RICHARD, and RAWLINGS, PHILIP, *Imprisonment in England and Wales: A Concise History* (London, Croom Helm, 1985).

HARRIS, JOSÉ, *Unemployment and Social Policy: A Study in English Social Policy 1886–1914* (Oxford, Clarendon Press, 1972).

HARRIS, RUTH, 'Murder under Hypnosis in the Case of Gabrielle Bompard: Psychiatry in the Courtroom in Belle Epoque Paris', in W. F. Bynum *et al.* (eds.), *The Anatomy of Madness*, ii (1985).

—— *Murders and Madness: Medicine, Law, Society in the Fin de Siècle* (Oxford, Clarendon Press, 1989).

HARRISON, BRIAN, 'Philanthropy and the Victorians', *Victorian Studies*, 9:4 (June 1966).

—— 'Drink and Sobriety in England 1815–1872: A Critical Bibliography', *International Review of Social History*, 12 (1967).

—— 'Underneath the Victorians', *Victorian Studies*, 10:3 (Mar. 1967).

—— *Drink and the Victorians: The Temperance Question in England 1815–1872* (London, Faber, 1971).

—— 'State Intervention and Moral Reform', in Patricia Hollis, *Pressure from Without in Early Victorian England* (London, Edward Arnold, 1974).

—— *Separate Spheres: The Opposition to Women's Suffrage in Britain* (London, Croom Helm, 1978).

—— *Peaceable Kingdom: Stability and Change in Modern Britain* (Oxford, Clarendon Press, 1982).

HARTMAN, MARY S., *Victorian Murderesses: A True History of Thirteen Respectable French and English Women Accused of Unspeakable Crimes* (London, Robson, 1985; first publ. 1977).

HARTMAN, MARY S., and BANNER, LOUIS (eds.), *Clio's Consciousness Raised* (New York, Harper Colophon, 1974).

HAY, DOUGLAS, 'Crime and Justice in Eighteenth and Nineteenth Century England', in M. Tonry and N. Morris (eds.), *Crime and Justice: An Annual Review of Research* (1983).

—— LIVEBAUGH, PETER, RULE, JOHN G., THOMPSON, E. P., and WINSLOW, C. (eds.), *Albion's Fatal Tree: Crime and Society in Eighteenth Century England* (London, Allen Lane, 1975).

HEENEY, BRIAN, *The Women's Movement in the Church of England 1850–1930* (Oxford, Clarendon Press, 1988).

HEFFERNAN, ESTHER, *Making it in Prison: The Square, the Cool, and the Life* (New York, Wiley-Interscience, 1972).

HEIDENSOHN, FRANCES, 'Women and the Penal System', in A. Morris and L. Gelsthorpe (eds.), *Women and Crime* (1981).

—— *Women and Crime* (Basingstoke, Macmillan, 1985).

HELSINGER, ELIZABETH K., SHEETS, ROBIN LAUTERBACH, and VEEDER, WILLIAM, *The Woman Question: Society and Literature in Britain and America 1837–1883*, i and iii (Manchester, Manchester University Press, 1983).

HENRIQUES, URSULA, 'The Rise and Decline of the Separate System of Prison Discipline', *Past and Present*, 54 (Feb. 1972).

—— *Before the Welfare State: Social Administration in Early Industrial Britain* (London, Longman, 1979).

HERVEY, N., 'A Slavish bowing down: The Lunacy Commission and the Psychiatric Profession 1845–1860', in W. F. Bynum *et al.* (eds.), *The Anatomy of Madness*, ii (1985).

HILTON, BOYD, *The Age of Atonement: The Influence of Evangelicalism on Social and Economic Thought 1795–1865* (Oxford, Clarendon Press, 1988).

HIMMELFARB, GERTRUDE, *Victorian Minds* (London, Weidenfeld & Nicolson, 1968; first publ. 1952).

—— *The Idea of Poverty: England in the Early Industrial Age* (London, Faber, 1984).

HINDE, R. S. E., *The British Penal System 1773–1950* (London, Duckworth, 1951).

HOFFER, PETER C., and HULL, N. E. H., *Murdering Mothers: Infanticide in England and New England 1558–1803* (New York, New York University Press, 1981).

HOUGHTON, WALTER E., *The Victorian Frame of Mind 1830–1870* (New Haven, Conn., Yale University Press, 1957).

HOUSDEN, LESLIE GEORGE, *The Prevention of Cruelty to Children* (London, Cape, 1955).

HUGHES, ROBERT, *The Fatal Shore: A History of the Transportation of Convicts to Australia 1787–1868* (London, Collins Harvill, 1987).

HUTTER, BRIDGET, and WILLIAMS, GILLIAN (eds.), *Controlling Women: The Normal and the Deviant* (London, Croom Helm, 1981).

IGNATIEFF, MICHAEL, *A Just Measure of Pain: The Penitentiary in the Industrial Revolution 1750–1850* (London, Macmillan, 1978).

—— 'State, Civil Society and Total Institutions: A Critique of Recent Social Histories of Punishment', originally in M. Tonry and N. Morris (eds.), *Crime and Justice: An Annual Review of Research* (1981), repr. in S. Cohen and A. Scull (eds.), *Social Control and the State* (1983).

—— 'Total Institutions and Working Classes: A Review Essay', *History Workshop*, 15 (Spring 1983).

INCIARDI, JAMES A., and FAUPEL, CHARLES E. (eds.), *History and Crime: Implications for Criminal Justice Policy* (Beverly Hills, Sage, 1980).

—— BLOCK, ALAN A., and HALLOWELL, LYLE A., *Historical Approaches to Crime: Research Strategies and Issues* (Beverly Hills, Sage, 1977).

JAMIESON, LYNN, 'Limited Resources and Limiting Conventions: Working Class Mothers and Daughters in Urban Scotland', in J. Lewis (ed.), *Labour and Love* (1986).

JEFFREYS, SHEILA, *The Spinster and Her Enemies: Feminism and Sexuality 1880–1930* (London, Pandora, 1985).

JONES, DAVID, *Crime, Protest, Community and Police in Nineteenth Century Britain* (London, Routledge and Kegan Paul, 1982).

JONES, KATHLEEN, *Mental Health and Social Policy 1845–1959* (London, Routledge and Kegan Paul, 1960).

KANNER, B., 'The Women of England in a Century of Social Change 1815–1914: A Select Bibliography', in M. Vicinus (ed.), *Suffer and Be Still* (1980).

KENT, RAYMOND A., *A History of British Empirical Sociology* (Aldershot, Gower, 1981).

KENT, SUSAN KINGSLEY, *Sex and Suffrage in Britain 1860–1914* (Princeton, NJ, Princeton University Press, 1987).

KLEIN, VIOLA, *The Feminine Character: History of an Ideology* (London, Kegan Paul, Trench Trubner, 1946).

KNIGHT, PATRICIA, 'Women and Abortion in Victorian and Edwardian England', *History Workshop*, 4 (Autumn 1977).

LANE, ROGER, 'Crime and the Industrial Revolution: British and American Views', *Journal of Social History*, 7:3 (Spring 1974).

LARNER, CHRISTINA, 'Crimen Exceptum? The Crime of Witchcraft in Europe', in V. A. C. Gatrell *et al.* (eds.), *Crime and the Law* (1980).

LEONARD, EILEEN, *Women, Crime and Society* (New York and London, Longman, 1982).

LEVIN, YALE, and LINDESMITH, ALFRED, 'English Ecology and Criminology of the Past Century', *Journal of Criminal Law, Criminology and Police Science*, 27 (1936–7).

LEWIS, E. O., 'Mental Deficiency and Criminal Behaviour', in *Mental Abnormality and Crime*, prefaced by P. H. Winfield (London, Macmillan, 1944).

LEWIS, JANE, *Women in England 1870–1950: Sexual Divisions and Social Change* (Sussex, Wheatsheaf, 1984).

—— (ed.), *Labour and Love: Women's Experience of Home and Family 1850–1940* (Oxford, Basil Blackwell, 1986).

LUCKIN, BILL, 'Towards a Social History of Institutionalisation', *Social History*, 8:1 (Jan. 1983).

MCCANDLESS, PETER, 'Liberty and Lunacy: The Victorians and Wrongful Confinement', *Journal of Social History*, 11:3 (Spring 1978).

MCCONVILLE, SEAN, *A History of English Prison Administration*, i, *1750–1877* (London, Routledge and Kegan Paul, 1981).

MCGREGOR, O. R., 'Social Research and Social Policy in the Nineteenth Century', *British Journal of Sociology*, 8:2 (June 1957).

MCHUGH, PAUL, *Prostitution and Victorian Social Reform* (London, Croom Helm, 1980).

MCLACHLAN, NOEL, 'Penal Reform and Penal History: Some Reflections', in L. Blom-Cooper (ed.), *Progress in Penal Reform* (1974).

MACLEOD, ROY M., 'The Edge of Hope: Social Policy and Chronic Alcoholism 1870–1900', *Journal of the History of Medicine*, 22:3 (1967).

MCMILLAN, JAMES, *Housewife or Harlot: The Place of Women in French Society 1870–1914* (Brighton, Harvester, 1981).

MAHOOD, L., *The Magdalenes: Prostitution in the Nineteenth Century* (London, Routledge, 1989).

MANNHEIM, HERMANN (ed.), *Pioneers in Criminology* (London, Stevens & Sons, 1960).

MARCUS, STEVEN, *The Other Victorians: A Study of Sexuality and Pornography in Mid-Nineteenth Century England* (London, Weidenfeld and Nicolson, 1966).

MAYER, JOHN A., 'Notes towards a Working Definition of Social Control in Historical Analysis', in S. Cohen and A. Scull (eds.), *Social Control and the State* (1983).

MELOSSI, DARIO, and PAVARINI, MASSIMO, *The Prison and the Factory: Origins of the Penitentiary System* (London, Macmillan, 1981; first publ. 1977).

MILLMAN, MARCIA, 'She Did It All For Love: A Feminist View of the Sociology of Deviance', in M. Millman and R. M. Kanter (eds.), *Another Voice* (1975).

—— and KANTER, ROSABETH MOSS (eds.), *Another Voice: Feminist Perspectives on Social Life and Social Science* (New York, Anchor, 1975).

MORRIS, ALLISON, *Women, Crime and Criminal Justice* (Oxford, Basil Blackwell, 1987).

—— and GELSTHORPE, L. (eds.), *Women and Crime* (Cambridge, Institute of Criminology, 1981).

—— 'False Clues and Female Crime', in A. Morris and L. Gelsthorpe (eds.), *Women and Crime* (1981).

MORT, FRANK, *Dangerous Sexualities: Medico-Moral Politics in England since 1830* (London and New York, Routledge and Kegan Paul, 1987).

NEWTON, JUDITH L., RYAN, MARY P., and WALKOWITZ, JUDITH R. (eds.), *Sex and Class in Women's History* (London, Routledge and Kegan Paul, 1983).

NYE, ROBERT A., 'Crime in Modern Societies: Some Research Strategies for Historians', *Journal of Social History*, 11:4 (Summer 1978).

—— *Crime, Madness and Politics in Modern France: The Medical Concept of National Decline* (Princeton, NJ, Princeton University Press, 1984).

OAKLEY, ANNE, and MITCHELL, JULIET (eds.), *The Rights and Wrongs of Women* (Harmondsworth, Penguin, 1976).

O'BRIEN, PATRICIA, 'Crime and Punishment as Historical Problem', *Journal of Social History*, 11:4 (Summer 1978).

—— *The Promise of Punishment: Prisons in Nineteenth Century France* (Princeton, NJ, Princeton University Press, 1982).

PARKER, JULIA, *Women and Welfare: Ten Victorian Women in Public Social Service* (Basingstoke, Macmillan, 1988).

PEARSON, GEOFFREY, *The Deviant Imagination: Psychiatry, Social Work and Social Change* (London, Macmillan, 1975).

—— *Hooligan: A History of Respectable Fears* (London and Basingstoke, Macmillan, 1983).

PERROT, MICHELLE, 'Délinquance et système pénitentiare en France au XIXe siècle', *Annales*, 30:1 (Jan.–Feb. 1975).

PLAYFAIR, GILES, *The Punitive Obsession: An Unvarnished History of the English Prison System* (London, Victor Gollancz, 1971).

POLLOCK, JOY, and CHESNEY-LIND, MEDA, 'Early Theories of Female Criminality', in L. H. Bowker (ed.), *Women, Crime, and the Criminal Justice System* (1978).

POPE, BARBARA CORRADO, 'Angels in the Devil's Workshop', in R. Bridenthal and C. Koonz (eds.), *Becoming Visible* (1977).

PREWER, R. R., 'The Contribution of Prison Medicine', in L. Blom-Cooper (ed.), *Progress in Penal Reform* (1974).

PRIESTLEY, PHILIP, *Victorian Prison Lives: English Prison Biography 1830–1914* (London, Methuen, 1985).

PROCEK, EVA, 'Psychiatry and the Social Control of Women', in A. Morris and L. Gelsthorpe (eds.), *Women and Crime* (1981).

PROCHASKA, F. K., *Women and Philanthropy in Nineteenth Century England* (Oxford, Clarendon Press, 1980).

RADZINOWICZ, LEON, *Ideology and Crime: A Study of Crime in its Social and Historical Context* (London, Heinemann, 1966).

RADZINOWICZ, Sir LEON, and HOOD, ROGER, 'Incapacitating the Habitual Criminal: The English Experience', *Michigan Law Review*, 78:8 (Aug. 1980).

—— —— *A History of English Criminal Law and its Administration from 1750*, v, *The Emergence of Penal Policy* (London, Stevens, 1986).

RAFTER, NICOLE HAHN, 'Chastizing the Unchaste: Social Control Functions of a Women's Reformatory 1894–1931', in S. Cohen and A. Scull (eds.), *Social Control and the State* (1983).

—— 'Prisons for Women 1790–1980', in M. Tonry and N. Morris (eds.), *Crime and Justice: An Annual Review of Research* (1983).

—— 'Gender, Prisons, and Prison History', *Social Science History*, 9:3 (Summer 1985).

—— *Partial Justice* (Boston, Northeastern University Press, 1985).

—— and STANKO, ELIZABETH ANNE, *Judge, Lawyer, Victim, Thief: Women, Gender Roles, and Criminal Justice* (Boston, Northeastern University Press, 1982).

RASCHE, CHRISTINE E., 'The Female Offender as an Object of Criminological Research', in A. M. Brodsky (ed.), *The Female Offender* (1975).

RATTRAY TAYLOR, G., *Sex in History* (London, Thames and Hudson, 1953).

ROBERTS, ELIZABETH, *A Woman's Place: An Oral History of Working Class Women 1890–1940* (Oxford, Blackwell, 1984).

ROBERTS, ROBERT, *The Classic Slum* (Manchester, Manchester University Press, 1971).

ROBSON, L. L., *The Convict Settlers of Australia* (London, Melbourne University Press, 1965).

ROCK, PAUL, *A History of British Criminology* (special edn. of the *British Journal of Criminology*, 28:2, Spring 1988 (Oxford, Clarendon Press, 1988).

ROGERS, JAMES ALLEN, 'Darwinism and Social Darwinism', *Journal of the History of Ideas*, 33 (Jan.–Mar. 1977).

ROSS, ELLEN, 'Fierce Questions and Taunts: Married Life in Working-Class London 1870–1914', *Feminist Studies*, 8:3 (Fall 1982).

—— 'Labour and Love: Rediscovering London's Working Class Mothers 1870–1918', in J. Lewis (ed.), *Labour and Love* (1986).

ROTHMAN, DAVID, *The Discovery of the Asylum: Social Order and Disorder in the New Republic* (Boston and Toronto, Little, Brown, 1971).
—— *Conscience and Convenience: the Asylum and its Alternatives in Progressive America* (Boston and Toronto, Little, Brown, 1980).
ROVER, CONSTANCE, *Love, Morals and the Feminists* (London, Routledge and Kegan Paul, 1970).
ROWETT, COLIN, and VAUGHAN, PHILLIP J., 'Women and Broadmoor: Treatment and Control in a Special Hospital', in B. Hutter and G. Williams (eds.), *Controlling Women* (1981).
RUDÉ, GEORGE, *Criminal and Victims: Crime and Society in Early Nineteenth Century England* (Oxford, Clarendon Press, 1985).
RUGGLES-BRISE, Sir EVELYN, *The English Prison System* (London, Macmillan, 1921).
RUSCHE, GEORG, and KIRCHHEIMER, OTTO, *Punishment and Social Structure* (New York, Columbia University Press, 1939).
SCULL, ANDREW, 'Madness and Segregative Control: The Rise of the Insane Asylum', *Social Problems*, 24:3 (Feb. 1977).
—— *Museums of Madness: The Social Organisation of Insanity in Nineteenth-Century England* (London, Allen Lane, 1979).
—— (ed.), *Madhouses, Mad-doctors and Madmen: The Social History of Psychiatry in the Victorian Era* (Philadelphia, University of Pennsylvania Press, 1981).
—— 'Moral Treatment Reconsidered: Some Sociological Comments on an Episode in the History of British Psychiatry', in A. Scull (ed.), *Madhouses, Mad-doctors and Madmen* (1981).
—— 'The Social History of Psychiatry in the Victorian Era', in A. Scull (ed.), *Madhouses, Mad-doctors and Madmen* (1981).
SEARLE, G. R., *Eugenics and Politics in Britain 1900–1914* (Noordoff, Noordoff Publishing International, 1976).
SHANLEY, MARY LYNDON, *Feminism, Marriage and the Law in Victorian England 1850–1895* (London, Tauris, 1989).
SHAW, A. G. L., *Convicts and the Colonies* (Faber, 1966).
SHORTER, EDWARD, *A History of Women's Bodies* (London, Allen Lane, 1983).
SHOWALTER, ELAINE, 'Victorian Women and Insanity', in A. Scull (ed.), *Madhouses, Mad-doctors and Madmen* (1981).
—— *The Female Malady: Women, Madness, and English Culture 1830–1980* (London, Virago, 1987).
SIMMONS, HARVEY G., 'Explaining Social Policy: The English Mental Deficiency Act of 1913', *Journal of Social History*, 11:3 (Spring 1978).
SKULTANS, VIEDA, *Madness and Morals: Ideas on Insanity in the Nineteenth Century* (London, Routledge and Kegan Paul, 1975).
SMART, CAROL, *Women, Crime and Criminology: A Feminist Critique* (London, Routledge and Kegan Paul, 1976).

SMART, CAROL, 'Criminological Theory: Its Ideology and Implications concerning Women', *British Journal of Sociology*, 28:1 (Mar. 1977).

―― and SMART, BARRY (eds.), *Women, Sexuality and Social Control* (London, Routledge and Kegan Paul, 1978).

SMITH, ANN, *Women in Prison* (London, Stevens, 1962).

―― 'The Woman Offender', in L. Blom-Cooper (ed.), *Progress in Penal Reform* (1974).

SMITH, F. BARRY, 'Sexuality in Britain 1800–1900', in M. Vicinus (ed.), *A Widening Sphere* (1977).

SMITH, ROGER, 'The boundary between Insanity and Criminal Responsibility in Nineteenth Century England', in A. Scull (ed.), *Madhouses, Mad-doctors and Madmen* (1981).

―― *Trial by Medicine: Insanity and Responsibility in Victorian Trials* (Edinburgh, Edinburgh University Press, 1981).

SOLOWAY, RICHARD, 'Counting the Degenerates: The Statistics of Race Deterioration in Edwardian England', *Journal of Contemporary History*, 17:1 (Jan. 1982).

SPATZ WIDOM, CATHY, 'Perspectives of Female Criminality', in A. Morris and L. Gelsthorpe (eds.), *Women and Crime* (1981).

STANSELL, CHRISTINE, 'Women, Children, and the Uses of the Streets: Class and Gender Conflicts in New York City, 1850–1860', *Feminist Studies*, 8:2 (Summer 1982).

STEDMAN JONES, GARETH, *Outcast London: A Study in the Relationship between Classes in Victorian Society* (Oxford, Clarendon Press, 1971).

STOCKDALE, ERIC, 'The Rise of Joshua Jebb 1837–1850', *British Journal of Criminology*, 16:2 (Apr. 1976).

―― *A Study of Bedford Prison, 1160–1877* (London and Chichester, Phillimore, 1977).

STURMA, MICHAEL, 'Eye of the Beholder: The Stereotype of Women Convicts 1788–1852', *Labour History*, 34 (May 1978).

SUMMERS, ANNE, *Damned Whores and God's Police: The Colonization of Women in Australia* (Victoria, Penguin, 1976).

SUTHERLAND, GILLIAN (ed.), *Studies in the Growth of Nineteenth Century Government* (London, Routledge and Kegan Paul, 1972).

SYKES, GRESHAM M., *The Society of Captives: A Study of a Maximum Security Prison* (Princeton, NJ, Princeton University Press, 1958).

THANE, PAT, 'Women and the Poor Law in Victorian and Edwardian England', *History Workshop*, 6 (Autumn 1978).

THOMAS, J. E., *The English Prison Officer Since 1850: A Study in Conflict* (London, Routledge and Kegan Paul, 1972).

―― 'Policy and Administration in Penal Establishments', in L. Blom-Cooper (ed.), *Progress in Penal Reform* (1974).

THOMAS, KEITH, 'The Double Standard', *Journal of the History of Ideas*, 20:2 (Apr. 1959).

THOMPSON, E. P., and YEO, EILEEN (eds.), *The Unknown Mayhew: Selections from the Morning Chronicle 1849–1850* (Merlin, 1971).

THOMPSON, F. M. L., 'Social Control in Victorian Britain', *Economic History Review*, 2nd ser., 34:2 (May 1981).

—— *The Rise of Respectable Society: A Social History of Victorian Britain 1830–1900* (London, Fontana, 1988).

TOBIAS, J. J., *Crime and Industrial Society in the Nineteenth Century* (London, Batsford, 1967).

—— *Nineteenth Century Crime: Prevention and Punishment* (Newton Abbot, David and Charles, 1972).

TOMES, NANCY, 'A "Torrent of Abuse": Crimes of Violence between Working-Class Men and Women in London 1840–1875', *Journal of Social History*, 11:3 (Spring 1978).

TOMLINSON, M. HEATHER, '"Prison Palaces": A Reappraisal of Early Victorian Prisons, 1835–1877', *Bulletin of the Institute of Historical Research*, 51:123 (May 1978).

—— 'Penal Servitude 1846–1865: A System in Evolution', in V. Bailey (ed.), *Policing and Punishment in Nineteenth Century Britain* (1981).

TONRY, MICHAEL, and MORRIS, NORVAL (eds.), *Crime and Justice: An Annual Review of Research* (Chicago, University of Chicago Press, 1979 onwards).

VICINUS, MARTHA (ed.), *A Widening Sphere: Changing Roles of Victorian Women* (Bloomington and London, Indiana University Press, 1977).

—— (ed.), *Suffer and Be Still: Women in the Victorian Age* (London, Methuen, 1980).

—— *Independent Women: Work and Community for Single Women 1850–1920* (London, Virago, 1985).

WALKER, NIGEL, *Crime and Insanity in England*, vol. 1, *The Historical Perspective* (Edinburgh, Edinburgh University Press, 1968).

—— and MCCABE, SARAH, *Crime and Insanity in England*, ii, *New Solutions and New Problems* (Edinburgh, Edinburgh University Press, 1973).

WALKOWITZ, JUDITH, 'The Making of an Outcast Group: Prostitutes and Working Women in Nineteenth Century Plymouth and Southampton', in M. Vicinus (ed.), *A Widening Sphere* (1977).

—— *Prostitution and Victorian Society: Women, Class and the State* (Cambridge, Cambridge University Press, 1980).

—— 'Male Vice and Feminist Virtue: Feminism and the Politics of Prostitution in Nineteenth Century Britain', *History Workshop*, 13 (Spring 1982).

358 *Bibliography*

WALKOWITZ, JUDITH, 'Jack the Ripper and the Myth of Male Violence', *Feminist Studies*, 8:3 (Fall 1982).
—— and WALKOWITZ, D.J., ' "We are not beasts of the field": Prostitution and the Poor in Plymouth and Southampton under the Contagious Diseases Act', in M. S. Hartman and L. Banner (eds.), *Clio's Consciousness Raised* (1974).
WALTON, J. K., 'Casting out and Bringing back in Victorian England: Pauper Lunatics 1840–60', in W. F. Bynum *et al.* (eds.), *The Anatomy of Madness*, vol. 2 (1985).
WALTON, RONALD G., *Women in Social Work* (London, Routledge and Kegan Paul, 1975).
WALVIN, JAMES, *Victorian Values* (London, Cardinal, 1988).
WARD, DAVID A., and KASSEBAUM, GENE G., *Women's Prison: Sex and Social Structure* (London, Weidenfeld and Nicolson, 1966).
WEEKS, JEFFREY, *Sex, Politics and Society: The Regulation of Sexuality since 1800* (London, Longman, 1981).
WHITING, J. R. S., *Prison Reform in Gloucester, 1775–1820* (Chichester, Phillimore, 1975).
WIMSHURST, KERRY, 'Control and Resistance: Reformatory School Girls in Late Nineteenth Century South Australia', *Journal of Social History* (Winter 1984).
WOHL, ANTHONY S. (ed.), *The Victorian Family: Structure and Stresses* (London, Croom Helm, 1978).
Women's National Liberal Federation, *Report of Temperance Sub-Committee on the Effects of Intemperance on the Home, the Woman and the Child* (Westminster, WNLF, n.d., *c.*1927).
WRIGLEY, E. A. (ed.), *Nineteenth Century Society: Essays in the Use of Quantitative Methods for the Study of Social Data* (Cambridge, Cambridge University Press, 1972).

INDEX

Newark State Custodial Home for
  Feeble-Minded Women 288
New South Wales 175, 177
'new woman', the 69–71
Nicolson, D. 79
Nye, R. 83

O'Brien, D. 166, 199–200, 209–10
O'Brien, P. 97, 206
Orme, E. 126, 230

Panopticon 102
Parkhurst Prison 182, 189
  see also convict prisons
Parramatta 175–6
  see also transportation
Paul, G. O. 108
Peel's Gaol Act (1823) 120, 133, 149
penal servitude, *see* convict prisons
Penal Servitude Act (1853) 178
  see also Commission into the Working
    of the Penal Servitude Acts
Perry, J. G. 141
Pinsent, H. 272, 273
Prevention of Crimes Act (1871) 148
Priestley, P. 100
prison chaplain 103, 111, 195–6
  see also convict prisons, religion; local
    prisons, religion; religion
Prison Commissioners 137
prison, deterrence in 104, 112–14
  see also convict prisons; local prisons
prison staff 120–2
  see also convict prisons; local prisons
prison historiography 93–100
prison, moral reform in 102–4, 108–12
  see also convict prisons; local prisons
prison, reform of 93–7, 100–4
prostitution 22, 32–3, 35, 37, 40, 60–1,
  65–6, 73, 75, 77, 80, 154, 246,
  261, 275
  see also fallen women; sexuality

Radzinowicz, L. 228
Rafter, N. H. 117
recidivism, *see* female criminality,
  habitual
refuges 181–2, 214–16
  see also Carlisle Refuge; Fulham
    Refuge; Winchester Refuge
religion 13
  evangelicalism 14, 102–3

'residuum' 78–9
respectability 12, 16–18
'revisionist' histories 94–7, 125, 191
Richardson, H. M. 75, 76
Roberts, E. 16
Roberts, R. 16
Rothman, D. 94–5
Royal Commission on the Feeble-
  Minded (1908) 262, 273, 277, 281,
  285, 286, 287–9
  see also feeble-mindedness
Royal Commission into the Penal
  Servitude Acts (1863) 186, 211,
  283–4
Royal Victoria Homes 232–3
  rural location 238–9
  see also inebriate reformatories
Ruggles-Brise, E. 124
Rusche, G. 132, 197
Russell, W. 105, 109–10, 115–16

Safford, A. H. 29
St Joseph's Reformatory 233, 234
  rural location 239
  see also inebriate reformatories
Sandford, E. 14
Sandlebridge Home for the Feeble-
  Minded 287
Sawyer, E. 276
Scull, A. 269
Select Committee on Habitual
  Drunkards (1872) 231
  see also drunkenness
Select Committee on Public Houses
  (1852–3) 225
  see also drunkenness
'separate system' 108–16, 125–6, 188
sexuality 18, 32, 48–9, 75, 87–8
  see also fallen women; prostitution
Showalter, E. 268–9
'silent system' 104–8
Simmons, H. G. 266, 268, 271, 273,
  295
Social Darwinism, *see* Eugenics
Smalley, H. 280
Smith, A. 98
Smith, R. 83–5, 86, 89–90
Snowden, P. 228
Social Science Association 19, 29, 44,
  47, 61, 64, 207, 220, 228
Society for the Improvement of Prison
  Discipline 103

Society for the Study and Cure of Inebriety 222
see also drunkenness
Somerset, Lady 244, 253
Spencer, H. 76
Stedman Jones, G. 78
Steer, M. 74
Stephen, J. 14
Stodart, M. A. 14
suffragettes 72, 76, 128–9
Sullivan, W. C. 60, 223
'surplus women' 64
Sykes, G. 148–9
Symons, J. C. 31, 47

Temperance Movement 220–1
see also drunkenness
Thomson, J. B. 79, 282
Thornton, H. 14
Tothill Fields House of Correction 107, 131, 137–8
accommodation 139
discipline and punishment 169
education 142–3
health 145–6
homosexual relations 161–2
inmate subculture 165–8
labour 141–2
nursery 147–8
numbers of women 151–3
prisoners' ages 156–7
prisoners' previous commitments 158–9
prisoners' previous occupation 153–5
prisoners' offences 156
religion 144
staff 149–51
staff–inmate relations 159–65
'trafficking' 163
violence 168
see also Fletcher, S. W. (prisoner); local prisons
transportation 173, 174–7, 178, 184, 213
see also convict ships; Parramatta
Tredgold, A. F. 273, 275, 276
Trevelyan, Sir C. 271
Twining, L. 121

Van Diemen's Land 175, 177

Waddington, H. 192
Warwick Reformatory 235–6
see also inebriate reformatories
Webb, S. 134
Wedgwood, J. 292
Wilberforce, W. 14
Wilson, G. 79
Winchester Memorial Refuge 214
see also refuges
Woking Prison 182, 212, 283
see also convict prisons
workhouses 56, 114–15, 171, 279
Wrench, M. 145, 171

Utilitarianism, *see* Bentham

Yonge, C. M. 41